*Environmental Science*

Series Editors: R. Allan U. Förstner W. Salomons

Springer
*Berlin*
*Heidelberg*
*New York*
*Barcelona*
*Budapest*
*Hong Kong*
*London*
*Milan*
*Paris*
*Tokyo*

W. Salomons W. M. Stigliani (Eds.)

# Biogeodynamics of Pollutants in Soils and Sediments

## Risk Assesment of Delayed and Non-Linear Responses

With 121 Figures

Springer

*Series Editors*

**Dr. Rod Allan**
Canada Centre for Inland Waters
867 Lakeshore Road, P.O.B. 5050, Burlington, Ontario L7R 4A6, Canada

**Prof. Dr. Ulrich Förstner**
Arbeitsbereich Umweltschutztechnik
Technische Universität Hamburg-Harburg
Eißendorfer Straße 40, D-21071 Hamburg, Germany

**Prof. Dr. Wim Salomons**
GKSS Research Center
Max Planck Straße, D-21502 Geesthacht, Germany

On leave of absence from:
Institute for Agrobiology and Soil Fertility (AB-DLO)
P.O. Box 129, N-9750 AC Haren, The Netherlands



*Volume Editors*

**Prof. Dr. Wim Salomons**
address see above

**Prof. Dr. William M. Stigliani**
Centre for Energy and Environmental Education (CEEE)
University of Northern Iowa
Cedar Falls, Iowa, 50614-0293

ISBN 3-530-58732-2 Springer-Verlag Berlin Heidelberg New York

Library of Congress Cataloging-in-Publication Data. Biogeodynamic of pollutants in soils and sediments : risk assesment of delayed and non-linear responses / W. Salomons, W.M. Stigliani (eds.). p. cm – (Environmental science) Includes bibliographical references and index. ISBN 0-387-58732-2. –ISBN 3-540-58732-2 1. Soil pollution. 2. Soil chemisty 3. Pollution–Risk assesment. I. Salomons, W. (William). 1945-. II. Stigliani, William M. III. Series: Environmental science (Berlin, Germany) TD878.B555 1995 628.5'5–dc2095-13009



Cover design: Struve & Partner, Heidelberg
Media conversion: Anne de Grosbois, Heidelberg
32/3136 – 5 4 3 2 1 0 - SPIN 10489955 - Printed on acid-free paper

# Preface

In the USA, and western and central Europe there are many large-scale polluted sites. These sites, however, are too large to be cleaned up with available technologies at reasonable costs. They are polluted by mining activities, smelters and other large industries, and the pollution extends not only to soils, but also to sediments in waterways and reservoirs. In addition, similar, new sites are being created with little or no concern for environmental protection. Since these areas are expected to remain polluted for many years, it is necessary to take a long-term view towards ensuring that the capacity to retain the contaminants is not diminished. Likewise, it will be necessary to understand the potential for large-scale contaminant mobilization at these sites, which is triggered by changing environmental conditions.

The mobility of contaminants is often determined by the pH, redox conditions and the presence of complexing agents such as dissolved organic matter and inorganic anions. These parameters, which control the balance between the retention and mobility of the contaminants, are termed *Capacity-Controlling Parameters (CCP)* and *master variables* in this volume.

For short-term risk assessment (5 to 10 years) it is sufficient to understand how the master variables determine contaminant mobility and hence their leaching and bioavailability. Much information is available in the literature on this subject already. Much less is known about the processes which regulate the dynamics of the master variables and CCP. This knowledge gap is not that important for short-term processes which determine the acute pollution status of sediments and soils and their immediate short-term impact on the environment. Over the long-term, however, changes in the CCP and master variables can cause mobilization of vast stores of pollutants manifested by behavior that is time-delayed and nonlinear – precisely the kind of response that catches policymakers, the public, and even the scientists by surprise. One example of this syndrome in recent decades is the advent of acid deposition as a new environmental stress. The large-scale acidification of the environment has led correspondingly to large- scale leaching of aluminum and other heavy metals from acid-impacted areas which is continuing to this day. For example, in Scandinavia, mercury levels in freshwater fish are increasing even though emissions and deposition of mercury have been decreasing for several decades. The contamination of fish has been attributed to the remobilization of mercury locked in watershed soils decades ago, which is now mobilized by soil acidification.

Changes in the CCP are of a long-term nature and are initiated by changes in the major element cycles in the soil-sediment system. The effects of these changes on concentrations of elements in the soil solution are in particular nonlinear for inorganic pollutants and may cause strong increases in concentrations over a short time period. This can result in increased leaching to ground- and surface waters, an impact on the ecosystem and the necessity to take control measures on short notice.

If one wants to control, for instance, diffuse pollution and prevent nonlinear responses, it is necessary to assess and manipulate those factors which determine the capacity-controlling parameters and master variables. This implies that it is necessary to study in detail the major biogeochemical cycles not only of sulphur, nitrogen, and carbon but also calcium, iron and manganese. The soil–sediment system with its stored pollutants has to be treated as a dynamic system in which changes take place over time frames in the order of years to decades to centuries. The soil–sediment system as a dynamic entity can be described as biogeodynamics, since a full understanding of long-term changes requires the combined efforts of scientists from the "geo-" and "bio-"sciences.

An overview of the biogeodynamics of the soil and sediment system is presented in the first chapter by Salomons. The various themes developed in this chapter are expanded upon in subsequent chapters.

In Chapters 2-6, the scientific basis for retention and mobilization of contaminants from the point of view of soil and aquatic science is discussed. However, unlike other treatments of this topic, these chapters place equal emphasis on factors controlling the soils' capacities for storing contaminants, and how these capacities could be altered over the long term. In fact, it is these capacity-altering processes that can trigger a soil or sediment to switch from a net retainer of toxic materials to a net emitter. The central role of the microbial processes is discussed in the Chapter 2 by Doelman, focusing on the microbial environment as the capacity-controlling factor. Geiger et al. document the importance of the dynamics of organic matter in soils and its role in both adsorbing and mobilizing heavy metals in soils. Chapter 4 by Bourg and Loch builds on the themes developed in the two previous chapters. They discuss the dynamics of pH and redox conditions in soil and sediment systems, and their effect on metal mobility. Kuhnt describes the long-term behavior of pesticides in agricultural soils.

In Chapter 7 ter Meulen-Smidt discusses the regional variability of inputs and the distribution of contaminants in Europe. Also discussed are regional differences in storage capacities and prospective region-specific problems and vulnerabilities related to contaminant mobilization.

A more detailed regional example, discussed in the paper by Smal, which highlights the combination of continuing soil acidification and metal inputs in specific "hot spots" in Poland. In a number of developing countries, a serious situation is developing with regard to mercury, which is finding increasing use in the extraction of gold in small-scale mining operations. Is the introduction of mercury into the tropical environment the precursor of a future Minimata-type disaster? This is the topic of discussion in the contribution by Lacerda et al. (Chap. 10).

It is not enough simply to describe the problem – solutions must also be sought. Förstner (Chap. 11) describes the approach of *geochemical engineering* as a means to control the flow of pollutants. Various methods to determine and manipulate CCP are described and illustrated by many examples in this chapter and in Chapter 12 by Eijsackers.

Finally, Stigliani, in a more global context, discusses the importance of harmonizing standards for air, soil, groundwaters, crops and human exposure. He also discusses the costs and benefits of reducing pollutant inputs to the soil and the particular problems associated with agricultural soils and the risks associated with the conversion of agricultural lands to other uses.

May 1995 Wim Salomons and William M. Stigliani

## Acknowledgments

This book originated from a long-term project, colloquially called "Chemical Time Bombs" to stress the often unexpected and sudden response of the environment to changing conditions. This project was financed by the Dutch Ministry of Housing, Physical Planning and the Environment with additional financial support of the European Commission, European Environmental Research Organization and the Umweltbundesamt in Germany. As part of the project, workshops were organized in various countries of Western, Eastern and Central Europe. We want to thank the numerous workshop participants who assisted in the development of ideas and concepts. Last but not least we want to thank Dr. G. Hekstra for his leading role in the CTB project and Dr. G.R.B. ter Meulen-Smidt of RIVM for project coordination.

# Table of Contents

# List of Contributors


Bourg, Prof. Dr. Alain C.M.
National Geological Survey
Water Group
Faculty of Sciences
University of Limoges
123 Avenue Albert-Thomas
F-87060 Limoges Cedex
France

Doelman, Dr. Peter
IWACO B.V.
P.O.Box 8520
NL-3009 AM Rotterdam
The Netherlands

Eijsackers, Dr. Herman
National Institute of Public
Health and Environmental
Protection
Laboratory for Ecotoxicology
P.O.Box 1
NL-3720 BA Bilthoven
The Netherlands

Förstner, Prof. Dr. Ulrich
Technische Universität
Hamburg-Harburg
Section of Environmental
Science & Technology
P.O. Box 90 10 52
D-21050 Hamburg
Germany

Furrer, Dr. Gerhard
ETH Zürich, Institute of
Terrestrial Ecology
Grabenstrasse 3 / 11a
CH-8952 Schlieren
Switzerland

Geiger, Dr. Gabriella
ETH Zürich, Institute of
Terrestrial Ecology
Grabenstrasse 3 / 11a
CH-8952 Schlieren
Switzerland

Gestel, Dr. C.A.M. van
Department of Ecology and
Ecotoxicology
Vakgroep Oecologie en
Oecotoxicologie
Vrije Universiteit Amsterdam
De Boelelaan 1087
NL-1081 HV Amsterdam
The Netherlands

Imeson, Dr. A.C.
University of Amsterdam
Faculteit der Ruimtelijke
Wetenschappen
Nieuwe Prinsegracht 130
NL-1018 VZ Amsterdam
The Netherlands

Kuhnt, Dr. Gerald
Department of Geography
University of Kiel
Ludewig Meyn Strasse 14
D-2300 Kiel 1
Germany

Lacerda, Dr. L.D.
Dept. Geoquímica,
Universidade Federal
Fluminense
Niteroi, 24020-007 RJ
Brazil

Loch, Dr. Dr. J.P. Gustav
Department Geochemistry
University of Utrecht
P.O. Box 80021
NL-3508 TA Utrecht
The Netherlands

Malm, Dr. Olaf
Laboratório de Radioisótopos
Instituto de Biofísica Carlos
Chagas Filho, CCS
Universidade Federal
do Rio de Janeiro
21949-900 Rio de Janeiro
Brazil

Meulen-Smidt, Dr. Gera R.B. ter
RIVM
P.O. Box 1
NL-3720 BA Bilthoven
The Netherlands

Rademaker, Dr. M.C.J.
Department of Ecology
and Ecotoxicology
Vrije Universiteit
De Boelelaan 1087
NL-1081 HV Amsterdam
The Netherlands

Straalen, Prof. Dr. N.M.van
Department of Ecology and
Ecotoxicology
Vrije Universiteit
De Boelelaan 1087
NL-1081 HV Amsterdam
The Netherlands

Salomons, Prof. Dr. Wim
GKSS
Institut für Chemie
Max Planck Strasse
D-21502 Geesthacht
Germany

Schulin, Prof. Dr. Rainer
ETH Zürich, Institute of
Terrestrial Ecology
Grabenstrasse 3
CH-8952 Schlieren
Switzerland

Smal, Dr. Halina
Institute of Soil Science and
Environment Management
University of Agriculture
Leszczynskiego 7
20-069 Lublin
Poland

Stigliani, Prof. Dr. William
Director
Centre for Energy and
Environmental Education CEE
University of Northern Iowa
Cedar Falls, Iowa, 50614-0293
U.S.A.

Wilken, Dr. Rolf-Dieter
GKSS Forschungszentrum
Max Planck Strasse
D-21502 Geesthacht
Germany

# 1 Long-term Strategies for Handling Contaminated Sites and Large-scale Areas

*W. Salomons*

## 1.1 Introduction

In the United States, western and central Europe there are many large-scale polluted sites. These sites are simply too large to be cleaned up with available technologies. They are polluted by mining activities, smelters and other large industries and the pollution extends not only to soils, but also to sediments in water-ways and reservoirs. In addition new large-scale polluted sites are being created under conditions of no or little environmental control. Since these large-scale polluted areas are expected to be present for many years, it is necessary to take a long-term view of the metal mobility and of the capacity of these areas to retain contaminants.

Also waste dumps are abundant and their number will increase in the future. The impact on the environment depends on the source strength of the pollutants at the dump site. To obtain information on the source strength, and in particular how it changes with time, it is necessary to assess the time-dependant processes.

Furthermore, large areas exist with low to moderate levels of contamination with heavy metals, pesticides and nutrients. This is in particular the case for those areas in Europe where intensive agriculture is practiced. A prime example is the Netherlands. In a number of cases impacts of these diffuse pollutants are already noticeable through pollution of groundwater resources. Other examples of low to moderately polluted areas are those affected by areal deposition of pollutants. In all cases the areas affected are large and not economical to clean up with conventional methods.

Common to all of these (at a first glance) non-related pollution cases is the important role of long-term processes. It is likely that both waste dumps and the diffuse polluted soils and sediments will exist for many years to come. Both have to be manipulated in such a way as to cause negligible long-term impacts on the environment.

## 1.2 Time Frames: Soils and Sediments as Part of the Hydrological Cycle

*Hydrological Cycle.* Soils and sediments are part of the hydrological cycle in which particulate material is transported from continents to oceans. Sediments and soils have a high storage capacity for contaminants. In any part of the hydrological cycle far less than 1% of these are actually dissolved in the water, more than 99.9% are stored in the sediments and soils. The dissolved fraction however, is the most mobile and most bio-available. Various interactions determine the actual concentrations of dissolved contaminants and cause changes in concentrations during transport. The pathway of contaminants in the hydrological cycle from soils to oceanic sediments is shown in Fig. 1.1.

The hydrological cycle is represented as three interlocking wheels. The inner one, the atmosphere, is fast running and has a short residence time, the water wheel is slower moving and has residence times ranging from weeks (rivers and estuaries) to years (lakes and coastal waters) and even longer (oceans). The outer wheel is the very slow moving particulates wheel, where the residence times exceed those of the water wheel by orders of magnitude. The ultimate storage of all materials on the continents, be it sediments or soils, is of course the ocean floor, but before the present-day pollutants will end up there and become part of the geological cycle it may take millions of years.

Figure 1.1A shows the situation in western Europe as it existed before about 1970 (no or little pollution control) and as it still exists in eastern/central Europe and increasingly in newly industrializing countries. Except for direct storage on land most contaminants enter the environment either through the air or are

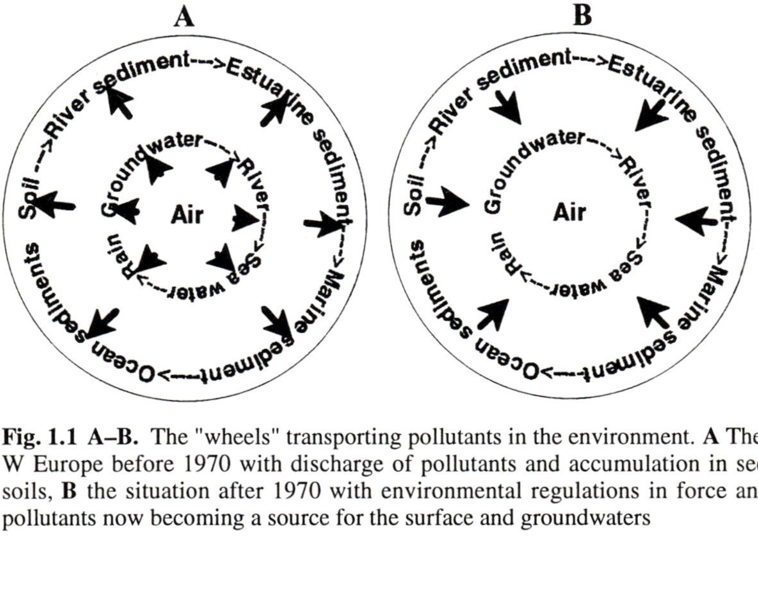


**Fig. 1.1 A–B.** The "wheels" transporting pollutants in the environment. **A** The situation in W Europe before 1970 with discharge of pollutants and accumulation in sediments and soils, **B** the situation after 1970 with environmental regulations in force and the stored pollutants now becoming a source for the surface and groundwaters

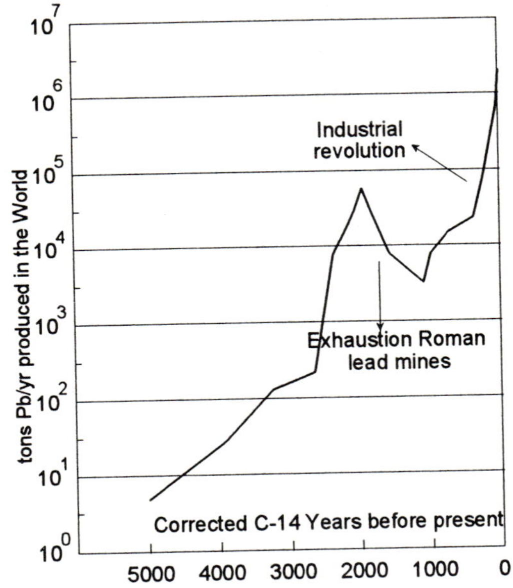


**Fig. 1.2.** Historical development in the use of lead during the last 5500 years. Redrawn from Settle and Patterson (1980)

discharged in the water-ways. For agricultural soils the input from direct application of pesticides and the presence contaminants in fertilizers, sewage sludge or compost have to be taken into account.

**Inputs.** Pollution by metals has a long history since it started with the extraction from ores in the Bronze Age. The estimated input of metals into the atmosphere since 6000 B.P. is shown in Fig. 1.2. The economic megacycles like the industrial revolution and exhaustion of mines are clearly shown by changes in the input into the atmosphere. The curve on production mimics the pattern found in dated cores of lake sediments in Sweden (Renberg et al. 1994), where lead was deposited via atmospheric deposition. The accumulation of organic micro-pollutants, on the other hand, has occurred for the past half century. Also intensive agriculture with its associated high application rate of fertilizers has occurred over the same time frame.

Generally the rate of accumulation of pollutants in western Europe has decreased. Pollution control has a fast response for concentrations in the water and in the air because of the short residence time. The contaminants stored in the sediments and soils will take a much longer clean-up time by natural means. For sediments this is expected to be faster than for soils. Especially in high sedimentation areas, the contaminated sediments will be covered with clean(er) sediments, thus preventing an exposure of the still contaminated sediments to surface waters. Soils are stationary and the stored pollutants are removed through leaching and surface run-off. For the hydrological cycle a new situation exists and the stored contaminants are now released from the reservoirs to the surface- and ground–waters and the atmosphere (Fig. 1.1B).

**Rate of Transfer.** Differences in the rate of transfer from soils to ground- and surface waters exist, caused by differences in properties of the chemicals and the soils. To illustrate these differences we calculated the time needed to remove different chemicals from a 1 m soil column by natural leaching. The soil chosen was a sandy soil due to its low retention properties. The results are shown in Fig. 1.3.

For strong hydrophobic compounds like PCB's and DDT, and some heavy metals, the time needed is on the order of thousands to tens of thousands of years. For chemicals which are less strongly adsorbed like nitrate and potassium the removal time is less than 10 years. This long retention time of hydrophobic chemicals puts severe boundaries on the allowable inputs in the soil from fertilizers and atmospheric deposition to prevent a further increase in the upper part of the soil. As an example consider the EC guidelines for organic micro-pollutants in groundwater. These guidelines are quite strict: 0.1 $\mu g\ l^{-1}$ of individual compounds regardless of their toxicological properties. If these guidelines are used for shallow groundwater, then the amount which can be leached from agricultural soils is less than 0.3 g $ha^{-1}$ of an individual compound. The prediction of these low leaching rates requires an extremely high accuracy of the sorption and transformation parameters of the mathematical models.

Human activities like civil engineering works, drainage of land, changing land use, disturb the equilibria between the toxic metals in sediments/soils and groundwater or surface water and can result in a increase in the rate of supply to the fast moving "water wheel". The change in the rate of supply is not necessarily linear but may exhibit non-linear characteristics. Due to these non-linear changes a

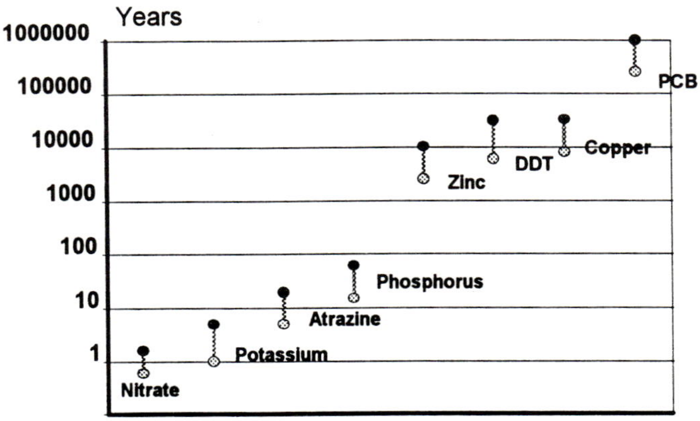


**Fig. 1.3.** Time needed to remove chemicals from 1 m of sandy soil by natural leaching (Salomons 1993)

sudden increase in pollutant levels in the "water wheel" may occur over short time periods in comparison with the time needed to build up the concentrations of the pollutants.

## 1.3 Conceptual Framework for Understanding Long-term Processes in Soils and Sediments

The capacity of soils and sediments to adsorb and retain contaminants depends on their composition. For organic micro-pollutants this is the organic matter since this component has the strongest binding capacity. A strong relationship is observed between the organic matter content of sediments and concentration of organic pollutants (Fig. 1.4). This is one of the reasons for including the organic matter content in Dutch soil standards (Vegter 1994).

For inorganic contaminants like heavy metals, the adsorbing capacity of both the organic and inorganic particles has to be taken into account. The adsorbing capacity is linked with the surface area and surface properties of the particles and hence small particles like clay minerals have a high adsorbing capacity. The clay minerals and other components like quartz and feldspar grains are usually coated with hydrous manganese and iron oxides, and by organic substances. These coatings additionally regulate concentrations in solution (Salomons and Förstner 1984, Hart and Hines 1994)

Several leaching methods have been developed to determine the binding of heavy metals in sediments and soils (Tessier et al. 1979, Förstner 1993).

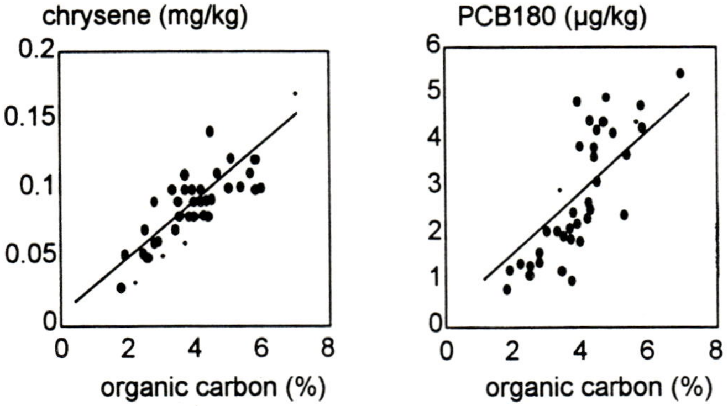


**Fig. 1.4.** Correlation between organic matter content and concentrations of organic micro-pollutants

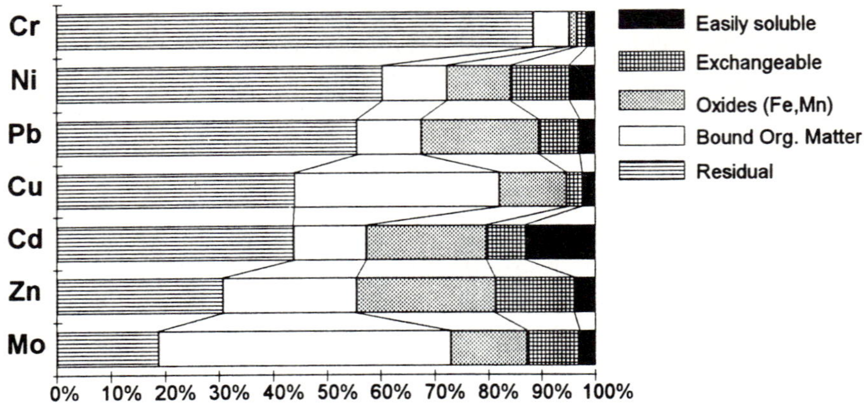


**Fig. 1.5.** Distribution of heavy metals over various phases in the soil. (Kabata-Pendias 1994)

Figure 1.5 shows the distribution of a number of heavy metals the main adsorbing phases in soils. In sediments similar distributions have been found (Salomons and Förstner 1980).

Important for retaining contaminants in soils and sediments are:

- The organic matter content
- The clay mineral content and nature of the clay minerals
- The iron and manganese content
- The carbonate content as a buffer against pH changes

These parameters which control the capacity to retain pollutants are shown schematically in the inner circle of Fig. 1.6. Also shown is the sum parameter "cation ion exchange capacity", which is determined by the surface area and nature of the particles.

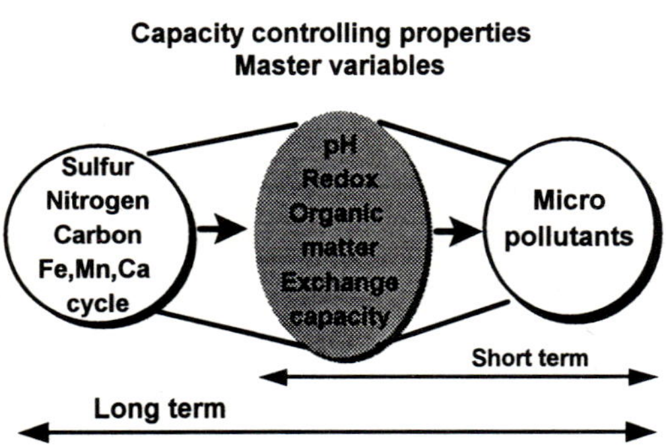


**Fig. 1.6.** The link between the capacity controlling parameters and the major element cycles

The actual mobility of contaminants is additionally determined by the pH, redox conditions and the presence of complexing agents like dissolved organic matter and inorganic anions. These parameters which control retention and mobility of contaminants are called *capacity controlling parameters* or *master variables.*

For short term risk assessment (5–10 years) it is sufficient to understand how the master variables determine the mobility (Fig. 1.6) and hence, leaching and bio-availability. Much information is available in the literature on this subject.

Less information is available on the processes which determine the master variables and the capacity controlling parameters. This is not that important for short term processes which determine the acute pollution status of sediments and soils and their immediate short term impact on the environment. In a number of cases the present impact may be small, however, it can become greater if the capacity of the soil for retention of pollutants changes or when the master variable controlling the interaction between the soil and the soil solution change. These changes are of a long-term nature and are initiated by changes in the major element cycles (Fig. 1.6) in the soil–sediment system. The effect of these changes on concentrations of elements in the soil solution are in particular non-linear for inorganic pollutants (Salomons 1993) and may cause strong increases in concentrations over a short time period. Changing land use, continued acid deposition and changes in hydrology are examples of how these changes can be brought about.

Although the discussion is focused on the chemical behaviour of chemicals in the soil, it is necessary to include other disciplines for a complete understanding of the system and to perform predictive long-term modelling. It is important to assess the importance of increased mobility on transport, plant uptake and impact on the soil ecosystem. Integration of these aspects will provide answers for the target areas like ecotoxicological guidelines, sustainable agriculture, changing land use and long-term protection of groundwater resources.

## 1.4 Long-term Changes: Capacity Controlling Parameters

### 1.4.1 The Organic Matter as a Capacity Controlling Parameter

Organic matter in the soils plays an important role in regulating the behaviour of soil pollutants. It provides a high storage capacity for metals and organic micropollutants and furthermore, it is concentrated at the interface between the soil and the atmosphere as well as between the soil and plants. Thus, it is directly exposed to atmospheric input of pollutants and through agricultural practices.

**Organic Matter Dynamics.** High organic matter content causes a high retention of micro-pollutants. This fact is recognized in the criteria for soil quality as they are applied in the Netherlands (Vegter 1994). Numerous models are available which describe the cycling of organic matter including living and dead material in the short term. Long-term models are hardly available. Nevertheless, for agricultural soil the change in organic matter content in the topsoil can be drastic. Changing arable land to pasture gives a threefold increase in organic matter content over a 100 year period. Whereas conversion of pasture to potato cultivation gives a decrease of organic matter content from 2.6 to 1.2% over a period of 7 years (Scheffer and Schachtschabel 1976). In Fig. 1.7 results of a model are shown describing the decrease of organic matter content after the soil is left barren. A fast decrease of organic matter content takes place over the first 3 years.

These examples show that organic matter is not a constant capacity controlling parameter in soils. Changes in the use of the soil will affect its capacity for retention of chemicals. In addition lower organic matter content will effect the distribution of chemicals over the soil and the soil solution. In contaminated sediments with often high organic matter concentrations, a large part of the organic matter is easily degraded. Reported degradation rates of organic matter vary by orders of magnitude.

Although organic matter is able to retain chemicals in soil and sediments, its breakdown products are able to mobilize heavy metals and organic micro-pollutants in particular (Fig. 1.8).

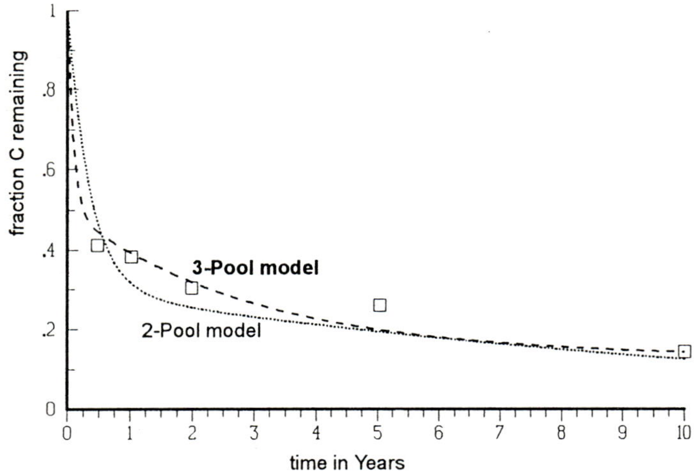


**Fig. 1.7.** Results of a model describing the organic matter in the soil with a 3-pool model. Results show the decrease in organic carbon when a soil becomes barren. Data points taken from Jenkinson and Rayner (1977); Model Bril (unpubl. results)

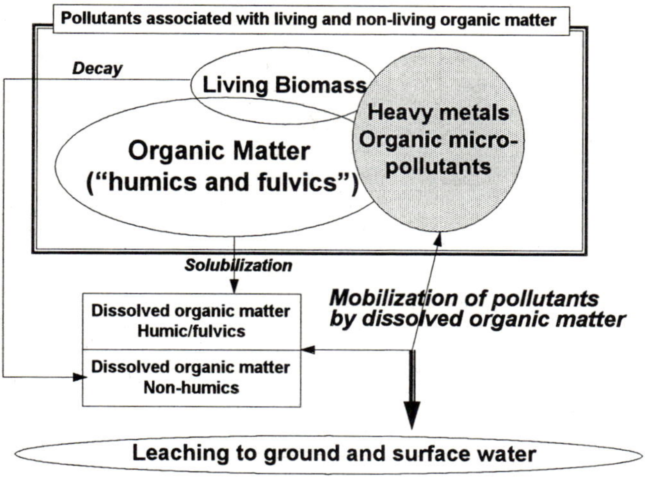


**Fig. 1.8.** The role of living and dead organic matter in the mobilization of contaminants

**Impact on Metals.** The organic matter cycle in soils changes due to acidification, changes in hydrology and changing land use. These changes are reflected in higher dissolved organic matter contents and increased leaching of dissolved organic matter to surface waters.

The changes in the organic matter dynamics in soils in Scandinavian countries has caused a release of mercury through its association with dissolved organic matter (Meili 1991). The mercury thus mobilized is transported to the many lakes which are now in large numbers are banned for fishing. Reservoirs in Canada have also shown elevated mercury levels due to the breakdown of organic matter causing a mobilization of mercury and subsequently increased levels of mercury in fish (Verdon et al. 1991).

Dissolved organic matter not only mobilizes heavy metals but also affects its bio-availability. It can be stated quite generally that the ionic form of the metal is the most toxic. If this toxic form associates itself with either the suspended matter or dissolved ligands the toxicity of the water or groundwater decreases (Kelly 1988).

Usually, the dissolved organic carbon levels in rivers vary in the range 1–10 mg $l^{-1}$, while solid or adsorbed organic carbon vary in the range 0.1–1 mg $l^{-1}$. Dissolved organic carbon is derived from the degradation of plant and animal tissues and has functional groups that can form chelates with metals. Humic and fulvic acids can form quite stable complexes (Tipping and Hurley 1992). Copper, due to its chemical properties, is a metal which is strongly associated with organic matter in surface waters. Figure 1.9 shows the importance of this interaction. With an increase in dissolved organic carbon concentrations,

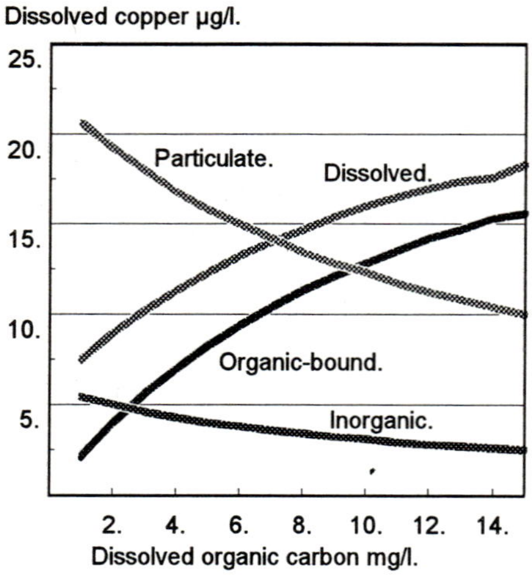


**Fig. 1.9.** The influence of changing concentrations of natural dissolved organic matter on the distribution of copper over the adsorbed and soluble phase. For the soluble phase a distinction is made between copper-organic and copper-inorganic complexes

the total dissolved copper concentration increases at the expense of the copper in the suspended matter (particulate). The increase in dissolved copper is due to the formation of dissolved organic-copper complexes. Although the total dissolved copper concentrations increase, the concentrations of the toxic inorganic copper ion decreases. This figure shows that an increase in total dissolved copper due to complexing agents does not necessarily imply an increase in toxicity.

**Impact on Organic Micro-pollutants.** For the soils, the pesticides are the most important organic micro-pollutants. They will be partly degraded in the soil environment and become biologically inactive. The degradation time (half-life) varies considerably and for substances like y-HCH it may be on the order of 1000 days. Information on the biochemically initiated breakdown has been obtained from laboratory simulations and from direct field evidence. Especially in the study of sediments it has been shown that under anaerobic conditions a biologically breakdown takes place (Beurskens et al. 1993). However, part of the organic micro-pollutants which are not mineralized form very strong associations with the organic matter or become chemically bound to it (bound residue). In those cases it is not possible to extract the pollutants with commonly used extraction liquids. The fate of organic micro-pollutants (pesticides) in the soil environment is shown schematically in Fig. 1.10.

Field studies (Baker et al. 1986, McCarthy and Zachara 1989) have shown the influence of colloidal and dissolved high molecular (humic) organic material on

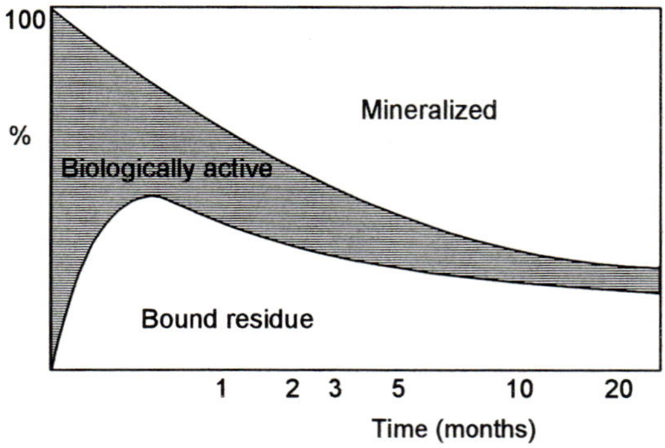


**Fig. 1.10.** Fate of pesticides in the soil environment. (Führ 1984)

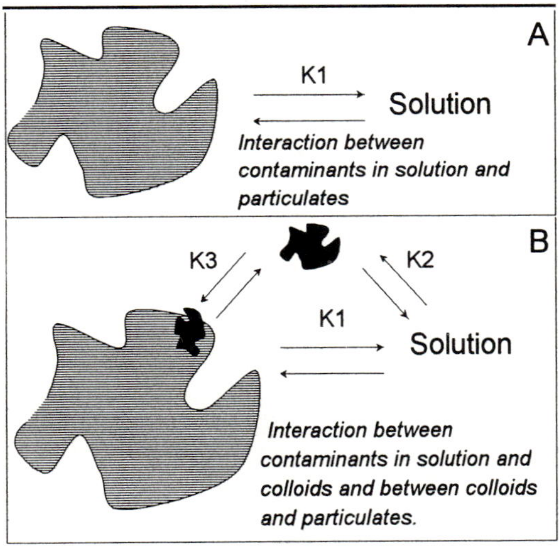


**Fig. 1.11 A–B.** Interaction between organic micro-pollutants in soils as conventionally described (**A**), and the description with the presence of colloids (**B**)

the behaviour of organic chemicals. A three phase model (Fig. 1.11) including not only solid material and water-dissolved contaminants, but also non-filterable micro particles and macromolecular organic matter showed that colloidal-associated contaminants might be the dominant species in equilibrium with the solid phase. In laboratory experiments with isolated non-filterable humic material this association was quantified (McCarthy et al. 1985, Hassett and Milicic 1985, Chiou et al. 1986).

Polycyclic aromatic hydrocarbons are very strongly bound to dissolved organic carbon in soil and pore waters. (Schlautman and Morgan 1993). Although in most cases it has been possible to describe the interactions in a three phase model, there is increasing evidence that the sorption also depends on the nature of the dissolved organic carbon (Chin et al. 1992) and on the major cation composition. The latter influences the morphological characteristics and whether the size of molecules allows them to fit into hydrophobic cavities in humic and fulvic material (Schlautman and Morgan 1993).

Changes in land use affect the organic carbon pool in the soil. Before a new equilibrium pool is established this may result in an increased formation of dissolved organic carbon. On the other hand different equilibrium pools will have different breakdown routes of the organic matter and different levels of dissolved organic carbon. A recent study on forests of different age in the northern part of the Netherlands shows that the shift from agriculture to forest and subsequent increase in the pool of organic carbon also causes a strong increase in dissolved organic carbon levels. As a result, an increase in mobility of hydrophobic organic chemicals is expected. Also heavy metals may increase their mobility due to organic complexing.

**Nutrients.** Although the discussion has been restricted to potentially toxic chemicals it should not be overlooked that nutrients are also associated with the organic matter (Qualls and Haines 1991) and increased breakdown cause an enhanced transport to the groundwater. For instance organic phosphorus may account for more than 50% of the total phosphorus in the soil solution.

### 1.4.2 Changes in Redox Conditions as Capacity and Intensity Controlling Factors

**Redox Sequences.** The oxygen availability (redox conditions) in soils and sediments determines the mobility of both contaminants and of capacity controlling parameters like the presence or absence of iron- and manganese (hydr)oxides. Redox conditions are largely controlled by the bacterial decomposition of organic matter. Shifting groundwater tables or the summer stratification in lakes cause strong changes in redox conditions. Redox changes occur in a number of predefined steps controlled by the biochemistry of the systems. Examples of redox processes – involving the elements C, N, O, S, H, Fe, and Mn – are arranged in Fig. 1.12 (Reddy et al. 1986) in the sequence of reactions observed in an aqueous system at various Eh-values.

Since the reactions considered (with the possible exception of the reduction of $MnO_2$ and FeOOH) are biologically mediated, the chemical reaction sequence is paralleled by an ecological succession of micro-organisms – aerobic heterotrophs, denitrifiers, fermenters, sulphate reducers, and methane bacteria (Stumm and Morgan 1981). The driving force of reduction processes is the decomposition of organic matter by non-photosynthetic organisms, thereby obtaining a source of energy for their metabolic needs. Reduction processes involving oxidation of

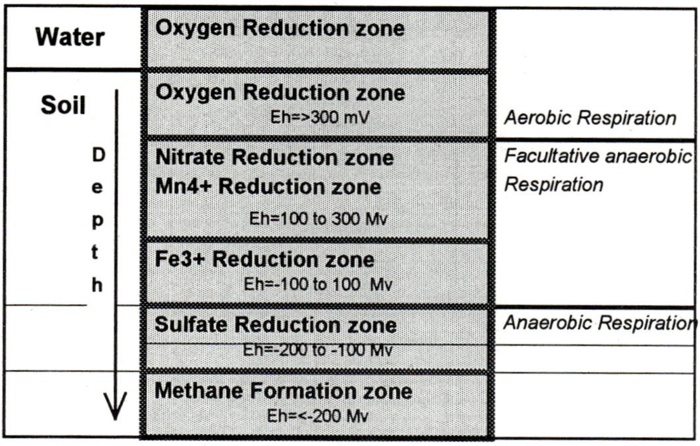


**Fig. 1.12.** Various zones of redox conditions in a waterlogged soil or sediment. (Reddy et al. 1986)

organic matter follow the sequence "aerobic transpiration", "denitrification", "nitrate reduction", "sulphate reduction" and "methane fermentation". In this succession the formation of soluble Mn(II) by reduction of Mn(III, IV)-oxides

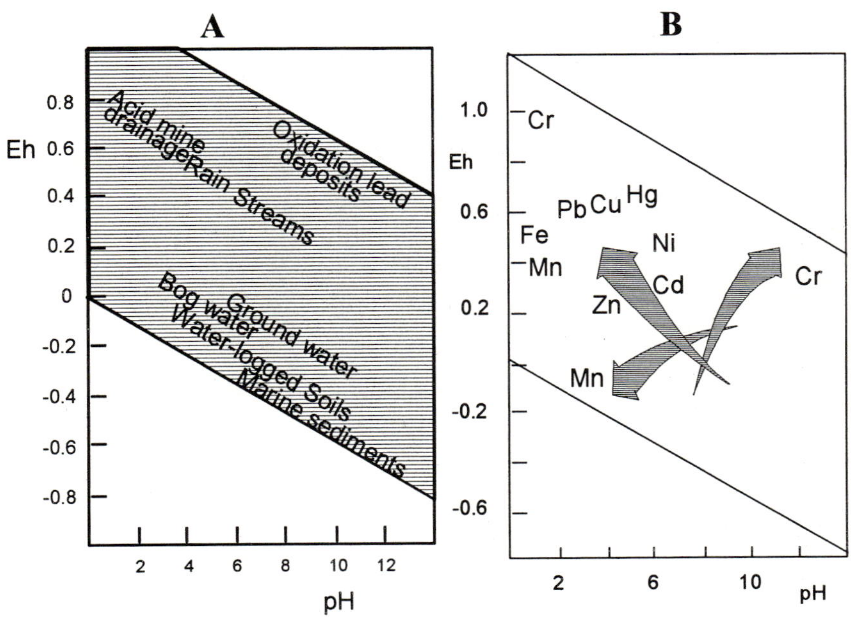


**Fig. 1.13 A–B.** Major environments in terms of their Eh and pH range. Based on Garrels and Christ (1965) and the shifts of increasing solubility with change in Eh and pH. (Bourg 1994)

parallels or follows nitrate reduction, whereas the formation of soluble Fe(II) by reducing Fe(III)-oxides takes place near the conditions for sulphate reduction; the combination of the latter constituents for insoluble iron sulphides is characteristic for more strongly reducing environments.

Apart from the redox conditions in soils and sediments, another important parameter is the pH. Chemical conditions as they represent rivers, soils, sediment etc. in the environment can conveniently be depicted in so-called Eh-pH diagrams (Fig. 1.13A). One of the impacts of mankind on the environment could be referred to as "messing-around" with the Eh-pH diagram. Draining of peat lands, flooding of soils or storage of dredged material on land are examples of man-induced changes in redox and pH conditions. These changes in conditions cause increased or decreased mobility of heavy metals. The changes in Eh and pH which cause increase in solubility are shown as arrows in Fig. 1.13B (see also Chap. 4 by Bourg and Loch).

**Indirect Effects of Redox Changes.** Redox conditions influence the mobility of metals in two different ways. Firstly, there are direct changes in the valences of certain metals. For example, under reducing conditions $Fe^{3+}$ is transformed to $Fe^{2+}$. Similarly, manganese and arsenic are subject to direct changes in valence. Since the reduced ions are more soluble, increased concentrations of these metals have been observed in reducing environments such as groundwaters and pore waters of sediments.

Redox changes also affect metals indirectly. Under reducing conditions (low redox) sulphate reduction will take place (see Fig. 1.14) and metal sulphides become a stable phase. Metal sulphides have an extremely low solubility and concentrations of metals in pore waters of sediments are generally low. The formation of metal sulphides is important for those depositional areas where sulphate is present: estuarine and marine environments. On the other hand, an increase in the redox will cause metal sulphides to become unstable with a subsequent rise in dissolved lead concentrations. This change is not gradual but rather sudden and an example of a non-linear response. This change occurs when polluted (anoxic) sediments are dredged from harbours and dumped either on land or in water-ways. In these oxygenated environments sulphides are not stable and will be oxidized.

Indirect effects also occur when chemicals are associated with components which are subject to redox changes. This is the case for phosphorus and most metals which are (at least partly) associated or adsorbed to iron and manganese hydroxides (Förstner 1986; Fig. 1.15). For long-term predictions it is therefore, not only necessary to consider variables like pH, competing ions, dissolved organic matter etc. which determine the initial adsorption but even more so to consider the stability of the adsorbing phases. Iron in particular is an important capacity controlling parameter. Two stability diagrams for iron are presented in Fig. 1.14 . The first diagram A deals with stability of iron in a freshwater environment; the second one B with a marine environment. The main difference is the presence of sulphate in the marine environment. Sulphate is used by bacteria and converted to

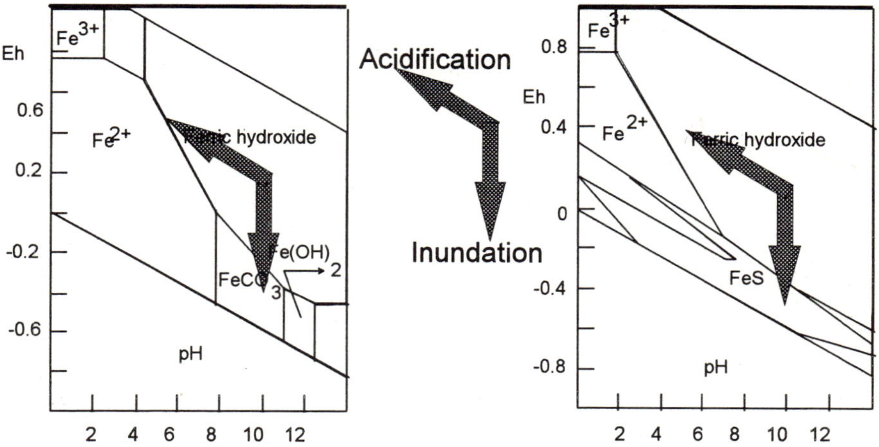


**Fig. 1.14.** The Eh-pH diagrams for iron in a freshwater environment and in a marine environment

sulphide (see Chap. 2 on sequence of bacterial processes in sediments). With the presence of sulphide, the iron hydroxides become unstable and are converted to iron sulphides. This type of compound has a far lower adsorbing capacity and the net result is a loss of adsorbing capacity from the system. Inundation and acidification are examples of impacts which will cause iron hydroxides to become unstable. When this occurs the metals associated with iron and manganese hydroxides are mobilized.

*Redox Changes and Groundwaters.* The sequence of redox changes which in a sediment system comprises a depth ranging from tenths to a few centimetres is also found for infiltrating river water in dune areas in the Netherlands (Stuyfzand 1991). The scale however, is quite different and is on the order of ten to hundred metres. The area where iron hydroxides become unstable shows a significant

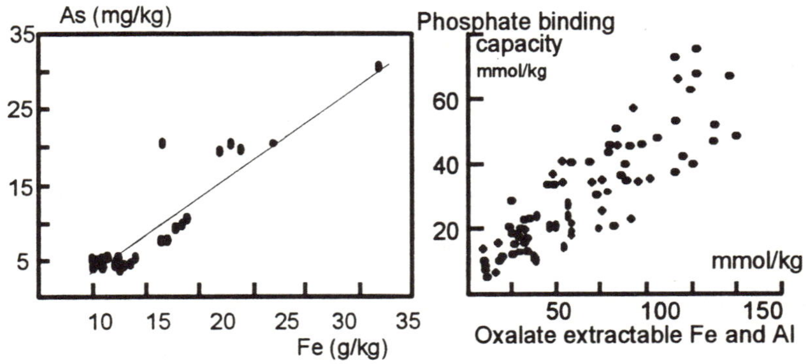


**Fig. 1.15.** The correlation between iron in soils and the arsenic and phosphorus contents

increase in arsenic concentrations (Stuyfzand 1991).

In plumes of landfill leacheates in Denmark similar redox zones have been observed. Parts of these plumes are important sites for the degradation of organic micro-pollutants. The main degradation of organic micro-pollutants takes place in the zone of iron reduction. The process is microbially mediated with the dissolved organic carbon in the leachate acting as a substrate (Lyngkilde and Christensen 1992).

*Redox Changes Following Disposal of Dredged Material and Erosion of Sediments.* The Eh-pH diagrams only refer to metal compounds. However, in a number of cases no compounds are formed, but the metals are absorbed to iron-manganese hydroxides, clay particles and organic matter. It is important to know whether dissolved metal concentrations are determined by precipitation-dissolution reactions (formation of metal compounds) or by adsorption processes. With precipitation-dissolution processes the concentrations will not depend on the total metal content in the sediment/soil, and an increased input in the system will not affect the dissolved concentrations (Fig. 1.16).

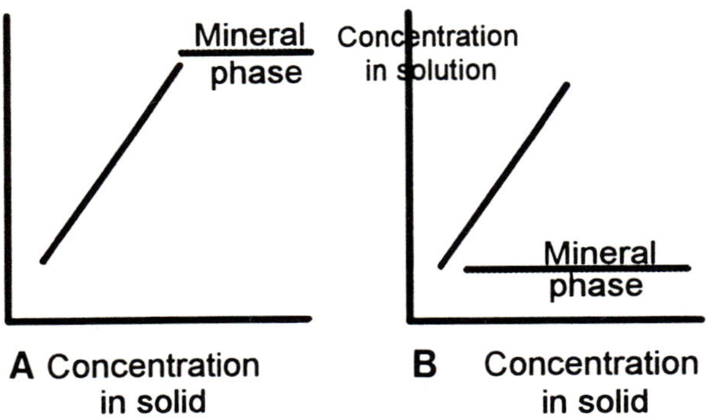


**Fig. 1.16.A–B.** Relationship between dissolved metal concentrations and the concentrations in the sediments. **A** The stable mineral phase has a high solubility compared with removal by adsorption; **B** the stable mineral phase has a low solubility

On the other hand, when adsorption is the main process for binding the metals, the increased input causes an increase in metal concentrations in the pore waters.

Figure 1.16 summarizes two possible situations of heavy metal concentrations in pore waters of sediments and soils in relation to the total concentrations:

- The metal compound has a higher solubility compared with the adsorbed form (Fig. 1.16A)
- The metal compound has a lower solubility compared with the adsorbed form (Fig. 1.16B)

Metal sulphides which occur under reduced conditions in sediments and soils have a low solubility corresponding to the situation in Fig. 1.16B.

Changes in redox conditions also take place when dredged material is removed from harbours and discharged in surface waters. Figure 1.17 shows the results of an experiment in which dredged material was suspended in river- and in sea water for extended periods of time to simulate the dumping of dredged material in a highly energetic environment. In both cases the dredged material contains cadmium sulphide which is unstable in surface water and a remobilization of Cd is expected. However, a drastic difference is observed between a freshwater and a marine environment. In both cases the cadmium sulphide decomposes. In the marine environment only a partial readsorption takes place because the chloride ions form very strong cadmium chloride complexes and thus prevent readsorption. Chloride is hardly present in the river water and a nearly complete readsorption and consequently low mobilization of cadmium takes place.

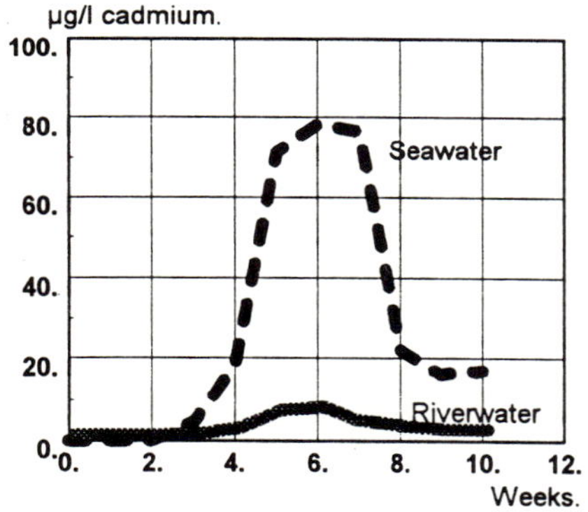


**Fig. 1.17.** Release of cadmium from dredged material suspended in sea water and in river water

This sequence of sulphide formation to adsorption may be cyclic when sediments are transported on the coastal shelf. In the Dutch Wadden Sea the concentrations of heavy metals decrease from west to east. This area is subject to frequent erosion and deposition of contaminated sediments. After each erosional episode the sulphides are unstable and heavy metals are partly released to the surface water and partly readsorbed on the resuspended particles. This cyclic process results in a natural clean-up of contaminated sediments (Fig. 1.18 )

Storage of contaminated dredged spoil on land causes an increase in the mobility of heavy metals in the surface layers of the dump site. Under reducing conditions most metals occur as sulphides with a low solubility. Under oxygenated conditions the sulphides become unstable and the metals become adsorbed to the sediment/soil particles with an increase in overall solubility (see Fig. 1.16B). As a result, when agricultural crops are grown in the disposal site an increased accumulation of metals in crops takes place (van Driel and Nijssen 1988).

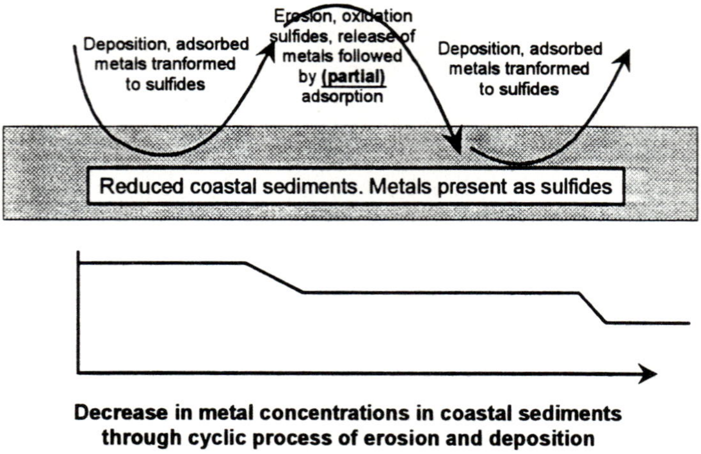


**Fig. 1.18.** Cyclic process of erosion and deposition of coastal sediments resulting in a decrease of metal levels in the sediments

### 1.4.3 Changes in pH as Capacity and Intensity Controlling Factors

Parameters which are relevant for adsorption in natural environments include the pH (e.g. algal blooms), ionic strength (e.g. estuaries) and natural and man-made complexing agents. Natural complexing agents are organic matter and inorganic anions. Examples of man-made complexing agents are substances like EDTA, NTA and DTPA. An overriding factor however, is the pH.

As shown in Fig. 1.19 the adsorption of metals is highly dependent on the pH

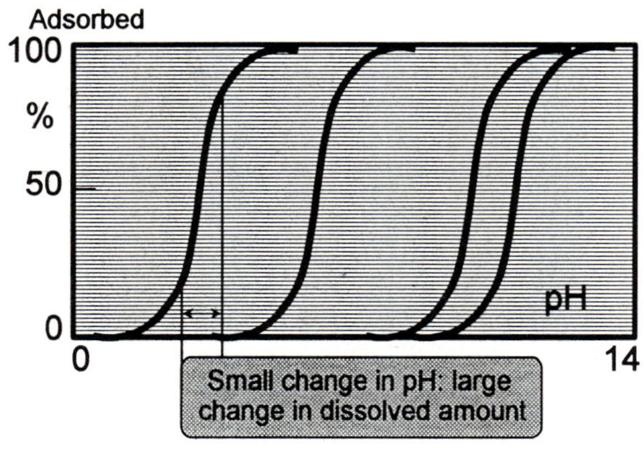


**Fig. 1.19.** The adsorption of metals on solid surfaces. Metals show different adsorption edges at different pH values

and non-linear. It increases from near nil to near 100% as pH increases through a critical range 1-2 units wide (Salomons and Förstner 1984).

This means that a small shift in pH in causes a sharp increase or decrease in dissolved metal levels. The pH depends to a large extent on the buffering capacity of the system. If the buffering capacity is slowly consumed as may happen through continuous input of acid from acid rain or acid mine drainage a certain threshold will be exceeded and metals will rather suddenly desorb from the sediments or the soils. The onset of the adsorption or the desorption is characteristic for each metal.

This non-linear response has also ecotoxicological significance. One example has been observed in the river Rhine and its distributary, the river IJssel (Salomons 1989). Travel time from the river Rhine to the measuring point in the river IJssel is only a few days. However, over this time period the pH increases about 0.5 units. As a result the distribution of cadmium over the suspended matter and the solution changes due to increased adsorption. The decrease in dissolved cadmium causes a decrease in uptake by the bivalve *Dreissena polymorpha* (Fig. 1.20). In this case it is a positive benefit (detoxifying), however, in a river or lake subject to acidic inputs the reverse would occur.

### 1.4.4 Redox and pH Changes

Redox conversions occur at many places in the hydrological cycle. In a number of cases this goes hand in hand with strong changes in the pH also. This is the case when pyrite is present. Pyrite is an important constituent of near shore sediments, dredged material, wetland and mining wastes. Negative impacts of both

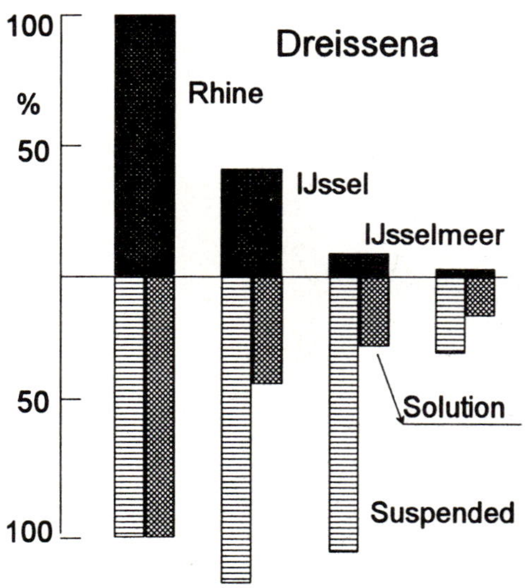


**Fig. 1.20.** Relative concentrations of cadmium in *Dreissena Polymorpha* in the river Rhine, in its distributary the river IJssel and in the IJsselmeer (a lake which is fed by the IJssel). The concentrations of cadmium in solution and in the suspended matter are shown below the X-axis. Concentrations in the river Rhine are given as 100%

pH and redox changes involving pyrite is well-documented for mining wastes (Axtmann and Luoma 1991) and for acid-sulphate soils (Dent 1986). In particular for metal mining, the oxidation of pyrite, cause environmental problems with regard to acid-mine drainage. The rate determining step in acid-mine drainage is the production of ferric sulphate, which is a powerful oxidant (Singer and Stumm 1970). At high pH pyrite is oxidized chemically, but at pH values between 3.5 and 4.5 the ferrous-ferric oxidation is accelerated by a factor of 200 by the bacterium Metallogenium. A further drop in pH (2–3.5) causes an increase in the iron oxidation by $10^5$ to $10^6$ by Thiobacillus ferroxidant.

The metals in mine waste are present as sulphides. In the mine waste dumps therefore, the remobilization of metals is the result of two processes:

- The oxidation of the metal sulphides
- The low pH caused by the oxidation of pyrite

The earlier case, where oxidation of metal sulphides (dredged material) also causes a release, is a minor problem compared with mine waste. Now the lowering of the pH and the much lower adsorbing capacity of the mine waste compared with sediments causes a much more significant increase in dissolved metal levels.

Oxidation of pyrite takes place almost immediately after the waste dumps have access to oxygen. Some results on the time frame involved in acidification of

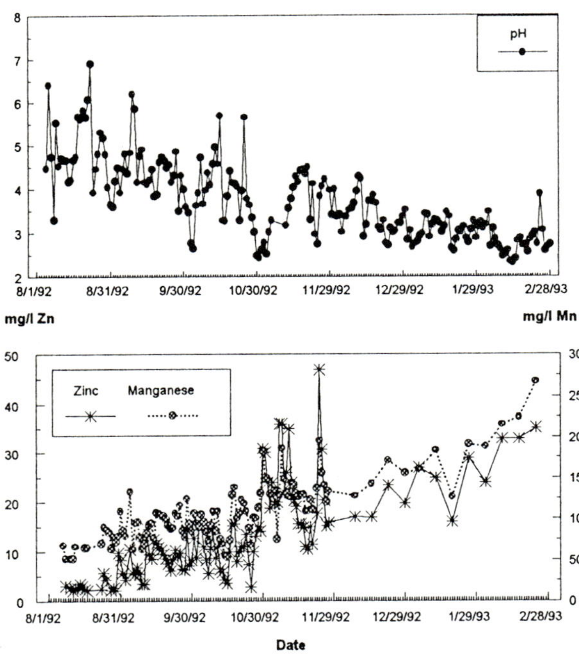


**Fig. 1.21.** Change in the pH and resulting concentrations in manganese of a stream draining a waste rock dump (Indonesia)

waste dumps and resulting drop in pH are shown in Fig. 1.21. These results refer to a waste dump of a gold mine in Indonesia, where the dump contained several percent of pyrite. Over a period of less than 1 year the waste dump began to generate acid with an increase in metal levels. In this particular case the waste rock contains abundant amounts of manganese and zinc.

In some cases, when the dredged material contains little or no carbonate and contains significant amounts of pyrite, a similar situation may arise (Fig. 1.22). During the first stage of metal sulphide oxidation the dissolved concentration increase because the insoluble metal sulphides are transformed to the adsorbed phase. The continuing oxidation causes a destruction of the buffering capacity by calcium carbonate and once this is completed, the drop in pH causes a further increase in dissolved concentrations.

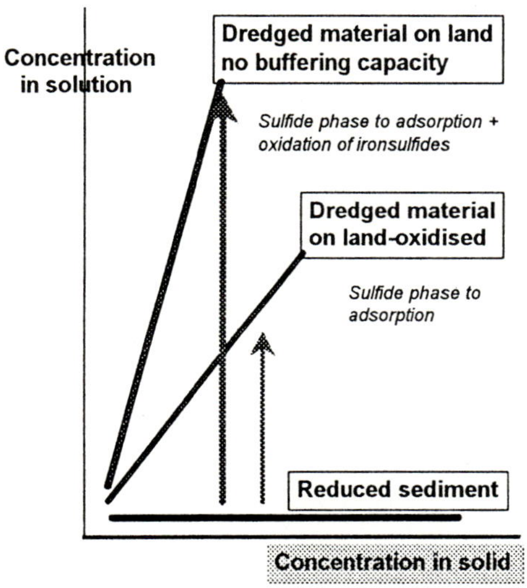


**Fig. 1.22.** The increase in concentrations when metal sulphides become unstable in an oxygen rich environment

Oxidation of pyrites also plays an important role in the negative environmental impact caused by draining of wetlands (or acid-sulphate soils). It is well documented that wetlands can serve to protect adjacent water bodies from inputs of sulphuric and nitric acids (acid rain). As these acids pass through the wetlands, sulphuric acid is reduced to insoluble sulphides and nitric acid is reduced to molecular nitrogen via the denitrification reaction. When wetlands dry out, the system may switch from one dominated by anaerobic chemical reactions to one

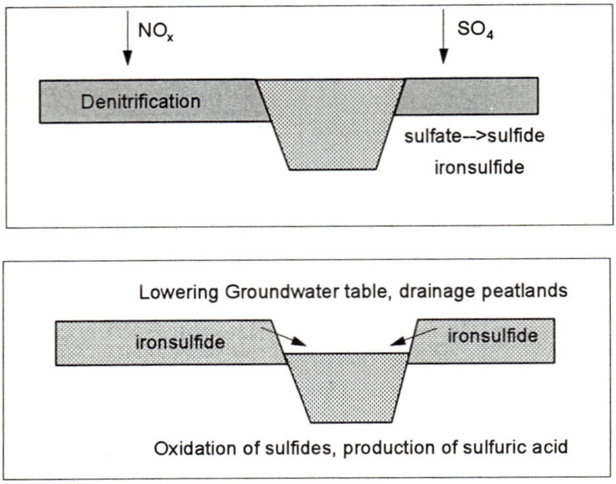


**Fig. 1.23.** Changes when wetlands are being drained

controlled by aerobic reactions (Fig. 1.23). Interestingly, Lake Blamissusjon in northern Sweden, known as "the most acidic lake in Sweden", was not acidified by acid deposition, but rather, as a result of drainage (for agricultural purposes) of the watershed soils surrounding the lake. Acidification occurred because the water-logged soils were rich in pyrite, a compound containing sulphur in the reduced chemical form ($S^{2-}$). After drainage, oxidizing conditions prevailed, causing oxidation of the sulphur to sulphuric acid, which subsequently washed into the lake and acidified it.

## 1.5 Case Studies

### 1.5.1 Acid Rain and Metal Uptake

The soil is subject to atmospheric deposition. Although the quantities may be small it is one of the contributors to the continued accumulation of chemicals in the soil. For heavy metals and for a number of organic chemicals the release to the groundwater and removal by crops is rather low. Therefore, even if the inputs decrease, the accumulation (although at a lower rate) will continue.

Calculation performed for cadmium accumulation in the agricultural soil in western Europe shows a continuous increase with some levelling off for later years (Stigliani et al. 1993). Soils, at least in western Europe, Canada and the United States, are also subject to acid inputs from the atmosphere. In interpreting these kind of predictions on metal accumulations and their impact on the environment, one should also consider the effects of acidification. Metal concentrations in the soil solution depend to a large extent on the pH (see also Fig. 1.24). The increase in dissolved cadmium and the change in speciation due to pH changes will both cause a greater bio-availability of cadmium. The two effects of increased cadmium in the soil and greater uptake due to pH change by crops can be combined to determine their potential impact on human intake (Stigliani and Salomons 1993). The results shown in Fig. 1.24 show that cadmium concentrations which at a certain time cause human uptake below the WHO limits will exceed the WHO limits when the acidity of the soil increases. These theoretical calculations show that it is necessary to take the whole system into account for predicting environmental impact. In this particular case however, it is known that farmers apply lime to raise the soil-pH for maintaining crop productivity (Mahler and McDole 1987). Thus the situation of increased uptake may not occur for human consumption in well limed soils.

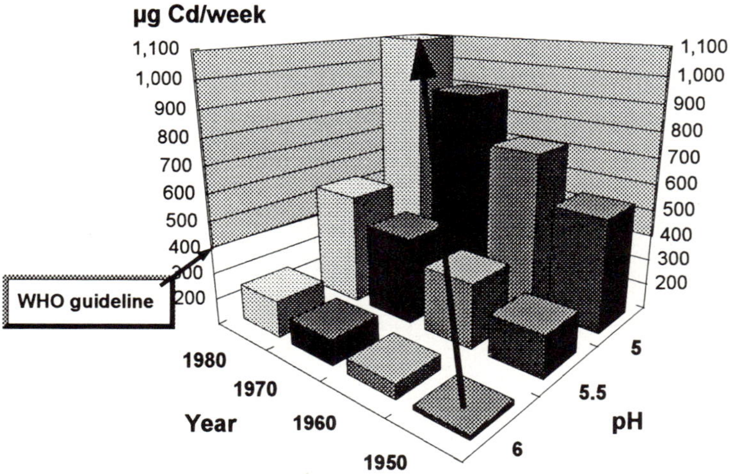


**Fig. 1.24.** Estimated cadmium uptake by humans as a function of pH for the years 1950-1990. (Stigliani and Salomons 1993)

In eastern and central Europe acidification of the soil is continuing as well as atmospheric deposition of heavy metals. The decreases in base cation deposition (Gorham 1994, Hedin et al. 1994), partly due to removal of particles from stacks, will reduce the buffering capacity of the atmospheric particles against $SO_x$ and $NO_x$. Economic constraints may in eastern and central Europe reduce regular liming of soils by farmers and cause an increase in soil acidification and metal uptake (Smal and Salomons 1993).

### 1.5.2 Changing Land Use

The situation of increased metal accumulation and loading of agricultural soils combined with acidification is highly relevant for the present practice of turning agricultural land into forest. Present agricultural soils have higher concentrations of heavy metals compared with forest soils (Fig. 1.25). These high concentrations are caused by agricultural practices such as the application of fertilizer or sewage. Despite the higher soil concentrations the concentrations in the soil solution are much lower.

On the other hand the forest soils have the highest concentrations of cadmium and zinc in the soil solution. These differences are simple due to differences in pH between the agricultural (limed) and the more acidic forest soils. When agricultural land is converted to forest we may expect even higher concentrations in the soil solution then in the present forest soils.

The mobility of contaminants in the new forest soil will also depend strongly on the organic matter content. In soils with an initial low organic matter content high concentrations of dissolved metals are expected to develop. This situation will occur when the acidification proceeds faster than the development of a new equilibrium pool of organic matter in the soil. This new pool will be higher for the forest compared with the former agricultural soil.

The litter of natural forest is known for its high accumulation of metals and accumulation in detritus feeders. No information is available on metal cycling in artificial forests where the soil contains increased levels of metals. The expected

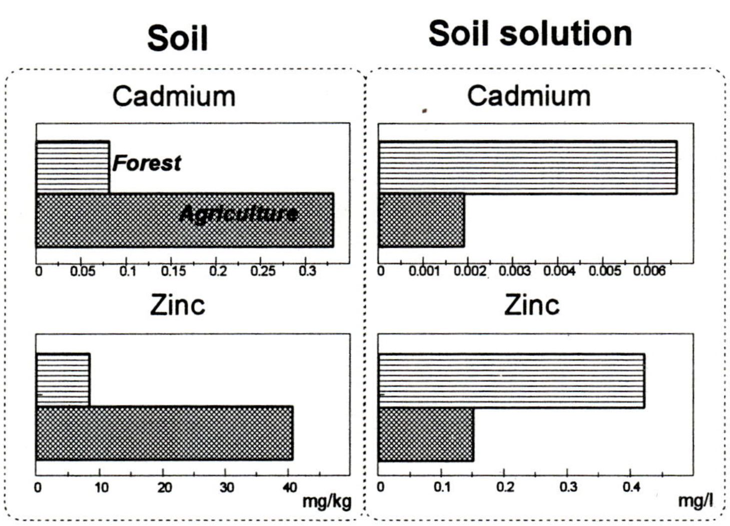


**Fig. 1.25.** The mean concentrations in the soil and in the soil solution of agricultural and forest soils in the Netherlands. (Römkens, unpubl. data)

higher concentrations in the most mobile and bio-available phase (soil solution) suggest increased accumulations.

Insufficient field information is available to speculate on the fate of accumulated organic chemicals in forest soils derived from agricultural soils. Their fate depends on a number of counteracting processes: transformation to the bound residue, microbial degradation and increased mobility due to association with increasing levels of dissolved organic carbon.

To illustrate these changes with actual data, some preliminary results of a survey of forest soils of different ages are shown in Fig. 1.26. A change in pH is observed over a period of 70 years for forest in the Netherlands derived from former arable soil. The first 20 years exhibits a large drop in pH of about 1.5 units. The dissolved organic carbon levels in the soil solution increased five times over a period of 70 years. The increased dissolved organic carbon levels together with the decrease in pH will increase the mobility of heavy metals and organic micropollutants (see also Fig. 1.27).

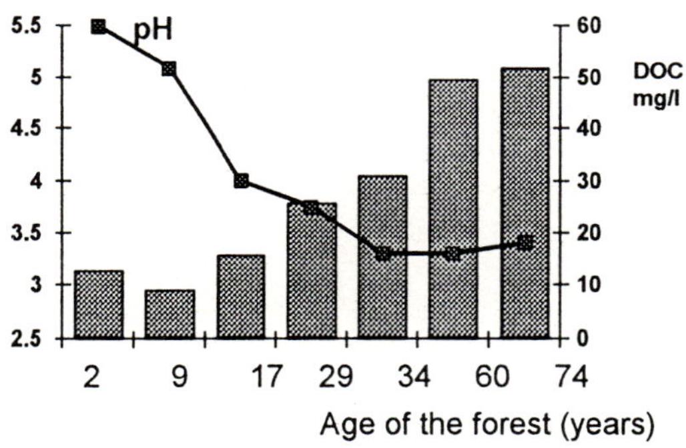


**Fig. 1.26.** The decrease in pH and increase in dissolved organic carbon for forests of different ages in the Netherlands

The relationship between pH and dissolved metal concentrations in a set of soils from western Europe is given as an example in Fig. 1.27. These soils had different degrees of metal contamination, however, Fig. 1.27A shows that pH is an overriding factor in determining the concentrations in solution. The non-linear response of the metals towards pH changes is also quite clear from these data. If buffering capacity in the form of calcium carbonate is present (e.g. liming by farmers) the pH will remain above about 6 and metals will have low concentrations in the pore waters of the soil. Once the buffering capacity is gone (in itself a non-linear phenomenon) the pH decreases and the metal concentrations in the soil solution increase rapidly.

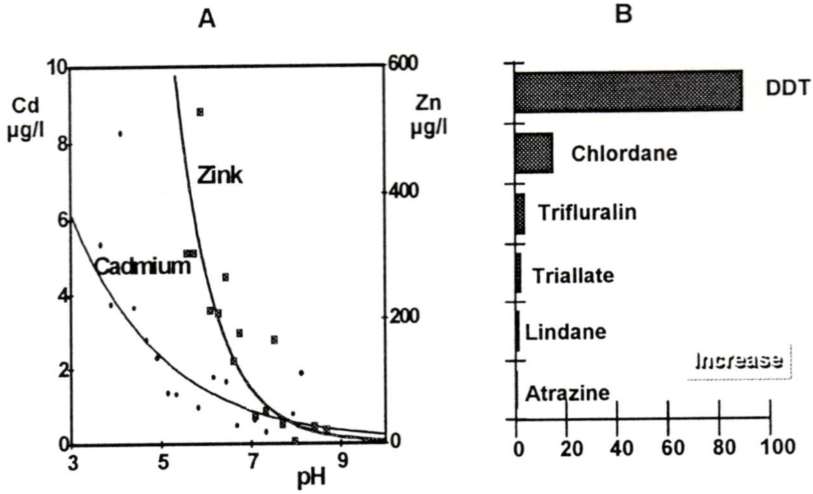


**Fig. 1.27.** **A** Relationship between pH and dissolved metal concentrations in the soil solution. **B** Increase of mobility of organic micro-pollutants due to the presence of dissolved organic carbon in the soil solution

## 1.6 Summary and Conclusions

The various issues discussed in the previous sections show that there is a need for long-term measures to deal with the accumulation of pollutants in soils and sediments. Long-term changes in mobility and the possible occurrence of non-linear responses are caused by changes in the master variables or capacity controlling properties of sediments and soils. A summary of the various capacity controlling parameters is given in Table 1.1.

Research into the behaviour of pollutants up to now was mainly confined to the effects of the master variables and capacity controlling parameters on mobility and bio-availability.

As a result most regulators and scientists have assumed in their decisions and research that these properties are constant and that a sediment or a soil is a static system. Of course this is true if we consider the short term, of about 1 to 5 years.

In the long term these capacity controlling parameters are not constant and are effected by changes in land use and changes in the hydrology. These changes effect cycles of elements like nitrogen, carbon and sulphur and hence, the capacity controlling parameters. Not only are these capacity controlling properties not constant in the dynamic soil and sediment system, but also the changes often exhibit a non-linear behaviour.

Table 1.1. Interlinkage between capacity controlling parameters (CCP) and major biogeochemical cycles

| CCP | Important parameters | Biogeochemical cycle |
|---|---|---|
| Cation exchange capacity | Clay minerals, organic matter | Carbon cycle |
| pH | Carbonate content | Calcium cycle |
| Redox potential | Iron, manganese oxides, sulphides | Iron, manganese and sulphur cycle |
| Organic matter content | Organic matter | Carbon cycle |
| Soil structure and texture | Organic matter and inorganic minerals | Carbon cycle |
| Salinity (complexing capacity of the soil solution) | | Carbon cycle (DOC) |

If one wants to control for instance diffuse pollution and prevent non-linear responses it is necessary to assess and manipulate those factors which determine the capacity controlling parameters and master variables. This implies that it is nec essary to study in detail the major biogeochemical cycles of sulphur, nitrogen, carbon and also of calcium, iron and manganese.

In fact, there is a strong contrast between geochemical engineering of large contaminated areas and engineering of hot spots of contamination. In hot spot clean-up the capacity controlling parameters are changed in such a way that they

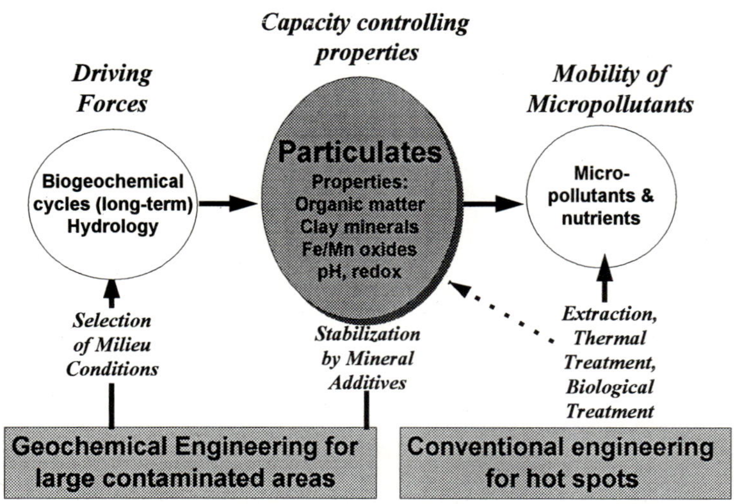


**Fig. 1.28.** The interlinkage between the major biogeochemical cycles, master variables and capacity controlling parameters (CCP) the mobility of micro-pollutants and the possibilities for manipulating (geochemical engineering) these interlinked processes and cycles

possibilities for manipulating (geochemical engineering) these interlinked processes and cycles
release contaminants for subsequent treatment. In geochemical engineering the reverse is attempted: influencing the capacity controlling parameters in such a way that they retain the contaminants, making them less mobile and decrease the bio-availability.

The various examples given show that a mono-disciplinary approach will not lead to a better understanding of the long-term behaviour of chemicals in the environment. On one side we have the scientists who know "nearly everything" about the intricate interactions of chemicals with organic matter, adsorbing surfaces in the soils etc. On the other side we have the geo- and biogeochemists who study the cycles of the major elements like carbon, nitrogen and sulphur. Element cycles which determine the capacity controlling parameters. It is necessary to bring these disciplines together with their own pieces of the puzzle to solve potential problems with continuous diffuse build-up of pollutants in soils and sediments and changes in major element cycles.

Figure 1.28 illustrates the various ways in which it is possible to manipulate the processes (more information is contributed by Förstner in Chap.11). The capacity controlling parameters can be changed by additives. In the case of pH changes this may be carbonates or slowly weathering silicates. To increase the adsorption capacity for phosphorus the addition of iron chloride to promote the precipitation of hydroxides has been suggested and tested. The capacity controlling parameters are also modified with changes in the biogeochemical cycles, changes which can be induced by changes in the hydrology or by changes in vegetation cover. Geochemical or ecological engineering of lake systems has been practiced for quite some time, however, for soils and their ecosystems such practices are just beginning.

**Acknowledgements.** We wish to thank the Dutch Integrated Soil Research Program for their support of this and related studies.

## References

Axtmann EV, Luoma SN (1991) Large scale distribution of metal contamination in fine-grained sediments of the Clark Fork river, Montana, USA. Appl Geochem 6:75-88

Baker JE, Capel PD, Eisenreich SJ (1986) Influence of colloids on sediment water partitioncoefficients of polychlorobiphenyl congeners in natural-waters. Env Sci Tech 20(11):1136-1143

Beurskens JEM, Mol GAJ, Barreveld HL, Munster B van, Winkels HJ (1993) Environ Toxicol Chem 12(9):1549-66

Bourg ACM (1994) Speciation of heavy metals in soils and groundwater and implications for their natural and provoked mobility. In: Salomons W, Mader P, Förstner U (eds) Pathways, impact and engineering aspects of metal polluted sites. Springer, Berlin Heidelberg New York

Chin YP, Gschwend PM (1992) Partitioning of polycyclic aromatic hydrocarbons to marine porewater organic colloids. Environ Sci Technol 26(8):1621-1626

Chiou CT, Malcolm RL, Brinton TI, Kile DE (1986) Water solubility enhancement of some organic pollutants and pesticides by dissolved humic and fulvic acids. Environ Sci Technol 20:502-508

Dent D (1986) Acid sulphate soils: a baseline for research and development. ILRI publ no 39 Wageningen, the Netherlands

Driel W Van, Nijssen JPJ (1988) Development of dredged material disposal sites: implications for soil, flora and food quality. In: Salomons W, Förstner U (eds) Chemistry and biology of soild waste: dredged material and mine tailings. Springer, Berlin Heidelberg New York

Förstner U (1986) Changes in metal mobilities in quatic and terrestrial cycles. In: Patterson JW (ed) Speciation, separation and recovery of metals. Lewis Publ Chelsea, Michigan, pp 3-26

Förstner U (1993) Metal speciation – general concepts and applications. Intern J Environ Analy Chem 51:5-23

Führ F (1984) Praxisnahe Tracerversuche zum Verbleib von Pflanzenschutzwirkstoffen in Agrarökosystemen. Rheinisch-Westfälische Akademie der Wissenschaften, Vorträge N326:5-35

Garrels RM, Christ CL (1965) Minerals and equilibria. Harper and Row, New York

Glässer W, Klapper H (1992) Stoffumsätze beim Füllprozeß von Tagebaurestseen. In: Boden, Wasser und Luft. Umweltvorsorge in der AGF, Bonn, W-Germany,

Gorham E (1994) Neutralizing acid rain. Nature 367:321

Hart BT, Hines T (1994) Trace elements in rivers. In: Salbu B, Steinnes E (eds) Trace elements in natural waters. CRC Press, Boca Raton (In press)

Hassett JP, Milicic E (1985) Determination of equilibrium and rate constants for binding of a polychlotinated biphenyl congener by dissolved humic substances. Env Sci Tech 19:638-643

Hedin LO, Geanat L, Likens GE, Buishand TA, Galloway JN, Butler TJ, Rohde H (1994) Steep decline in atmospheric base cations in regions of Europe and North America. Nature 367:351-354

Jenkinson DS, Rayner JH (1977) The turnover of soil organic matter in some of the Rothamsted classical experiments. Soil Science 123:298-303

Johnston AE, Goulding KWT , Poulton PR (1986) Soil acidification during more than 100 years under permanent grassland and woodland at Rothamsted. Soil Use and Management 2:3-10

Kabata-Pendias A (1994) Agricultural problems related to excessive trace metal contents of soils. In: Salomons W, Mader P , Förstner U (eds): Pathways, impact and engineering aspects of metal polluted sites. Springer, Berlin Heidelberg New York

Kelly M (1988) Mining and the freshwater environment. Elsevier, Amersterdam, 231 pp

Leistra M, Boesten JJTI (1989) Pesticide contamination of groundwater in Western Europe. Agric Ecosyst Env 26:369-389

Lyngkilde J, Christensen TH (1992) Fate of organic contaminants in redox zones of a landfill leachate pollute plume (Vejen, Denmark). J Contam Hydrol 10:291-307

Mahler RL , McDole RE (1987) Effect of pH on crop yield in northern Idaho. Soil Sci Soc Am J 79:751-755

McCarthy JF, Zachara JM (1989) Subsurface transport of contaminants. Env Sci Technol 23(5):496-502

McCarthy JF, Jimenez D, Barbee T (1985) Effect of dissolved humic material on accumulation of polycyclic aromatic hydrocarbons; structure activity relationships. Aquatic Toxicology 7(1-2):15-24

Meili M (1991) The coupling of mercury and organic matter in the biogeochemical cycle – towards a mechanistic model for the boreal forest zone. Water, Air and Soil Pollution 56:333-347

Qualls RG, Haines BL (1991) Geochemistry of dissolved organic nutrients in water percolating through a forest ecosystem. Soil Sci Soc Am J 55:1112-1123

Reddy KR, Feijtel TC, Patrick WH Jr (1986) Effect of soil redox conditions on microbial oxidation of organic matter. In: Chen Y, Evnimelech Y (eds) The role of organic matter in modern agriculture. Martinus Nijhoff, The Hague

Renberg I, Persson MO, Emteryd (1994) Pre-industrial atmospheric lead contamination detected in Swedish lake sediments. Nature 368(6469):323-326

Salomons W (1993) Non-linear and delayed responses of toxic chemicals in the environment. In: Arendt F, Annokkée GJ, Bosman R, van den Brink WJ (eds) Contaminated soil '93. Kluwer, Dordrecht, pp 225-238

Salomons W (1989) Fate and behavior of trace metals in a shallow eutrophic lake. In: Boudou A, Ribeyre F (eds) Aquatic ecotoxicology fundamental concepts and methodologies. CRC Press, Boca Raton, pp 185-200

Salomons W, Förstner U (1980) Trace metal analysis on polluted sediments. Part II. Evaluation of environmental impact. Environ Technol Lett 1:506-517

Salomons W, Förstner U (1984) Metals in the hydrocycle. Springer, Berlin Heidelberg New York

Salomons W, Stol BEM (1993) Diffuse pollutant sources in agriculture (in Dutch). Report IB-DLO

Scheffer P, Schachtschabel P (1976) Lehrbuch der Bodenkunde. Ferdinand Enke Verlag, Stuttgart, 394 pp

Schloutman MA, Morgan JJ (1993) Binding of a fluorescent hydrophobic organic probe by dissolved humic substances and organically-coated aluminum oxide surfaces. Env Sci Technol 12:2523-2532

Settle DM, Patterson CC (1980) Lead in Albacore: guide to lead pollution in Americans. Science 207:1167-1176

Singer PC, Stumm W (1970) Acidic mine drainage: the rate determination step. Science 167:1121-1123, 1969

Smal H, Salomons W (1993) Heavy metals and acidification in Poland. Chemical time bombs. In: Meulen GRB ter, Stigliani WM, Salomons W, Bridges EM, Imeson AC (eds) Proceedings of the state of the art conference on delayed effects of chemicals in soil and sediments

Stigliani WM, Jaffe PR, Anderberg S (1993) Heavy metal pollution in the Rhine Basin. Env Sci Technol 27:786-793

Stigliani W, Salomons W (1993) Our fathers' toxic sins. New Scientist 140:38-42

Stumm W, Morgan J (1981) Aquatic chemistry. Wiley, New York

Stuyfzand J (1991) Sporenelementen in duinwater en kunstmatig geinflitreerd opperlvaktewater in de kuststreek. In: Beek GEM van, Stuyfzand PJ (eds) Sporenelementen in grondwater. Kiwa Mededeling nummer 118

Tessier A, Campbell PGC, Bisson M (1979) Sequential extraction procedure for the speciation of particulate metals. Anal Chem 51:844-851

Tipping E, Hurley MA (1992) A unifying model of cation binding by humic substances. Geochim Cosmochim Acta 56:3627-3641

Vegter J (1994) Soil protection in the Netherlands. In: Salomons W, Mader P, Förstner U (eds) Heavy metals: problems and solutions. Springer, Berlin Heidelberg New York

Verdon R, Brouard D, Lalumiere R, Laperle M, Schetagne R (1991) Mercury evolution (1978-1988) in fishes of the La Grande hydroelectric complex, Quebec, Canada. Water, Air and Soil Pollution 56:405-417

# 2 Microbiology of Soil and Sediments

*P. Doelman*

## 2.1 Introduction

Recycling of elements is the function of micro-organisms. This has been possible because of their ubiquity, their abundance, their shape and small size and therefore their extremely high surface area to volume ratio and their versatility.

The capacity of the planet Earth to produce organic matter is due to photosynthetic activity of green plants and algae. Annually approximately 13 x $10^{10}$ tons of carbon, as $CO_2$, are reduced to natural organic matter. This primary production process is counteracted by the capacity of microbial degradation processes in which organic material is recycled to inorganic material. More than 90% of this biological recycling is performed by $O_2$ producing or consuming organisms (Zehnder 1982).

The perfection of recycling, in which for any organic component a suitable biochemical conversion pathway exists, has evolved during three and a half billion years of evolution of microbial life. This phenomenon has been called "the principle of biological infallibility". This biotic potential can also be called the microbial carrying capacity and it is the storehouse or the matrix for the transformation and mineralization of unnatural or xenobiotic compounds. Xenobiotic compounds have entered the environment in large amounts, particularly in the last 50 years. The quantitatively most important contaminants are mineral oil, aromatic hydrocarbons as benzene, toluene, ethylbenzene and xylene, polyaromatic hydrocarbons, chlorinated organic compounds as hexachlorocyclohexane, hexachlorobenzene, polychlorinated biphenyls, aldrin, dieldrin, solvents as trichloroethylene, tetrachloroethylene and dichloromethane, organic waste, cyanide and heavy metals. Soil and sediment are the contaminants' ultimate sinks. Once "trapped" they are subject to various kinds of dynamic processes such as microbial transformations.

A risk assessment due to microbial transformations of pollutants stored in soil and sediment is the ultimate aim of this contribution. Therefore, we focus on the microbial environment, as the capacity controlling unit, and briefly on classical principles of degradation of organic matter in soil and sediment. It will be emphasized that environmental conditions with regard to the presence of oxygen or sulphate determine the microbial degradation pathways and rates. Particular examples will be given to show how alterations of environmental conditions can hamper as well as stimulate degradation rates. Furthermore, the interaction between heavy metals and micro-organisms in soil and sediment will be discussed with special attention to microbial methylation of heavy metals, since this preferably occurs along the sulphur cycle.

## 2.2 Microbial Environment of Soil and Sediment

Historically, recycling of elements in soil has been investigated for agricultural reasons: fertility due to the nitrogen and the phosphorus cycle. The presence and activity of micro-organisms under the rhizosphere region have long been neglected. Microbiology in deeper layers was believed to be either extremely limited or non-existent on account of poor or oligotrophic nutrient conditions (Waksman 1916). Lack of easily applicable techniques and lack of scientific and practical need and interest left deeper soil layers and all sediments unstudied for a long time.

Newly developed techniques in microbial ecology indicated that micro-organisms are present and active at considerable depths in the subsurface (Webster et al. 1985). Inspired by the discovery of pollutants at greater depths, it was discovered that deeper regions within the unsaturated soil and shallow water-table aquifers could contain populations of around a million organisms per gram, of which 1–10% was metabolically active. Similar numbers of diverse metabolically active micro-organisms exist in subsurface sediments, sometimes even at a depth of 500 m (Flierman 1989).

Microbial features correlate with such environmental conditions as pH, redox-potential, electron acceptors and temperature. So techniques to determine those conditions are important to understand recycling and dynamics. The introduction of polarographic oxygen sensors (Teal and Kanwisher 1961) and the ion-specific silver silver sulphide electrode (Berner 1963) as additional tools for characterization of the environmental conditions of a sediment was an important step. It became explicable that in the top millimetres of marine sediments the dynamics and fluctuations in pH/Eh conditions are extremely high (Baas Becking 1934), while at deeper layers more stable conditions prevail. Some sediment ecosystems turned out to be not only carbon- but also sulphur-dominated (Baas Becking 1925).

The golden triangle (summarized in Fig. 2.1) for understanding the microbial degradation of any natural or xenobiotic organic compound consists of knowledge of the microbial community, environmental conditions and knowledge of the

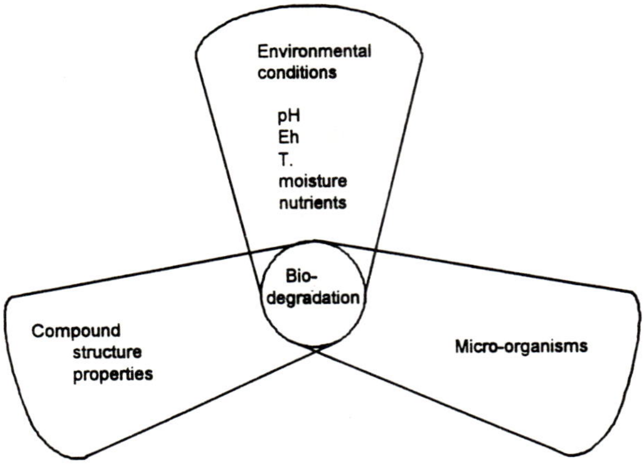


**Fig. 2.1.** The golden triangle for microbial degradation

structure and physico-chemical characteristics of the organic compound to be degraded.

**Microbial Degradation of Natural Organic Material.** Although stated earlier we repeat that the natural microbial versatility of soil and sediment is the biological basis or the biological carrying capacity from which the degradation of pollutants has to occur. In nature the primary substrates for micro-organisms, such as bacteria, actinomycetes and fungi, are mainly complex polymers. Microbial degradation, or recycling of organic material, consists of a sequence of many transformation steps carried out by either one population or a supplementary consortium, ultimately resulting in complete mineralization. An enormous variety of micro-organisms, with an average size of 1 mm, metabolize and use directly those reduced forms of organic matter as growth material and an energy source. This source is called the "electron donor". The beneficial energy for the micro-organism can be released and used by reaction with "electron acceptors", of which oxygen is extremely important. So in principal, microbial degradation is based on connecting or plugging electron donors with electron acceptors. In most reactions, molecules of reduced organic carbon represent the electron-donor, whereas oxygen or oxygen containing compounds of nitrogen, sulphur, manganese or iron, act as electron acceptors.

Besides serving as an electron acceptor $O_2$ also has a role as reactant for primary attack of compounds as a result of which oxygen molecules are inserted into the substrate molecule. In aerobic respiration, $O_2$ acts as a terminal electron acceptor for the electrons temporarily stored in co-enzymes during the oxidation of reduced energy substrates. When all the oxygen has been utilized, the system becomes anoxic or anaerobic. The microbial community must look for another kind of respiration, for other electron acceptors. One of the consequences is the generation of less energy, expressed as adenosine triphosphate (ATP) and of less biomass. From an energetic point of view nitrate is the next best acceptor. For the biodegradation mechanism of the organic substrate $(CH_2O)_n$ the mode of action and the sequence of effective electron acceptors has been presented in Fig. 2.2 (Bouwer 1992).

Depending on the electron acceptor, other populations of the microbial community become active (Zehnder and Stumm 1988). Indirectly there are many natural as well xenobiotic organic compounds, that are transformed by micro-organisms without being either a source of energy or producing any growth. In this case, the organic compound becomes co-metabolized in the obligate presence of a growth substrate or another transformable compound (Dalton and Stirling 1982).

There is a huge difference between transformation and mineralization. The latter means a complete breakdown to carbon dioxide, water and minerals available for plants. The former is just one degradation step. At the time when chemical determination was difficult, this first step was the difference between detection or absence. This first or primary biodegradation step causes a discrete

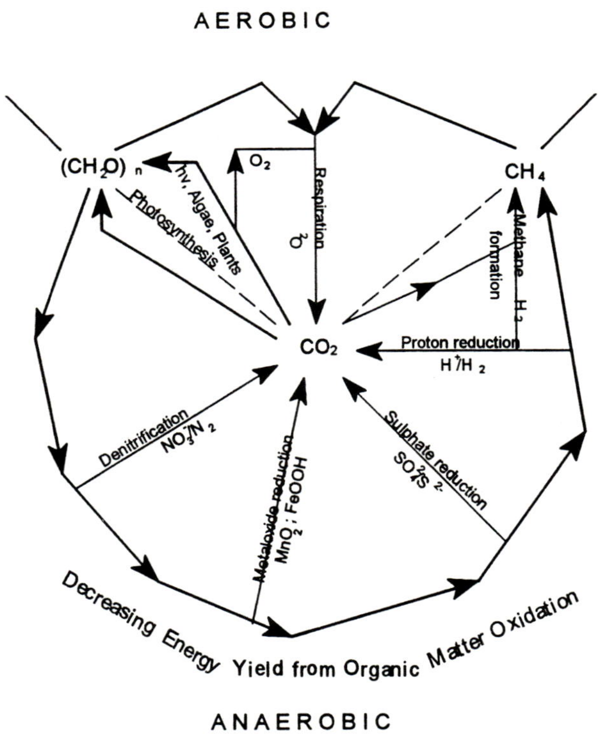


**Fig. 2.2.** Biodegradation mechanism of organic compounds showing the various modes of action of oxygen and other electron acceptors

alteration in the structure of the component in such a way that basic physical–chemical properties are lost and chemical detection becomes very difficult. For many large organic molecules, such as lignin and polyphenols, polyaromatic hydrocarbons (PAH), and polychlorinated biphenyls (PCB) it is difficult, without the help of advanced techniques such as GC-MS, to distinguish between the first and the second or between the eleventh or twelfth transformation steps.

In general aerobic processes proceed faster than anaerobic ones. With respect to the variety of available biochemical pathways anaerobes are not inferior to aerobes; on the contrary, anaerobes have had more time to evolve (their evolution is older) and the various potential electron acceptors have provided these organisms with the possibility of creating a large number of metabolic routes. In principle aerobic organisms have evolved from anaerobes, and represent only a specialized group of organisms which have adapted to a "highly toxic" compound oxygen. From a practical point of view biological decomposition processes historically were mainly focused on aerobic conversions. Therefore our knowledge of aerobic microbiology is older. Our knowledge, however, about the biodegradative potential of anaerobic microflora has tremendously expanded over the last 20 years.

**Carrying Capacity Potential of Soil And Sediment.** Upper layers of soils and sediments are exposed to strong fluctuating conditions, for example due to bioturbation, water-erosion deposition, diurnal light and dark cycles, and changing water content. In deeper layers the environment may be more stable with predictable redox conditions. All these environments contain generally a large variety of organisms: the storehouse for various degradation pathways. The carrying capacity potential (c.c.p.) of soils can differ remarkably from the c.c.p. of sediments. The differences are mainly caused by oxygen penetration depth, stratification, salinity and nutrient content of the water column above the sediment, or in the soil, pH and temperature.

**Oxygen Penetration Depth**. Oxygen penetration to a certain depth causes the most striking difference between soil and sediment with respect to microbial transformation processes. In marine and freshwater sediments oxygen can penetrate from less than 1 mm in systems with high microbial activity, to approximately 10 cm where vertical stratification is disturbed by animal activity (Gerritse 1993). Soils can range from being totally oxic, as well-drained soils to anoxic such as waterlogged swamps and marshes. Here we will mainly concentrate on drained arable soils, composed of conglomerates of sand-clay and organic matter with pores containing water and gas.

**Stratification.** Sediments are far more stratified than arable soils. In soils deep-rooting vegetation and burrowing biota such as earthworms can cause an aggregated structure. These aggregates can be very nutrient rich and therefore, heavily colonized by various groups of micro-organisms.

**Salinity and Nutrient Content of the Water Column above the Sediment or in the Soil.** Besides aerobic activities such as respiration and nitrification, denitrification, sulphate reduction and methanogenesis may occur within a distance of a few millimetres. So in arable soils dominated by aerobic degradation processes, anaerobic processes also occur, for example the well known denitrification reactions. Marine and freshwaters differ in salinity and sulphur. This influences the chemical, physical and microbial composition of the underlying sediment. Also the nutrient status in the water column above sediments varies from oligotrophic to eutrophic and determines the oxygen content.

The upper layer of a sediment is colonized by many aerobic and facultative anaerobic micro-organisms attached to particles. The thickness of these aerobic layers may range from 0.05 mm, for example in a Beggiatoa mat in a meso-eutrophic lake, to more than 4.8 mm in a oligotrophic lake (Sweerts 1990). Much of the sediment is anoxic and generally colonized by four distinct groups of bacteria:

1. *Fermentative bacteria*; these bacteria convert primary polymeric organic matter into a variety of metabolites, including alcohols, fatty acids, acetate, molecular hydrogen and dioxide.
2. *Acetogenic bacteria*; in freshwater sediments, mainly low in sulphate, acetogenic bacteria convert the reduced fermentation products to acetate, molecular hydrogen and $CO_2$.
3. *Methanogenic bacteria*; these bacteria convert the substrates produced by the former bacteria to methane and $CO_2$.
4. *Sulphate reducing bacteria*; in sulphate rich marine sediments sulphate reducing bacteria have a competitive advantage and oxidize the fermentation products via low molecular weight organic acids to $CO_2$ with concomitant reduction of sulphate to sulphide (Ward and Winfrey 1985).

**pH**. The pH ranges from 3 to 10 in soils and sediments. This has consequences for the solubility of the compound to be degraded and for the composition of the microbial community. Bacteria are mainly dominant and active in neutral and alkaline ecosystems. Fungi are active in lower pH ecosystems. The pH of the marine environment is uniform, steady and slightly alkaline. The pH of various soils encompasses a wide range, however, most of them are somewhat acidic. The marine environment is well buffered against acidification by the carbonate-bicarbonate system, as are some soils. But by no means all soils are well buffered. In soils low in carbonate organic or mineral acids from various metabolic processes can lower the soil pH to rather extreme values such as pH 3.

**Temperature**. In the very upper layers of soils the temperature can strongly vary within short periods of time and within short distances. Nevertheless, average soil temperatures are 5–15°C during summer in temperate climates. For sediments similar ranges can be mentioned. It should be kept in mind that a temperature rise can also be caused by microbial activity of which composting is the most extreme example.

In Fig. 2.3 generalized soil and sediment profiles are presented and quantitatively compared in relation to environmental factors and the numbers of specific micro-organisms.

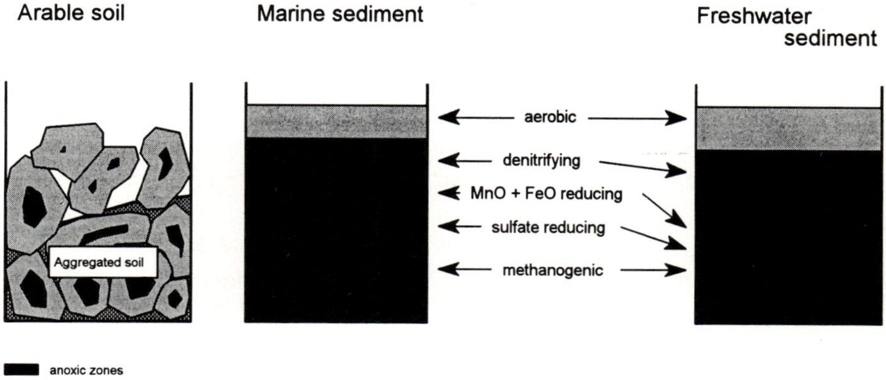


| | | Arable soil | Marine sediment | | Freshwater sediment |
|---|---|---|---|---|---|
| pH | | 3 - 10 | 6 - 9 | | 6 - 9 |
| organic matter% | 1 - 40 | | | | |
| salinity (‰) | | 1 | 20 - 40 | | 1 - 5 |
| number micro-org. (/g) with electron acceptor | | | | | |
| $O_2$ | | $10^8$ | $10^8$ | only in upper mm | $10^8$ |
| $NO_3^-$ | $10^4$ | $10^4$ | | $10^4$ | |
| MeO | 10 | $10^2$ | | $10^4$ | |
| $SO_4^{2-}$ | 10 | $10^6$ | | $10^4$ | |

**Fig. 2.3.** Environmental factors and microbial potention of soil- and sediment layers

## 2.3 Degradation of Xenobiotics in Soil and Sediment

In principle many of the most important pollutants in soils and sediments can become completely eliminated by biodegradation: mineral oil under aerobic conditions (Vanloocke et al. 1975; Atlas 1981; Bewley et al. 1989; Bossert and Bartha 1984), polyaromatic hydrocarbons mainly under aerobic conditions but the smaller ones also with $NO_3$ of FeO as electron acceptor (Cerniglia 1992; Weissenfels et al. 1992; Warith et al. 1992), alpha-, gamma- and delta- hexachlorocyclohexane mainly under aerobic conditions (Bachmann et al. 1988; Doelman et al. 1990; Loonen 1987), polychlorinated biphenyls under sequential anaerobic and aerobic conditions (Abramowizc 1990; Harkness et al. 1993; Quensen et al. 1988; Rhee et al. 1989), aliphatic chlorinated solvents as dichloroethene, vinyl chloride, trichloroethylene and tetrachloroethylene sometimes under aerobic and sometimes under extreme anaerobic conditions (Kleopfer et al. 1985; Ensley 1991; de Bruin et al. 1992; Holliger 1992), mono-aromatics as benzene, toluene, ethylbenzene and xylene with various electron acceptors (Aronstein et al. 1991; Hutchins 1991; Beller et al. 1992; Evans et al. 1991; Grbic-Galic and Vogel 1987; Lovley and Lonergan 1990), free cyanide (Knowles 1976; Ingvorsen et al. 1991; Fallon et al. 1991) and complexed cyanide (Fallon 1992; Rollinson et al. 1987) under aerobic and anaerobic conditions.

For reasons of classification a differentiation has been made into three groups of organic substances: (1) natural organic matter; (2) mono- and polyaromatic hydrocarbons; and (3) chlorinated organics.

For groups (2) and (3), rates and pathways of the microbial degradation in soils and sediments will be discussed in relation to the risk assessment of delayed and non-linear responses, either caused by environmental changes or by microbial phenomena.

**Biodegradation Kinetics and Rates**. Biodegradation of xenobiotics, their pathways and the kinetics of defined enzymatic degradation steps, has generally been determined in well defined optimal laboratory conditions such as aqueous systems or shake flask experiments using water–soil/sediment suspensions or batch experiments. Mainly those data have been used for model approaches, but they are hardly relevant for biodegradation rates in situ. Half-life periods as a parameter and first-order kinetics as a function are most commonly used for describing degradation of pollutants in soil or sediments. Degradation however strongly depends on the specific ecosystem investigated and the environmental conditions. Measuring half-life is rather easy, since it is based on disappearance, but it does not take into account the difference between one transformation step and complete mineralization.

Kinetic models for microbial degradation are based on substrate concentration and biomass (Simkins and Alexander 1984). This leaves in principle three types of kinetic order. When the substrate is completely available the degradation only depends on the activity of the micro-organisms following logarithmic growth. The degradation follows 0-order kinetics: a logarithmic disappearance. A process follows first order kinetics when the rate of biodegradation of a compound is directly proportional to its concentration. The second-order approach, in which the first-order kinetics is related to the population density, is the most realistic one. Lack of detailed stepwise degradation information may be one of the reasons why the occurrence of non-linear reactions is presumed.

The response of the microbial community towards natural as well as xenobiotic organic compounds does not depend on total concentration, but mainly on the water-soluble concentration. This soluble fraction is relatively small for mineral oil, PAH, PCB and many chlorinated organic compounds such as HCH and HCB; but not for TRI and PER. The sorbed or bounded fractions of xenobiotica are not available for microbial interference. Microbial activity takes place in the soluble phase. The soluble concentration may be too low to support measurable growth of the microbial population. However, even without growth, micro-organisms may completely degrade pollutants. In principle a minimum substrate mass, or threshold value, necessary to start degradation does not exist for micro-organisms. Recent research (Tros 1994) has shown that a low threshold for degradation may exist for some micro-organisms, but for other micro-organisms it does not exist. So when the substrate is physically available, micro-organisms can "clean their plate completely". Sorbed portions or bound residues are not available on their plate.

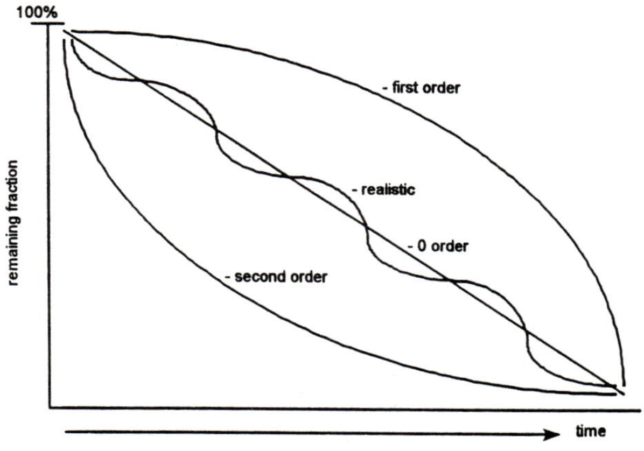


**Fig. 2.4.** The order kinetics

When two or more different sequential populations are demanded for complete degradation, adaptation time may be involved. For large molecules this is rather a rule than an exception. Different kinetics of the various degradation steps and this adaptation time are the reasons why the overall disappearance seems to follow a more or less repeatable second order (shown in Fig. 2.4). From this it can be concluded that non-linear responses in reality are rather a rule than an exception.

Adaptation to a new substrate and/or new electron acceptors may take a long period of time. For example, in a creosote-contaminated anaerobic sediment, an adaptation time of 100–120 days was measured before toluene was degraded under methanogenic conditions (Edwards and Grbic-Galic 1994). For oxylene this period was 200–250 days. Since this research was carried out at 35°C the adaptation at lower temperature may be even longer. On the other hand other

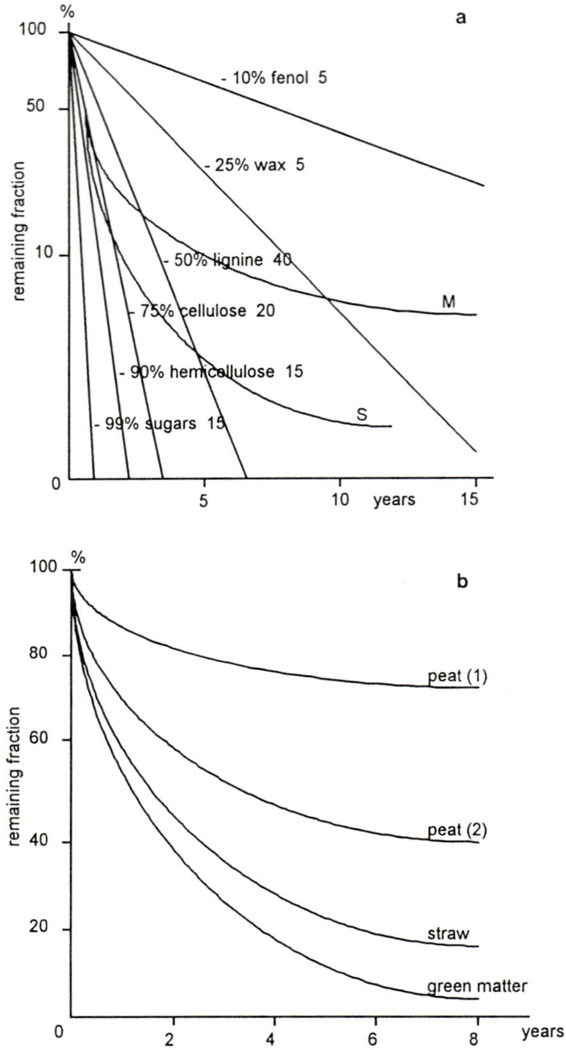


**Fig. 2.5 a–b.** Degradation rates in temporate soils

organisms may adapt at lower temperature with other sensitivities towards electron acceptors and organic matter. For defined organic matter the degradation rates in soil are given in Fig. 2.5a (Minderman 1968).

For various complex organics the degradation rate in soils are given in Fig. 2.5b (Janssen 1992). The numbers in Fig. 2.5 indicate that for the organic material such as cellulose, 75% of the original 15% is degraded within 1 year. It is clear that in a forest, soil degradation is a slow process. So xenobiotic

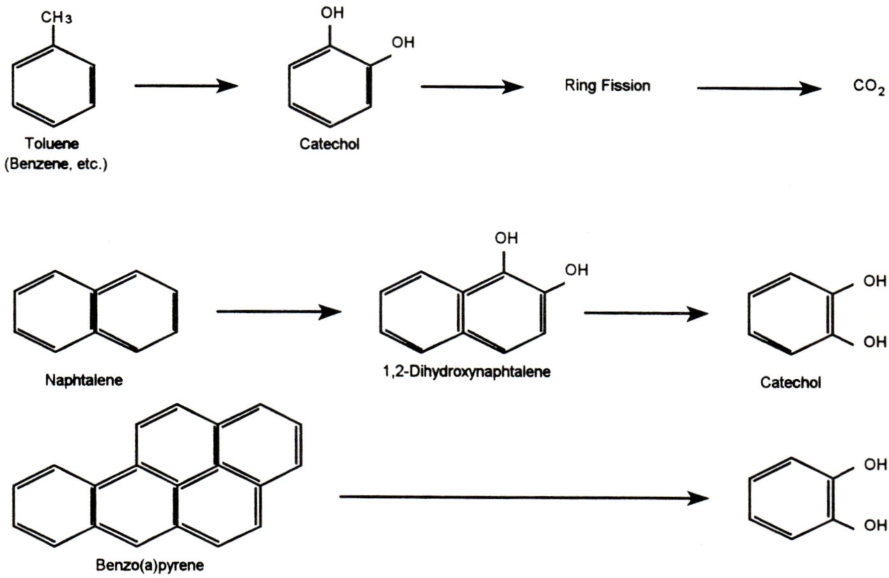


**Fig. 2.6.** Aerobic degradation pathway of aromatic hydrocarbons

compounds such as aromatics and polyaromatic hydrocarbons will have, similar to natural compounds, degradation rates of the same order of magnitude.

**Degradation Pathway and Rates of Aromatic Hydrocarbons.** In Fig. 2.6 the aerobic microbial degradation pathways of mono and polyaromatic hydrocarbons have been summarized. In principle, the degradation follows the dioxygenase route, which means the insertion of two oxygen groups. Next, the molecule is transformed to a smaller size, gradually "breaking off" $CO_2$ units. For naphthalene (two benzene rings) the circle has to be followed twice, and for pyrene four times. For PAH degradation it has been proven that a lower rate occurs at a lower redox-potential. For naphthalene and benzo(a)pyrene degradation rates are presented in detail in Table 2.1.

When a micro-organism can not completely mineralize the original molecule, the reaction product or intermediate will be transported from inside the cell to the outside and becomes available for another type of micro-organism. During degradation the intermediates become generally more water-soluble due to the insertion of hydroxyl groups. So outside the cell wall of the micro-organism the intermediate may be more mobile in soil and sediment than the original pollutant. Whether this mobility of the intermediate poses an additional risk to soil/sediment ecosystem, strongly depends on the intermediates toxicity and the ability of other organisms to "neutralize" the compound.

**Table 2.1.** Degradation rates ($t½$) of Napthalene and Benzo(a)pyrene at different redox potentials

| PAH | Redox | Conc [mg $l^{-1}$] | pH | T (°C) | Mineralizations | $t½$(d) |
|---|---|---|---|---|---|---|
| Naphthalene | | | | | | |
| 500 | mV-ox | 1 | 6.5 | 30 | 59% in 31d | 24 |
| 500 | mV-ox | 1 | 8.0 | 30 | 48% in 32d | 33 |
| 250 | mV-ox | 1 | 6.5 | 30 | 57% in 31d | 25 |
| 250 | mV-ox | 1 | 8.0 | 30 | 63% in 32d | 22 |
| 0 | mV-den | 1 | 6.5 | 30 | 48% in 31d | 32 |
| 0 | mV-den | 1 | 8.0 | 30 | 30% in 32d | 61 |
| -250 | mV-red | 1 | 6.5 | 30 | 0.29% in 31d | 7400 |
| -250 | mV-red | 1 | 8.0 | 30 | 0.39% in 32d | 5700 |
| Benzo(a)pyrene | | | | | | |
| 500 | mV-ox | 1.25 | 6.5 | 30 | 4% in 37d | 640 |
| 500 | mV-ox | 1.25 | 8.0 | 30 | 6% in 37d | 390 |
| 250 | mV-ox | 1.25 | 6.5 | 30 | 0.8% in 37d | 2970 |
| 250 | mV-ox | 1.25 | 8.0 | 30 | 1.89% in 37d | 1340 |
| 0 | mV-den | 1.25 | 6.5 | 30 | 0.14% in 37d | 8300 |
| 0 | mV-den | 1.25 | 8.0 | 30 | 0.15% in 37d | 7100 |
| -250 | mV-red | 1.25 | 6.5 | 30 | 0.11% in 37d | 23300 |
| -250 | mV-red | 1.25 | 8.0 | 30 | 0.09% in 37d | 28500 |

**Degradation of Chlorinated Organics.** The microbial degradation of relevant chlorinated organic contaminants such as hexachlorobenzene (HCB), polychlorinated biphenyls (PCB) and the solvents trichloroethylene (TRI) and tetrachloroethylene (PER) are based on similar principles.

When three or four chlorides are closely linked together in an organic molecule as for example in PER, HCB and highly chlorinated PCB, the primary degradation demands anaerobic conditions. After this reductive dechlorination, further degradation may occur under aerobic conditions. So PER and sometimes TRI need an initial anaerobic degradation step. For TRI a whole range of anaerobic as well as aerobic degradation pathways are known (Ensley 1991). For those xenobiotics a qualitative structural degradability relation (QSBR) may exist. The microbial degradation pathways are summarized in Fig. 2.7.

If the dechlorination of PER or TRI produces vinyl chloride (VC) as an end-product, it is an undesirable process because VC is carcinogenic. The literature hints that VC can be further dechlorinated to ethylene and even simpler compounds. However, the final dechlorination step from VC to ethylene may be rate limiting, and significant amounts of VC can exist for a long time (Cobb 1989, DiStefano et al. 1991).

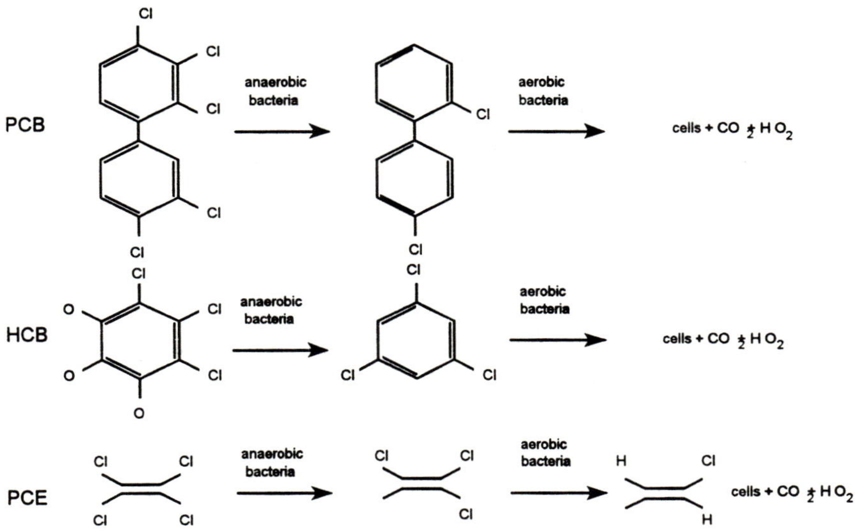


**Fig. 2.7.** Summarized degradation pathways of chlorinated organics

Under anaerobic conditions at an optimal temperature HCB can be dechlorinated to tri- and dichlorobenzene within 3 weeks. No further dechlorination of 1,3,5 -trichlorobenzene occurred under anaerobic conditions (Fig. 2.7; Fathepure et al. 1988). For further mineralization, aerobic conditions are necessary. In activated or dynamic soil systems the degradation rate is extremely high (0.1–10 mg $kg^{-1}$ $day^{-1}$). However, in undisturbed soils and sediments the rate is extremely low with half-life periods varying between 1 and 25 years, due to their unavailability to micro-organisms. In general, the degradation rates of the rather insoluble components such as PCB are very low and the half-life periods vary between 2 and 25 years (Beurskens et al. 1993).

During microbial transformation of PCB, chlorobenzoacids and chlorophenol are major intermediates. Their solubility and mobility, their toxicity and their degradation rate differ completely from those of the original compound. Degradation rates of these intermediates formed under various environmental conditions are extremely variable from days to months. For many chlorinated phenols it is shown that decreasing biodegradation rates correlate with an increasing octanol-water partition coefficient. This may result in quantitative structure–activity relationships (Parsons and Govers 1990).

**Microbial Consortia Controlling Carrying Capacity**. The degradation capacity depends on the number of micro-organisms present and their ability to degrade the specific substrate. The size of the specific biomass in relation to the natural biomass is generalized and given in Table 2.2.

In the aerobic part of a soil or sediment the number of hydrocarbon degrading

**Table 2.2.** Number of degraders of various pollutants given as a percentage of the natural population

| | | No.[a] | Bacteria | Fungi | Metabolism | Co-metabolism |
|---|---|---|---|---|---|---|
| Mineral oil | C12-20 | ±1-5% | ++ | + | + | - |
| | C21-40 | ±1-5% | ++ | + | + | - |
| PAH | 2-3 rings | ±0.1% | + | + | + | + |
| | 4 rings | ±0.1% | + | + | + | + |
| | 3 5rings | ? | + | + | + | + |
| PCB | 1-2 chloro | 0.1% | ++ | + | + | + |
| | 3-4 chloro | specific | + | + | + | + |
| | ±5 chloro | specific | + | + | + | + |
| HCH | alpha | 0.1% | ++ | - | + | - |
| | beta | - | - | - | - | - |
| | gamma | 0.1% | ++ | - | + | - |
| | delta | 0.1% | + | - | + | - |
| Cl-ethene | PER | specific | + | - | - | + |
| | TRI | 0.01% | + | - | + | + |
| | DCE | 0.01% | + | - | + | + |
| Cyanide | free | 1% | + | + | + | - |
| | complex | specific | + | - | - | + |

[a] number of microorganisms

organisms is relatively high. This may be expected since these compounds closely reflect common natural compounds such as fatty acids, lignin and polyphenols. The number of chlorinated organics degrading organisms is extremely small.

**Two Examples of Consequences of Environmental Changes.** Man-induced changes of the groundwater level and eutrophication of the soil by nitrogen and sulphur deposition, may affect redox conditions and pH. Information on the influence of pH on the degradation of chlorinated organic compounds is scarce. An example is the effect of pH on the reductive dechlorination of 1,2,3 trichlorobenzene and on tetrachloroethylene as published by Holliger (1992; Fig. 2.8).

From those data it is clear that acidification from pH 7 towards pH 6 may completely stop the degradation. Although it should be kept in mind that these data are derived from laboratory studies, it nevertheless, illustrates how strong the effect of a small change in pH can be. For temperature similar strong barriers can be given as shown in Fig. 2.9.

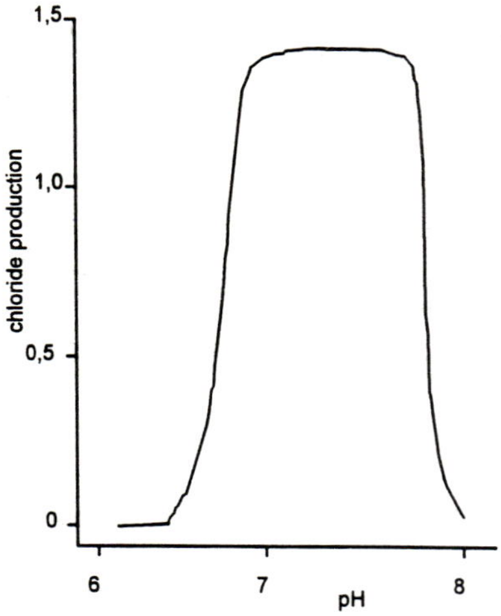


**Fig. 2.8.** pH-dependence of reductive dechlorination of PER and of trichlorobenzene

The data, as presented in Figs. 2.8 and 2.9, have been derived from enrichment cultures from Rhine River sediment. In this experiment dechlorination activity could only be maintained with extra nutrients. So in nature, where less extra nutrients may be available, these decreases within small pH or temperature margins may be less sharp. The examples are simply given to show how small environmental changes may have drastic effects on microbial transformations.

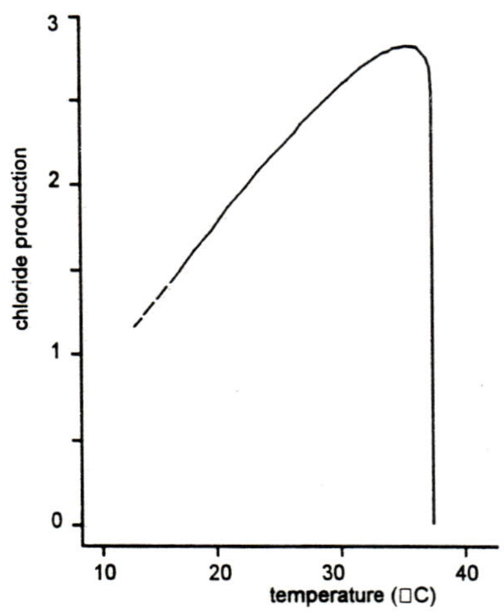


**Fig. 2.9.** Temperature dependence of reductive dechlorination

## 2.4 Interaction of Heavy Metals and Micro-organisms

Heavy metals are elements with a mass density above five. The most important micropollutants are Pb, Cr, Cd, Hg, Ni, Cu and Zn. Only the last two are essential for many biochemical processes, but at very low concentrations. Those pollutants are often toxic depending on their availability and speciation, depending on Eh-pH, and the nutrient status of the environment. The direct inhibitory effect of heavy metals on microbial decomposition processes have been widely covered in literature (Doelman 1985, Baath 1990). Here we pay attention to consequences of microbial shifts due to heavy metal pollution and to microbially induced speciation changes such as methylation.

**Adaptation to Heavy Metals and Consequences for Degradation**. Gradual and diffuse pollution of soils causes a shift within the microbial community. In general the different groups of micro-organisms show a difference in sensitivity towards heavy metals. Fungi are more resistant to heavy metals than are actinomycetes, the latter are more resistant than are Gram-negative bacteria, and Gram-negative bacteria are more resistant than Gram-positives (Doelman 1985). Hence, the ratio fungi/bacteria increases in soils stressed by heavy metal pollution. This shift will in turn cause a change in the potential of the soils to degrade harmful chemicals to harmless inorganic end-products.

Another shift within the bacterial community is an increase in the ratio of strains resistant to heavy metals over sensitive strains (Doelman and Haanstra 1979; Duxbury and Bicknell 1983). Resistance has been defined as growth in the presence of high concentrations of a heavy metal (e.g. 10 mg Hg/kg in the growth medium). Sensitivity has been defined as dying off in the presence of low concentrations of heavy metals (e.g. 0.1 mg Hg/kg in the growth medium).

Sensitive bacteria proved to have a greater ability to grow on a complex nutrition substrate like aromatic compounds than did the resistant ones (Doelman et al. 1994). Hence, the alteration of the environment favours the predominance of a heavy-metal-resistant microbial community with subsequent loss of the ability of soil micro-organisms to degrade certain persistent organic compounds. When this chain of events is followed by pollution of the ecosystem by chemicals such as polycyclic aromatic hydrocarbons or chlorinated organic compounds (e.g. anthracene, PCB's and dibenzofurans), their persistence will increase significantly relative to what would be anticipated in a system not contaminated by heavy metals. Subsequently, the concentrations of these toxic chemicals and their bioaccumulation in the food-chain could be greatly enhanced.

**Microbial Transformations of Heavy Metal**. The chemical oxidation of sulphide to sulphate may be become accelerated by a factor of 10 000 when the bacterial population *Thiobacilli* is present. They easily oxidize the insoluble metal sulphides to more soluble metal sulphates (Silver and Torma 1974). This

**Table 2.3.** Micro-organism-mediated heavy metal oxidation-reduction reactions

| Transformation | Metal |
|---|---|
| Reduction | $As^{5+}$, $Cr^{6+}$, $Fe^{3+}$, $Hg^{+}$, $Hg^{2+}$, $Mn^{4+}$, $Se^{4+}$, $Te^{4+}$ |
| Oxidation | $As^{3+}$, $Cr^{3+}$, $Fe^{0}$, $Fe^{2+}$, $Mn^{2+}$, $Sb^{3+}$ |
| Methylation | $As^{5+}$, $Cd^{2+}$, $Hg^{2+}$, $Se^{4+}$, $Sn^{2+}$, $Te^{4+}$ |
| Demethylation | R $Hg^{2+}$ |

microbially induced mobilization affects the metal leaching from the dump site to the underlying soil, groundwater and surface waters. In Table 2.3 the various oxidation reduction processes haven been summarized. As can be seen from Table 2.3 several heavy metals may be subject of various oxidation reduction processes.

**Methylation**. Studies on the microbial methylation of lead, arsenic and mercury are available, of which the latter is the most striking. Due to the unforeseen capability of bacteria to convert Hg to highly toxic mercury, the Minamata disease become world famous. The formation of methylmercury by bacteria and its release to the aquatic environment is a microbial defence mechanism protecting the microbes from mercury poisoning. Methylmercury is at least 100 times as toxic to fish as is metallic Hg. Similar phenomenon may occur now in the gold-mining area of the Amazon Basin in Brazil. Mercury is used for concentrating the fine gold particles through amalgamation. The gold is separated from the amalgam by application of heat to drive off the mercury which is volatilized to the atmosphere. Mercury, through atmospheric transport and deposition, is contaminating a vast area of the Amazon River (Malm et al. 1990). Currently about 200 tons of mercury per year are being deposited. All of the conditions exist for a Minimata type event, but on a large scale. In the 1970s and early 1980s it was thought that mercury methylation was highly pH-dependent and mainly occurred in the narrow pH range of 5.5–6.5. So mercury methylating micro-organisms were thought to be highly sensitive to changes in pH and also to changes in salinity and nutrient level. On the contrary, arsenic methylation was supposed to occur over a wide pH range 3.5 to 7.5 (Summers and Silver 1978). It can be argued that arsenic, antimony and mercury are also subject to abiotic methylation under specific physico-chemical conditions. Recently the literature has revealed that biotic sediments can become mayor sites for methylation of inorganic mercury, where sulphate reducing bacteria contribute considerably (Weber 1993; Gilmour et al. 1992; Kerry et al. 1991; Gilmour and Henry 1991; Oremland et al. 1991; Compeau and Bartha 1987). So increased sulphate deposition may ultimately cause an increased toxicity. Several model studies have shown that humic matter effectively methylates mercury forming methylmercury (Weber 1993). Experimental addition of sulphate to anoxic sediment resulted in increased microbial production of methylmercury from added inorganic mercury (Gilmour et al. 1992; Kerry et al. 1991). A special review focused on microbial and biogeochemical changes leading to increased levels of methylmercury (Gilmour and Henry 1991). Demethylation may occur under aerobic as well as anaerobic conditions (Oremland et al. 1991). The interference of salinity with methylation has been refuted (Compeau and Bartha 1987).

## 2.5 Concluding Considerations in Relation to Risk Assessment

The capacity of microflora in soils and sediments to recycle organic matter is controlled by environmental conditions. This capacity also comprises the storehouse for degradation of xenobiotics and chemical transformation of heavy metals. For degradation of xenobiotic electron donors such as hydrocarbons and chlorinated organics electron acceptors such as oxygen, nitrate, iron- and manganese-oxide, sulphate and carbon dioxide are important capacity controlling factors. Hence, pH, temperature and eutrophication with readily degradable organic matter and macro-elements such as nitrogen and sulphur play a role. The latter two factors may strongly change the oxygen supply and consequently the redox potential. Acid rain, involving nitrogen and sulphur pollution, ultimately result in increased $NO_3^-$ and $SO_4^{2-}$ levels. This affects the pH and provide new electron acceptors, thereby influencing the microbial ability under different redox conditions. These phenomena and consequences of environmental pollution, have hardly been investigated.

In this selected short overview, it has been indicated that on human time-scales non-linear reactions should be considered as normal, since complete degradation sometimes demands several transformation steps under different redox conditions and by various sequential changes in micro-organism populations. There are some classical examples of concern due to microbial transformation processes. The formation of vinyl chloride as an intermediate in the complete degradation of PER and TRI shows that some compounds may degrade to forms that are far more toxic and carcinogenic than the original contaminant. The microbial methylation of mercury, predominantly under sulphate reducing conditions, produces a compound at least 100 times as toxic for fish, while for micro-organisms this process of methylation is an detoxification mechanism. Furthermore, gaining resistance of micro-organisms against heavy metal pollution may decrease their potential to degrade persistent organic contaminants.

From a human and ecotoxicological point of view contamination of soil and sediment is extremely paradoxical. Due to toxicity and sometimes accumulation in the food-chain, the risk for higher animals and humans increases with increasing contamination. For micro-organisms, toxicity of pollutants may be one side of the coin. Exploring new ways, by adaptation, mutation and natural gene-transfer, within the principle of infallibility, is the other creative side of the coin. So contamination may cause many concerns for mankind but provides a "paradise regained" for micro-organisms.

## References

Abramowicz DA (1990) Aerobic and anaerobic biodegradation of PCB: a review. Crit Rev Biotechnol 10:241-251

Aronstein BN, Calvillo YM, Alexander M (1991) Effect of surfactants at low concentrations on the desorption and bio-degradation of sorbed aromatic compounds in soil. Environ Sci Technol 25:1728-1731

Atlas RM (1981) Microbial degradation of petroleum hydrocarbons: an environmental perspective. Microbial Reviews 45:180-209

Baas Becking LGM (1925) Studies on the sulphur bacteria. Anal Bot 39: 613-651

Baas Becking LGM (1934) Geobiologie. Stockum and Zoon,. Den Haag, 263 pp

Baath E (1990) Effects of heavy metals on microorganisms and microbially mediated processes in soil. An overview. Workshop Atvidaberg, Sweden

Bachmann A, Walet P, Bruin W de, Huntjens JLM, Roelofsen W, Zehnder AJB, (1988) Biodegradation of alpha and beta hexachlorocyclohexane in a soil slurry under different redox conditions. Appl Environ Microbiol 54:143-149

Beller HR, Grbic-Galic D, Reinhard M (1992) Microbial degradation of toluene under sulfate-reducing conditions and the influence of iron on the process. Appl Environ Microbiol 58:786-793

Berner RA (1963) Electrode studies of hydrogen sulphide in marine sediments. Geochim Cosmochim Acta 27:563-575

Beurskens JEM, Dekker CGC, Jonkhoff J, Pompstra L (1993) Microbial dechlorination of hexachlorobenzene in a sedimentation area of the Rhine river. Biogeochemistry 19:61-81

Bewley R, Ellis B, Thiele P, Vibey Th, Rees J (1989) Microbial cleanup of contaminated soil. Chemistry and Industry 4 dec:778-784

Bossert I, Bartha R (1984) The fate of petroleum in soil ecosystems. In: Atlas RM (ed) Petroleum microbiology. Macmillan Publ Comp, New York, pp 435-474

Bouwer EJ (1992) Bioremediation of organic contaminants in the subsurface. In: Mitchell J (ed) New concepts in environmental microbiology. John Wiley and Sons, New York, pp 285-316

Bruin WP de, Kotterman MJJ, Posthumus MA, Schraa G, Zehnder AJB (1992) Complete biological reductive transformation of tetrachloroethene to ethane. Appl Environ Microbiol 58:1996-2000

Cerniglia CE (1992) Degradation of PAK's in soil. State of art of US-research work. Dechema, 6-9 dec 1992, Karlsruhe

Cobb DD (1989) Modelling and experimental simulations of organic contaminant biotransformation in subsurface environments. Dissertation, The John Hopkins University, Baltimore, Maryland

Compeau GC, Bartha R (1987) Effect of salinity on mercury methylation activity of sulfate reducing bacteria in estuarine sediments. Appl Environ Microbiol 53:261-265

Dalton H, Stirling DI (1982) Co-metabolism. Phil Trans Roy Soc London 297:481-496

DiStefano TD, Gosselt JM, Zinder SH (1991) Reductive dechlorination of high concentrations of tetrachloro-ethylene to ethene by an anaerobic enrichment culture in the absence of methanogenesis. Appl Environ Microbiol 57:2287-2292

Doelman P (1985) Resistance of soil microbial communities to heavy metals. In Jensen et al. (eds) Microbial communities in soil. Elsevier Appl Sci Publish London

Doelman P, Haanstra L (1979) Effects of lead on the soil bacterial microflora. Soil Biol Biochem 11:487-491

Doelman P, Haanstra L, Loonen H, Vos A (1990) Decomposition of alpha and beta hexachlorocyclohexane in soil under field conditions in a temporate climate. Soil Biol Biochem 22:629-634

Doelman P, Jansen E, Michels M, Til M van (1994) Effects of heavy metals in soil on microbial diversity and activity as shown by the sensitivity-resistance index, an ecologically relevant parameter. Biol Fertil Soils 17:177-184

Duxbury T, Bicknell B (1983) Metal-tolerant bacterial populations from natural and metal-polluted soils. Soil Biol Biochem 15:243-250

Edwards EA, Grbic-Galic D (1994) Anaerobic degradation of toluene and o-xylene by a methangenic consortium. Appl Environ Microbiol 60:313-322

Ensley BD (1991) Biochemical diversity of trichloroethylene metabolism. Annual Rev Microbiol 45:283-299

Evans PJ, Mang DT, Young LY (1991) Degradation of toluene and m-xylene and transformation of o-xylene by denitrifying enrichment cultures. Appl Environ Microbiol 57:450-454

Fallon RD (1992) Evidence of hydrolytic route for anaerobic cyanide degradation. Appl Environ Microbiol 58:3163-3164

Fallon RD, Cooper DA, Speece R, Henson M (1991) Anaerobic biodegradation of cyanide under methanogenic conditions. Appl Environ Micobiol 57:1656-1662

Fathepure BZ, Tiedje JM, Boyd SA (1988) Reductive dechlorination of hexachlorobenzene to tri- and di-chlorobenzene in anaerobic sewage sludge. Appl Environ Microbiol 54:327-330

Flierman CB (1989) Microbial life in terrestrial subsurface of southeastern coastal plain sediments. Haz Waste Haz Materials 6:155-171

Gerritse J (1993) Growth of bacteria at low oxygen concentrations. PhD thesis. Rijks University, Groningen

Gilmour CC, Henry EA (1991) Mercury methylation in aquatic systems affected by acid deposition. Environ Pollut 71:131-170

Gilmour CC, Henry EA, Mitchell R (1992) Sulfate stimulation of mercury methylation in freshwater sediments. Environ Sci Technol 26:2281-2287

Grbic-Galic D, Vogel TM (1987) Transformation of toluene and benzene by mixed methanogenic cultures. Appl Environ Microbiol 53:254-260

Harkness MR, McDermott JB, Abramowicz DA, Salvo JJ, Flanager WP, Stephens ML, Mondello FJ, May RJ, Lobos JH, Carroll KM, Brennan MJ, Bracco AA, Fish KM, Warner GL, Wilson PR, Dietrich DK, Lin DT, Morgan CB, Gately WL (1993) In situ stimulation of aerobic PCB biodegradation in Hudson River sediments. Science 259:503-507

Holliger C (1992) Reductive dehalogenation by anaerobic bacteria. PhD Thesis LUW

Hutchins SR (1991) Biodegradation of mono aromatic hydrocarbons by aquifer micro-organisms using oxygen, nitrate or nitrons oxide as terminal electron acceptor. Appl Environ Microbiol 57:2403-2407

Ingvorsen K, Hojer-Pedersen B, Godtfredsen SE (1991) Novel cyanide hydrolyzing enzyme from Alcaligenes xylosoxidans subs denitrificans. Appl Env Microbiol 57:1783-1789

Janssen BH (1992) Organic matter and soil fertility. Agricultural University Wageningen, LUW J 100-225

Kerry A, Welbourn PM, Prucha B, Mierle G (1991) Mercury methylation by sulfate reducing bacteria from sediments of an acid stressed lake. Water, Air, Soil Pollut 56:565-575

Kleopfer RD, Ensley BD, Haas BB, Deihl TG (1985) Anaerobic degradation of trichloroethene in soil. Environ Sci and Technol 19:277-280

Knowles CJ (1976) Microorganisms and cyanide. Bacteriological Reviews 40:652-680

Loonen H (1987) Zelfreinigend vermogen van hexachloorcyclohexaan verontreinigde grond voor alpha-HCH onder landfarming condities. RIN-report, 44 pp

Lovley DR, Lonergan DJ (1990) Anaerobic oxidation of toluene, phenol and p-cresol by the dissimilatory iron-reducing organisms GS-15. Appl Environ Microbiol 56:1858-1864

Malm O, Pfiefer WC, Souza CCM, Reuther R (1990) Mercury pollution due to goldmining in the Madeira river basin, Brazil. Ambio 19:11-15

Minderman G (1968) Addition, decomposition and accumulation of organic matter in forests. J Ecology 56:355-362

Oremland RS, Culbertson CW, Winfrey MR (1991) Methylmercury decomposition in sediments and bacterial cultures: involvement of methanogens and sulfate reducers in oxidative demethylation. Appl Environ Microbiol 57:130-137

Parsons JR, Govers HAJ (1990) Quantitative Structure-Activity Relationships for Biodegradation. Ecotox Environ Safety 19:212-227

Quensen JF, Tiedje JM, Boyd SA (1988) Reductive dechlorination of polychlorinated biphenyls by anaerobic micro-organisms from sediments. Science 242:752-754

Rhee GY, Bush B, Brown MP, Viane M, Shane L (1989) Anaerobic biodegradation of polychlorinated biphenyls in Hudson river sediments and dredged sediments in clay encapsulated. Water Res 23:957-964

Rollinson G, Jones R, Meadows MP, Harris RE, Knowles CJ (1987) The growth of a cyanide utilising strain of Pseudomonas florescence in liquid culture on nickel cyanide as source of nitrogen. Fems Microbiol Lett 40:199-205

Silver M, Torma AE (1974) Oxidation of metal sulfides by *Thiobacillus ferro-oxidans* grown on different substrates. Can J Microbiol 20:141-147

Simkins SM, Alexander M (1984) Models for mineralization kinetics with the variables of substrate concentration and population density. Appl Environ Microbiol 47:1299-1306

Summers AO, Silver S (1978) Microbial transformation of metals. Annual Rev Microbiol 32:637-672

Sweerts JRRA (1990) Oxygen consumption processes, mineralization and nitrogen cycling at the sediment-water interface of north temporate lakes. PhD Thesis, Rijks University, Groningen

Teal JM, Kanwisher J (1961) Gas exchange in a Georgia salt marsh. Limn Oceanogr 6:388-399

Tros M (1994) Transformation of low concentrations of trichlorobenzoate by Pseudomonas sp B13; kinetics and residual concentrations. in prep

Vanloocke R, De Borger JP, Voets J, Verstraete W (1975) Soil and groundwater contamination by spills; problems and remedies. The Int J of Environ Studies 8:99-111

Waksman SA (1916) Bacterial numbers in soils, at different depths and in different seasons of the year. Soil science 1:363-374

Ward EJ, Winfrey MR (1985) Interactions between methanogenic and sulphate educing bacteria in sediments. Adv Aquatic Microbiol 3:141-151

Warith MA, Ferehner R, Fernands L (1992) Bioremediation of organic contaminated soil. Hazardous Waste and Haz Mat 9:137-177

Weber JH (1993) Review of possible paths for abiotic methylation of mercury (II) in the aquatic environment. Chemosphere 26:2063-2077

Webster JJ, Hampton GJ, Wilson JT, Ghiorse WC, Leach FR (1985) Determination of microbial numbers in subsurface samples. Groundwater 23:17-25

Weissenfels WD, Klewer HJ, Langhoff J (1992) Adsorption of polycyclic aromatic hydrocarbons (PAH's) by soil particles: influence on biodegradability and potential risk of contaminated sites. Appl Microbiol Biotechn 36:689-696

Zehnder AJB (1982) The carbon cycle. In: Hutzinger O (ed) Handbook of environmental chemistry. Springer, Berlin Heidelberg NewYork, vol. 1, Part B
Zehnder AJB, Stumm W (1988) Geochemistry and biochemistry of anaerobic habitats. In: Zehnder AJB (ed) Biology of anaerobic microorganisms. John Wiley and Sons, New York

# 3 Heavy Metal Retention by Soil Organic Matter under Changing Environmental Conditions

*R. Schulin, G. Geiger and G. Furrer*

## 3.1 Introduction

Impacts of human activities, which lead to a sudden, catastrophic event after a long period of delay, range among the most critical environmental problems. The algal blooms in the Adriatic Sea, or the stratospheric ozone losses are only two examples of such events for which the term "chemical time bomb" (CTB) has been coined. Continuous accumulation of pollutants in the environment from anthropogenic sources has to be recognized as the cause. With respect to soils and sediments, Stigliani et al. (1991) have defined a CTB as "a concept that refers to a chain of events, resulting in the delayed and sudden occurrence of harmful effects due to the mobilization of chemicals stored in soils and sediments in response to slow alterations to the environment". It should be noted that chemicals which are mobilized are not necessarily man-made, but can also have a natural source. Aluminum e.g., the third most abundant element in the earth's crust, is mobilized in soil only under acidic conditions. It is supposed, however, that the environmental alterations resulting in the mobilization of these chemicals are driven by human activities. Finally, the term "sudden" has to be understood with respect to the time scales of environmental changes and of possible political or operational reactions and countermeasures.

The unexpectedness, suddenness, irreversibility and hazardousness of CTBs effects urge for prophylactic measures. It is of utmost importance to detect potential CTBs as early as possible. The first and most valuable step is to review the state-of-the-art knowledge of environmental processes and to pinpoint latent CTB risks. As a direct consequence of the taken approach, the identification of potential CTBs is a big challenge to environmental research. This is contrary to the traditional philosophy of reacting to environmental hazards only after they have emerged. Anticipation of potential CTB effects in soils and sediments involves the identification of:

- Harmful chemicals accumulated or accumulating in soils and sediments
- Storage and release mechanisms
- Factors which can trigger a rapid release upon environmental change

The purpose of this chapter is to evaluate the potential role of heavy metal retention by soil organic matter (SOM) in the context of the CTB concept. Fig. 3.1 schematically shows the main reservoirs and the transfer processes affecting the biogeochemical cycling of carbon and heavy metals in soil systems. Heavy metals are known to be widespread soil pollutants which are continuously emitted into the environment by various human activities. Due to strong sorption to solid matter they are accumulated in soils and sediments. As toxic levels are eventually reached with respect to plants and soil organisms, this gradual accumulation already constitutes a serious environmental problem by itself. An unexpected rapid mobilization of accumulated heavy metals, however, could result in an environmental catastrophe. Thus environmental alterations triggering such

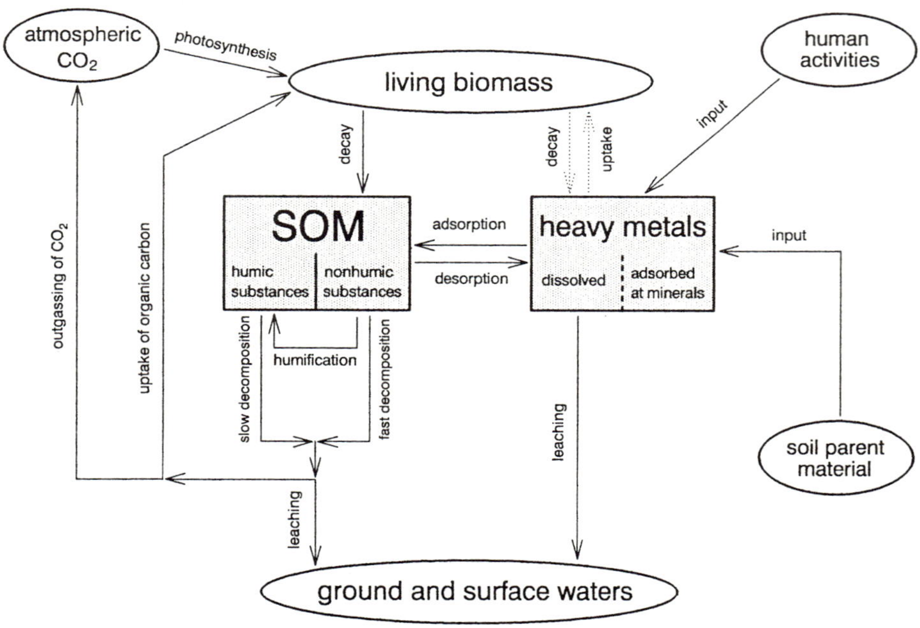


**Fig. 3.1.** The main reservoirs and the transfer processes affecting the biogeochemical cycling of carbon and heavy metals in soil systems

mechanisms represent high CTB risks menacing human health and welfare by poisoning water and food resources.

SOM qualifies as a prime CTB candidate with respect to storage and release of heavy metals because of various reasons:

- It has a high specific storage capacity for heavy metals and, furthermore, is concentrated at the interfaces between soil and atmosphere as well as between soil and plants. It is, therefore, most directly exposed to heavy metal inputs due to pollution. Heavy metals released from SOM are in immediate contact to plant roots, from where they may enter the human food chain.
- Heavy metal binding to other soil compounds is known to be sensitively dependent on pH and the presence of organic chelators. Soil acidity as well as the composition and mobility of SOM are subject to considerable spatial and temporal variations. Such examples suggest that there are various possible mechanisms by which the retention of heavy metals can drastically change.
- The nature and dynamics of SOM strongly depend on the biological soil activity as well as on above-ground organisms including plants, which in turn are closely linked to processes of environmental changes (e.g., global climate change) and to effects of land use changes (e.g., deforestation or drainage of wetlands).

Whereas on one hand, SOM substantially affects the fate of heavy metals in soils, heavy metals may, on the other hand, significantly change the nature and dynamics of SOM if they accumulate with time and affect negatively the growth conditions of soil organisms and plants. Feedback mechanisms, which are involved in these mutual effects, may have a buffering, damping or a self-amplifying influence on the release of heavy metals which are bound to organic matter. The complexity of such feedback mechanisms is the major obstacle in the identification of CTB risks. Based on the rudimentary knowledge about many important mechanisms, only some very rough speculations can be described in this chapter.

This chapter gives a brief introduction to the composition, dynamics and ecological role of SOM. Mechanisms of heavy metal retention by SOM and the toxic effects of heavy metals to soil microbial activity are reviewed. Finally potential CTB scenarios are discussed.

## 3.2 The Cycling of SOM

**SOM in the Global Carbon Cycle.** SOM is a crucial compartment in the global C and N cycles. Most estimates of the total mass of organic carbon in soils range from 1.4 to 1.7 $10^{18}$ grams worldwide (Schlesinger 1986). This is slightly more than twice the present-day atmospheric carbon pool. The annual net release of carbon from soils by agricultural clearing has been estimated to be about 0.05% of the global soil organic carbon pool (Schlesinger 1986). The carbon pool in the SOM compartment is particularly important in temperate and frigid zones. In Great Britain, for example, 200 times more carbon is stored in soils (especially in Scotland's peat bogs) than in the vegetation (Pearce 1994). Through its dependence on climate and as a source or sink for atmospheric gases, SOM is closely linked to the greenhouse effect (Bouwman 1990).

**The Soil Ecological Role of SOM.** For the growth of higher plants SOM is an important factor. It serves as an irreplaceable energy and nutrient source for the soil biosphere and thus forms an indispensable basis for soil fertility (Stevenson 1982; Follet et al. 1987). Soil biological activity entails the decomposition and assimilation of organic matter. The mineralization of residual organic matter and the recycling of nutrients by plants is central in the coupling of soil to the above-ground biosphere.

Composition and turnover of organic matter has a profound bearing also on the physical and chemical properties of soils. The relationship between organic matter and soil structure is one of the most important functions in this context. SOM forms strong complexes between organic polymers and soil minerals, which significantly contribute to the stabilization of soil structure. In addition, indirect contributions to the formation and maintenance of the soil structure derive from the promotion of biological activity and plant growth (e.g., stabilization of aggregates by fine roots and fungal hyphae, protection of aggregates against rain drop splash by the vegetation canopy). Through its influences on soil structure as well as on pore surface wettability, organic matter also affects the water and air regime of a soil. As SOM is concentrated at the soil surface and in the rhizosphere, these effects are particularly important for the control of energy and matter transfer (e.g., heat exchange, water infiltration and evapotranspiration) across the interfaces soil–atmosphere and soil–plants, and last not least for the resistance of soil surfaces against erosion (Smith and Wischmeier 1957).

The role of SOM with respect to chemical soil functions is in no way inferior to its role in physical soil functions. Due to the oxidation of organic carbon, decomposition processes may strongly affect the redox potential and, by producing organic acids and carbon dioxide, also the pH of the soil solution. The negative charge of humic matter is often a major factor of the cation exchange capacity. Equally important is the high specific buffer capacity of soil humus.

Paramount for the chemical reactivity of SOM is the high density and diversity of functional groups. This polyfunctionality provides the basis to sorb most kinds of chemicals and to catalyze the degradation of many organic pollutants (Greenland and Hayes 1981; Schwarzenbach and Westall 1981; Huang and Schnitzer 1986; Sposito 1986). The ability of organic matter to bind cations is not only effective in binding pollutants, but also plays a substantial role in the dynamics of undisturbed soils. The formation of surface complexes by organic ligands may significantly promote the weathering of minerals (Stumm and Furrer 1987). Furthermore, the translocation of dissolved organo-metallic complexes is considered to be a key process in the formation of podzolic soils.

**Difficulties in SOM Research.** Because of its central role with respect to soil fertility (Follett et al. 1987; MacCarthy et al. 1990), SOM has been studied intensively since the beginning of the discipline of soil science. However, the complexity of its composition, structure and turnover dynamics has as yet defeated many efforts to reach a complete understanding. This complexity is characterized in particular by (1) an extreme heterogeneity of chemical composition and structure; (2) myriads of possible interactions between biotic and abiotic soil components; and (3) an extreme diversity of temporal and spatial scales of turnover processes. Furthermore, the study of soil processes is subject to considerable technical problems. Soils are very difficult to access for in situ observations and measurements, and, because many essential soil processes respond sensitively to changes of the soil structure, artefacts arise easily in sample preparation.

Nonetheless, due to the increasing power of analytical tools such as NMR (nuclear magnetic resonance), ESR (electron spin resonance) and FTIR (Fourier transform infrared) spectroscopy and due to the development of computer modeling, considerable progress has been made in the understanding of the nature of organic matter and its role in ecosystem dynamics, in particular during the last two decades, when growing concern about the impacts of increasing soil pollution and degradation has led to a new interest in SOM (Schnitzer 1991).

**Composition of SOM.** SOM in the widest sense includes all living and nonliving carbon-containing organic materials in soil, excluding charcoal (Oades 1988). The living fraction comprises the biomass of soil organisms, including plant roots. The nonliving fraction or necromass may be subdivided operationally into macroorganic matter and humus. The term *humus* however, is sometimes also used in the wider sense of soil necromass (Gisi et al. 1990) Macroorganic matter is the residual nonliving organic matter that still preserves identifiable morphological structures of its origin. In particular it includes plant and root litter and their fragments of visible size, representing the primary resources for soil metabolism. Humus is the unstructured nonliving material that remains after removal of macroorganic matter (Oades 1988). Its traditional classification into humic and nonhumic substances is based on chemical characteristics.

*Non-humic substances* comprise all compounds which are also found as biopolymers or intermediary metabolites in the living matter. Quantitatively most important are carbohydrates such as cellulose, hemicellulose and mucopolysaccharides, aromatic polymers such as lignins, tannins and microbial melanins, proteins including extracellular enzymes (exo-enzymes), and aliphatic biomacromolecules such as cell wall lipids, suberins, and cutins (Kögel-Knabner 1992). Monomers are quantitatively less important since they are in general rapidly metabolized by microbes and, thus, have only very short life times. They comprise in particular degradation products of humic and nonhumic polymers, prevalently carbonic acids, which may occur in soil solutions at concentrations up to millimolar, but also amino acids, phenolic compounds (e.g., quinones), and sugars. Many of these substances are strong metal chelators and highly soluble. Despite their low concentrations, they may thus contribute significantly to the mobilization of metals by desorption and mineral dissolution and to their transfer into deeper soil horizons.

Non-humic substances can get into the humus fraction on various pathways: by active excretion (e.g., exoenzymes, microbial mucus, siderophores) or passive leakage through cell walls (small metabolites), by release through cell decay, and by extracellular degradation of larger units in the course of the decomposition process.

Due to their origin, nonhumic polymers are characterized by subunits with well-defined and easily recognizable chemical structures. *Humic substances* on the contrary are characterized by extremely irregular structures. The related concepts so far have only partially been revealed (Frimmel and Christman 1988; Oades 1989; Sposito 1989). Humic substances account for 70 to 80% of the humic material in most inorganic soils (Schnitzer 1986). They are traditionally fractionated and accordingly classified on the basis of their solubility in acids and bases into (1) completely insoluble material, termed "humin"; (2) acid-insoluble, but base-soluble "humic acids"; and (3) the remaining soluble fraction of "fulvic acids".

Humic and fulvic acids represent a heterogeneous mixture of interacting colloidal polymers with the properties of high polyfunctionality, macromolecular charge, hydrophilicity, and structural lability (Sposito 1989). The ability of humic and fulvic acids to form stable complexes with metal ions is based on their high content of electron-donor functional groups, in particular -COOH and phenolic -OH (Gamble et al. 1970), but also $-OCH_3$, $-NH_2$, =N, -SH, and -C=O (Schnitzer and Khan 1972; Stevenson 1982). While fulvic acids are fairly strong acids (about half of the carboxyl groups are dissociated at pH = 3.5), humic acids are dissociated to the extent of 50% only at pH = 6 (Buffle 1988). The humic matter may contribute substantially to the cation exchange capacity of a soil.

**Decomposition and Humus Formation.** Decomposition is the gradual breakdown and final mineralization of nonliving macroorganic matter by soil organisms. It involves an iterative sequence of catabolism, comminution and leaching processes and is controlled by "site-specific environment", "resource quality", and "decomposer community" (Swift et al. 1979). The breakdown starts with the death of an organism by autolytic processes. As the cell membranes lose their semipermeability, solute leaching leads to a selective preservation of refractory structures. The leachates serve as a readily available substrate for microbial activity and growth. Microbial attack by excreted enzymes will promote the breakdown process. Soil fauna feeding on microbes accelerates the release of mineral nutrients which are taken up and temporarily stored as microbial biomass.

Undigested residues can again serve as a substrate for microbes or other saprovoric organisms. The rate of the decomposition process is significantly dependent on the structural integrity of the macroorganic substrates. Comminution and production of feces by the soil meso- and macrofauna facilitates further microbial breakdown, as it increases the accessible surface area for further attack and mixes the substrate fragments with the decomposing microorganisms present in the gut. As a portion of the degraded organic matter is used to build up new biomass, a new decomposition cycle is initiated. Eijsackers and Zehnder (1990) compared the recursive structure of these processes with a series of Russian dolls fitted into another.

Reactions with passively or actively released substances, mechanical homogenization, and selective extraction of metabolites by organisms result in a large manifold of changes in the chemical composition and physical structure of the organic substrate. These may be conceived rather as a displacement within a continuous space of chemical and physical quality than as a cascade of transitions between discrete organic matter pools (Bosatta and Ågren 1985 1991).

The continuous transformation of primary sources by the iteration between direct transformation, selective preservation and microbial resynthesis is tantamount to the humification of organic matter or humus formation (Hatcher and Spiker 1988; Kögel-Knabner 1992).

**Turnover of SOM.** Size, composition, and turnover rate of the SOM pool are essential factors for the retention and transfer of heavy metals. Being directly dependent on the biological activity of the soil, these factors are governed by state variables such as soil temperature, moisture content, pH, oxygen availability, mineral nutrition status, and stress by toxic chemicals. These factors in turn are influenced by site conditions such as the above-ground climatic conditions, type of vegetation, stage of growth, fertilization regime, and the input of pollutants into the soil (Swift et al. 1979). An example for the importance of the plant nutrition status is the production of organic acids by plant roots in periods of phosphorous shortage.

**Modeling of Organic Matter Cycling.** Since the study of SOM cycling is attended by many difficulties, indirect approaches employing mathematical modeling and computer simulation have become increasingly important (van Veen 1987). Among the large variety of such models the single-state variable model of Jenny (1941) was one of the earliest. The effects of changes in environmental conditions on the kinetics of organic matter decomposition were often accounted for by lumping the effects of soil climatic variables in empirical modifications of rate parameters (e.g., Bunnell et al. 1976).

Whereas black-box lumped-parameter models or statistical models like the regression model of Meentemeyer (1978) may be useful for data reduction or extrapolation purposes, mechanistic models like those by Browder and Volk (1978), Smith (1979), McGill et al. (1981), Knapp et al. (1983), van Veen et al.

(1984), or Verberne et al. (1990) have given substantially more insights into the dynamics of SOM turnover as they are conceptually designed to consider the structure of the investigated systems. Mechanistic models are particularly useful because they help (1) to integrate fragmentary knowledge and to improve the understanding of the system's behavior in its entirety: (2) to formulate hypotheses and to derive verifiable predictions; (3) to establish the relative importance of the various system parameters through sensitivity analysis; and (4) to identify and estimate parameter values based on experimental and field observations.

A model explicitly designed to synthesize knowledge on organic matter decomposition has been developed by Smith (1979). It is based on a comprehensive compilation of published experimental data of that time. The model is composed of four submodels, each one accounting for the nitrogen, phosphorus, potassium and carbon turnover of a different fraction of organic matter. In addition to the explicit simulation of microbe dynamics, it treats physical and chemical processes of precipitation, input of fertilizers, dissolution of soil parent material, nutrient leaching, sorption of organic and inorganic chemicals onto soil colloids, chemical reactions between inorganic nitrogen and aromatic compounds, with process rates being functionally dependent on temperature and moisture. However, like the model of Knapp et al. (1983), it focuses on short time scales and does not account for the long-term dynamics of humic substances.

**Different Fractions of SOM.** Studies on the long-term dynamics of SOM using $^{14}C$-dating and $^{14}C$-tracer methods have shown that the life-spans of different fractions vary between less than one and several thousand years (Jenkinson 1977; Jenkinson and Rayner 1977). Based on these findings, mechanistic long-term models designate SOM with different stabilities to separated compartments (van Veen and Paul 1981; van Veen et al. 1985; Parton et al. 1987; Verberne et al. 1990). The differences in the turnover rates of the various carbon pools are not only determined by their stabilities, but also by physical and chemical inhibition mechanisms (Jenkinson and Rayner 1977; Ladd et al. 1981; Anderson and Paul 1984; van Veen et al. 1985).

Since they are applicable in a similar way to nonliving as well as living fractions of SOM, multi-region concepts considering physical and chemical protection may provide a general clue to the understanding of SOM dynamics (van Veen et al. 1984; Alexander and Scow 1989). Van Veen et al. (1984) explained contrasting carbon and nitrogen metabolism in a sand soil and in a clay soil by their different preservation capacities for microorganisms as well as differences in physical protection of organic matter against biological breakdown.

Various mechanisms are known which could serve to protect living or nonliving organic matter against biotic or abiotic decomposition. Purely physical protection results from differential accessibility depending on the location in the pore space (Elliott et al. 1980). Other important physical or chemical stabilization of SOM may result from association with mineral surfaces or recalcitrant humic substances (Martin 1971; Tiessen et al. 1984; Huang and Schnitzer 1986).

**Decomposition, Substrate Quality and Physical Availability.** The decomposition of organic matter during its entry into soils is directly related to the metabolic rate of the decomposer populations. These in turn depend on the simultaneous availability of nutrients and energy. Carbon utilization is closely coupled to the availability and quality of N sources. It has been reported, for example, by Knapp et al. (1983), that excess C is apparently immobilized as polysaccharides when N is growth limiting, whereas excess N is partially immobilized by uptake into the biomass when available C is limiting. Fairly detailed models were developed in order to represent mechanistically the observed C and N transformation behavior during decomposition (e.g., Parnas 1975; Frissel and van Veen 1981; Knapp et al. 1983; van Veen et al. 1985; Verberne et al. 1990).

Besides the C:N ratio, the lignin content of the substrates is also an important parameter influencing decomposition and humification rates (Fogel and Cromack 1977; Meentemeyer 1978; Melillo et al. 1982). Parton et al. (1987) succeeded in predicting regional trends of steady-state contents of SOM using four site-specific variables: temperature, moisture, soil texture, and plant lignin content. Due to very slow degradation, the intermediates of lignin biodegradation contribute decidedly more to humification than the more easily degradable plant components (Hatcher and Spiker 1988).

The formation of stable organo-mineral complexes leads to the aggregation of soil particles. Further protection of the organic matter, which is bound inside dense aggregates, results from the poor accessibility of these locations to the metabolic activity of soil microorganisms. This is firstly due to the fact that organisms are excluded from pores by their size. Secondly, also enzymes are bound by mineral surfaces and thus partially inactivated. Further, diffusion is generally by orders of magnitude slower in fine pores than in a bulk solution. Thus, also indirect metabolization by exoenzymes is very much reduced.

**Climatic Factors.** The influence of substrate quality on organic matter decomposition can be strongly modified by climatic variables, in particular soil temperature and moisture regime (Birch 1958; Sørensen 1974; Meentemeyer 1978; Clark and Gilmour 1983; Parton et al. 1987). The knowledge of the responsible mechanisms for the observed effects of climatic factors seems to be very limited, however. Since the way how these environmental variables are incorporated into the models profoundly affects the obtained results (van Veen 1987). Inadequate understanding of the impact of climatic conditions on organic matter turnover may be one of the major reasons for the poor predictive capabilities of mechanistic models when applied to field conditions. Not only conditions of average temperature and moisture are affecting decomposition, but sudden changes (rewetting of a dried soil, thawing of frozen soils, etc.) are equally important to the dynamics of SOM.

**Trophic Interactions.** Because of the lack of adequate data, trophic relationships between different groups of organisms have usually been neglected in organic matter models. One exception is the model of Paustian (1985), which distinguishes between bacteria and fungi. However, new information on food webs in soil ecosystems (e.g., Ingham et al. 1985; Ingham et al. 1986a,b; Hunt et al. 1987; Freckman 1988; Brussaard et al. 1990; Moore et al. 1990) now allows the improvement of the models with respect to trophic interactions.

The turnover of SOM may depend in several ways on the composition and dynamics of the food web. An effect which has received much attention is the acceleration of nutrient recycling by predation (Coleman et al. 1978; Hunt et al. 1984; Coleman 1985). Thus, grazing of bacteria by protozoa and bacteriophagous nematodes may enhance nutrient availability through release from microbial biomass. This in turn can stimulate microbial growth and organic matter decomposition. On the other hand, excessive grazing reduces bacterial numbers (Freckman 1988).

In a similar way as substrates can be protected from biodegradation by physical exclusion of organisms from fine pores, organisms can be sheltered against

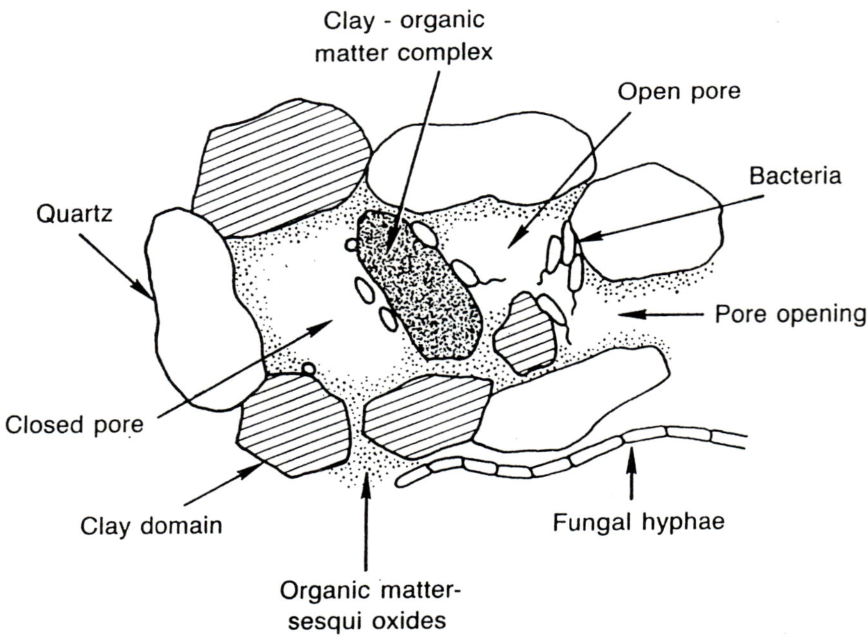


**Fig. 3.2.** Microscopic associations between organisms, organic matter and mineral particles (from Paul and Clark 1989)

predators and other harmful impacts (Fig. 3.2). Particular attention has been given in this respect to interactions between particles of the clay fraction and soil bacteria (cf. Huang and Schnitzer 1986; Oades 1988). In their SOM turnover model van Veen et al. (1984) accounted for such protection effects by partitioning the microbial biomass into a protected and a nonprotected fraction and by assigning the protected fraction a lower death rate than the nonprotected one.

Hunt et al. (1987) calculated the nitrogen flux through the detrital food web of a short-grass prairie and found that fauna directly account for more than one-third of N mineralization. Using an approach similar to Hunt et al. (1987) based on the same food web, Moore et al. (1990) obtained higher N turnover rates for integrated than for conventional agricultural management practices. In the integrated plot, N turnover was significantly faster in the upper 10 cm than in the 10 to 25 cm layer beneath, whereas no such difference was found in the conventional plot.

Additionally, the rhizosphere is a favorable habitat for acid-producing bacteria. Therefore, the concentrations of organic acids in the rhizosphere are significantly higher than in the bulk soil.

## 3.3 Interactions between SOM and Heavy Metals

The mobility and bioavailability of heavy metals depend crucially on their chemical speciation. Metals can exist in solution either as free ions or inorganic and organic complexes. They can also be associated with SOM and minerals by adsorption, precipitation, occlusion, or by uptake and incorporation into organisms (Fig. 3.3). Some heavy metals such as mercury, chromium, iron, and manganese occur at different oxidation states depending on the redox potential. Additionally, if slow redox processes are involved (e.g., reductive dissolution of iron oxides, oxidation of dissolved Cr(III)), the actual distribution of a metal between the various oxidation states can deviate quite far from thermodynamic equilibrium, depending on the boundary conditions and on the reaction kinetics. In soils, the speciation of metals is dominantly affected by the composition of the organic matter as well as of the mineral surfaces.

**Chemical Association of Metallic Cations with Organic Matter.** Chemical binding mechanisms may be classified into "passive" and "active" bonds. Binding is considered to be passive when it is primarily driven by entropy, without a significant contribution of the enthalpy term to the free energy (Hassett and Banwart 1989). Hydrophobic sorption of nonpolar organics onto organic surfaces surrounded by an aqueous phase belongs to this category. The strong internal forces of the liquid water phase are the main cause of the hydrophobic effect. Nonionic and nonpolar solutes are only sparingly soluble in water because they disturb the internal structuring of the solvent. Whereas direct interactions are

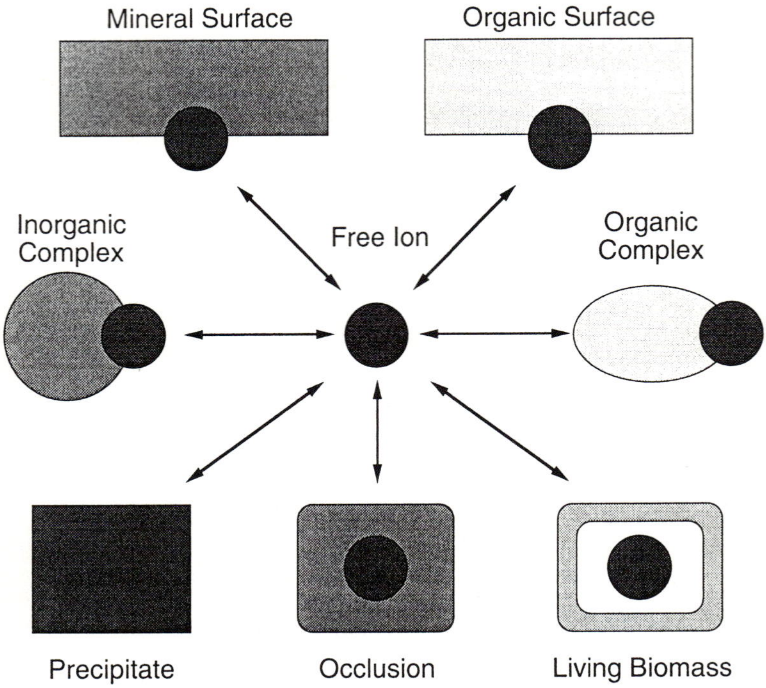


**Fig. 3.3.** Heavy metal interactions in soil

very weak in passive binding, active bonds involve preferentially forces between the binding partners and thus are primarily enthalpy driven. Active binding forces may be due to classical covalent and ionic bonds, p bonds, complex formation, ion–dipole and dipole–dipole interactions, including London-van der Waals forces, as well as purely electrostatic attraction due to Coulomb forces. A special chemical interaction, which does not fit into the above classification scheme, is the hydrogen bonding. It results from binding interactions of a hydrogen atom with the lone electron pairs of two or more other atoms.

The spectrum of metal bonding types in soils ranges between purely electrostatic attraction, ion–dipole interactions, outer-sphere and inner-sphere complexes. Inner-sphere complexes are formed between an electron-accepting cation and one or more electron-donating ligands (Stumm and Morgan 1981) and involve direct contact between cation and ligands. Special types of inner-sphere complexes are p-electron-donor systems in which cations react with the p orbitals of unsaturated and aromatic compounds. Outer-sphere complexes are ion pairs which form stable units. However, in contrast to inner-sphere complexes, the ions remain separated from each other by at least one solvent molecule (Sposito 1984).

Inner-sphere complexes are the dominant binding type between heavy metals and organic ligands. Most reactive are carboxyl and phenolic hydroxyl groups. Organic molecules with more than one electron-donating functional group may form polydentate complexes with the same cation. Such complexes are called chelates. Due to high stability and compensation of positive charge, dissolved chelate complexes can greatly increase the mobility of metal ions. On the other hand, complexes with nondissolved organic matter lead to immobilization of metal ions.

Complexes may also involve more than one ligand molecule. In soil, this results in a wide variety of interactions between different organic or inorganic ligands with cations playing the role of bridges. These are termed mixed or ternary complexes. They are of particular importance in providing an effective mechanism for the sorption of negatively charged organic molecules on equally charged mineral or organic matrix surfaces (Schindler 1990). Multiple complexation can lead to an entire network of ligands bound by cations if the ligands can bind to more than one cation through different functional groups. The stability of this network increases with the number of participating ligand groups per molecular or mineral entity. Large polymers with a high density of carboxyl groups such as humic and fulvic acids, thus provide particularly effective bridging structures. Furthermore, polyvalent cations play a crucial role for the aggregation and coagulation of organic matter and mineral colloids.

Coordination between cations and polar molecules is also possible without covalent bonding through ion–dipole interactions. These interactions result in the coordination of cations with surrounding water molecules. On the basis of mass action competition for coordination positions, other polar molecules such as alcohols may replace the water molecules of the solvation shell.

**Formation Kinetics of Heavy Metal Complexes.** The rate of complex formation with simple ligands is independent of the ligand's structure and only limited by the dehydration rate of the metal cations. However, with tangled ligands, complex formation is highly dependent on the nature of organic matter. The reaction is slower if a change in conformation or a diffusion step is required prior to the formation of organo-metallic complexes than without such complications. Moreover, adsorption is faster if the selectivity of the ion exchanger for the cations is low. The half-times for adsorption and desorption of $Pb^{2+}$, $Cu^{2+}$, $Cd^{2+}$, $Zn^{2+}$ and $Ca^{2+}$ with organic matter in vigorously stirred aqueous suspensions of peat were observed to be in the range of 5 to 15 seconds (Bunzl et al. 1976a).

The sorption kinetics can be also affected by the metal/ligand ratio. Bunzl et al. (1976b) found that half-times for complex formation with humic acids in stirred solutions changed with total Zn(II) concentration, however, they changed very little in the case of Cu(II) and Cd(II). The half-times for the sorption of all three metal ions were around 30 seconds. Similar reaction half-times (typically 60 seconds) were found for ligand exchange reactions of Cu(II)-humate complexes in stirred solutions by Hering and Morel (1990).

These results suggest that the kinetics of complex formation between bivalent metal cations and SOM are not limiting in soil environments and that the approach of thermodynamic equilibria is fairly accurate for this type of reaction.

**Thermodynamic Equilibria of Metal Complexes**. *Stability of complexes.* Whereas strongly hydrated monovalent cations form outer-sphere complexes with organic anions, bivalent metal cations generally form inner-sphere complexes. The stability of metal-organic complexes is mainly determined by the number and chemical nature of atoms that form a bond with the metal ion, the number of rings that are formed, the number of atoms in the rings, the nature of metal ions, the concentrations of metals and ligands, pH and ionic strength. Since the stability of organo-metallic complexes varies with the different metal cations, it is appropriate to classify them in a stability sequence. However, stability series depend critically on the pH and the chemical characteristics of the organic material (Schnitzer and Khan 1972; Stevenson 1982). $Pb^{2+}$ and $Cu^{2+}$ form stronger complexes with SOM than $Cd^{2+}$ and $Zn^{2+}$ (Fig. 3.4). The stability of metal-organic complexes increases with pH due to increasing dissociation of functional groups, especially carboxyls. Furthermore, the stability of complexes between metal ions and fulvic acids decreases with increasing ionic strength (Schnitzer and Hanson 1970).

**Retention and Translocation of Heavy Metals in Soils.** Metal adsorption on soil matrix surfaces is approximately a linear function of the total metal concentration in solution at low concentration, but approaches a saturation limit at high concentrations (Davies et al. 1969; Marinsky et al.1982; Schmitt and Sticher 1991). The sorption capacity of organic matter for metals is given in general by the amount of negatively charged functional groups. To a first approximation, these correspond to the number of dissociated carboxyl and phenolic hydroxyl groups. The exchange capacity of humic substances is typically in the range of 1.5 to 5.0 mEq./g. Bloomfield and Sanders (1977) found that the specific Cu-binding capacity of organic matter extracts increased with decreasing molecular size, which may be attributed to the higher concentration of phenolic and carboxylic groups of fulvic acids as compared to humic acids (Stevenson 1976).

The transport of metals has manifold effects on the overall biogeochemical processes such as changing the steady-state equilibrium as well as microbial toxicity. Both small organic acids (aliphatic acids, amino acids, polyphenols) and fulvic acid-type constituents are involved in the formation of mobile complexes in soil. $Pb^{2+}$ and $Cu^{2+}$ are generally bound to the high-molecular weight fractions of dissolved organic carbon which results in the almost complete removal of these metals from the soil solution by their coprecipitation in the B horizon. $Zn^{2+}$ and $Cd^{2+}$ tend to be associated with medium-molecular-weight dissolved organic carbon and therefore, are translocated easily in the soil profile. Waller and Pickering (1990) studied the distribution of $Cu^{2+}$ between "fixed" and "soluble"

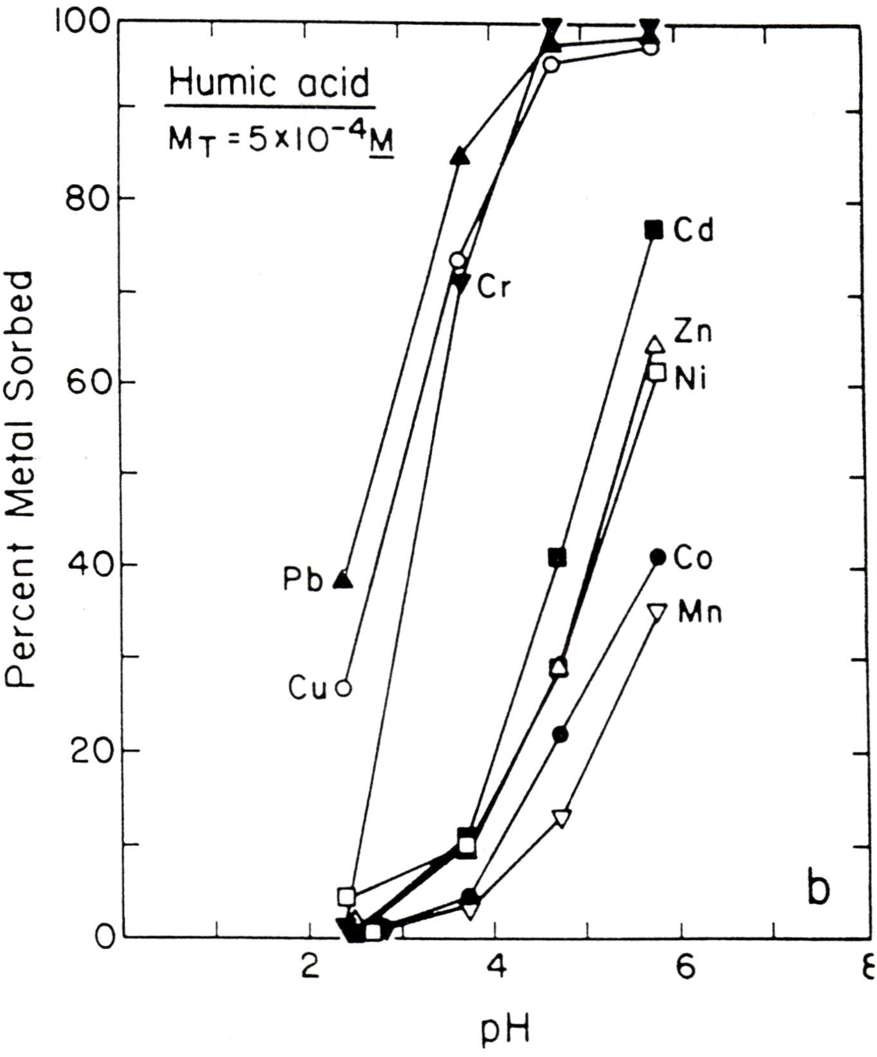


**Fig. 3.4.** The effect of pH on the sorption of heavy metals by humic acids (Sposito 1984)

$Cu^{2+}$-humic acid complexes at different pH values from 5 to 8.5. They found an increasing soluble fraction with increasing pH affecting the transport behavior of copper.

Soils with high contents of solid organic matter and/or clay minerals can accumulate large amounts of heavy metals. In general, their sorption capacities are used up only after decades or centuries. Changing conditions such as decreasing pH, increasing concentrations of organic acids, or decreasing redox potential, however, can drastically decrease the binding strength for heavy metals and concomitantly increase their mobility and bioavailability.

## 3.4 Heavy Metal Interactions with Soil Biota

Whereas heavy metals are harmless in trace concentrations or may even be essential as micronutrients, they are severely toxic for organisms at higher concentrations (Tate 1987). The interference of excess amounts of heavy metals with essential biotic soil functions such as organic matter decomposition, nitrogen fixation and nitrification must be expected to have substantial impacts on the entire ecological dynamics of soil. In particular the high metal sensitivity of functions such as nitrogen fixation which can be performed by only a few species in soil must be considered as very critical. Simple cause-effect mechanisms are, however, by far not the whole story of metal ecotoxicity, but only the pixels of a complex picture. Involving the mechanisms which have been outlined in the preceding sections, the bioavailability of metals in soil will in turn strongly depend on biotic processes. A further complication for the disentanglement of the intricate system of feedback relationships results from the fact that processes with very different time scales are involved and that in the long run, evolutionary processes gain more importance for the reaction of ecosystems under stress. With respect to soil pollution by heavy metals, the development of genetically metal-insensitive species is found.

**Metal toxicity.** The toxicity of metals derives primarily from their capability to form stable complexes with proteins which may denature the conformation of active sites or displace essential metal cofactors. Extracellular (exogenous) enzyme activity is affected in much the same way as intracellular enzyme activity. Enzymes containing metals as a structure element are very metal specific and substitution by another metal may induce the complete loss of the enzyme activity. Most enzymes belong to the class of metal activated enzymes, where the metal specificity is small. Thus, the substitution by another metal produces no, or very small reduction of activity. As reported, enzyme synthesis decreased drastically in response to Pb(II) addition, whereas the enzyme activity itself was very little affected (Cole 1977). Thus low enzyme activity can also be due to low concentration of the enzyme. The decrease in enzyme activity found in several investigations is probably mainly an effect of a decreased enzyme synthesis associated with inhibited microbial growth than of direct enzyme inhibition by metals (Bååth 1989).

Boudot and Bel Hadj Brahim (1989) reported that the biodegradation of metal-organic complexes was slower than of the free forms of the corresponding organic compounds. The protective effect appeared primarily in complexes involving highly charged metal ions ($Al^{3+}$, $Fe^{3+}$) and was the result of flocculation of initially soluble organo-metal complexes. These complexes reduce the exposure of organic matter to extracellular soil enzymes and, thus, degradation and mineralization (Boudot 1992).

Toxic effects may also be caused by specific metal complexation at the cell wall which may lead to disturbance of the cell functions (Collins and Stotzky 1989). Rich and Horsfall (1963) suggested that metallic ions may destroy the integrity of semipermeable membranes, allowing metabolites to leach. Metal toxicity can also be caused by the competition of metal cations with essential nutrients for available exchange sites at the cell wall.

**Phenomenological Effects of Heavy Metals on Microorganisms.** Manifestations of toxic effects on microorganisms due to heavy metals have been found in a variety of life functions, including growth (Bååth 1989), sporulation (Doelman and Haanstra 1979b), adhesion, morphology and biochemical activities (Collins and Stotzky 1989), respiration (Gabteni and Gallali 1988; Monteil 1989; Schindler et al. 1991), nitrogen fixation and nitrification (Morrissey 1975), mineralization of nitrogen (Liang and Tabatabai 1977) and phosphorous (Tyler 1981).

In general, soil microorganisms are most sensitive to metals during spore germination. Küster and Grün (1984) showed in tests on agar, that Cd(II) additions of more than 5 ppm resulted in a significant decrease in the germ production of bacteria, fungi and actinomycetes. Doelman and Haanstra (1979b) found similar results for Pb. The addition of 5 to 10 ppm Cd(II) lengthened the lag phase of *A. tumefaciens* from 3 to 6 h (Babich and Stotzky 1977). Blundell and Wild (1969) found that Cd, Cu, Zn, Hg, and Ni caused a reduction in the synthesis of RNA and protein in *E. coli*, whereas the addition of 5 to 10 ppm stimulated growth of *E. coli* (Doyle et al. 1975).

**Soil Chemical Factors of Heavy Metal Toxicity.** It has long been known that metal toxicity depends on the chemical speciation (Piscator 1986). In general, the free metal ion is the most toxic form, in the sense that the uptake of metals by living cells requires that complexing ligands have to be shed (Sposito 1989). The important exception are special "scavenger compounds", which are produced and released by organisms. Since these scavengers themselves have to compete with other ligands for the free ions, this uptake mechanism is also directly related to the free metal concentration, especially at low metal concentrations.

Contrary to the free concentration, the total concentration of a metal has no direct bearing on the toxicity. The higher the sorption capacity of the soil matrix, the lower is the bioavailability and thus the toxicity of a heavy metal. This relationship explains the results of Doelman and Haanstra (1979a) who found that soil respiration and dehydrogenase activity was much less inhibited at the same total lead concentration in a clay soil than in a sand soil and even less in a peat soil. The prediction of free dissolved metal cations, however, is far from being trivial because metal speciation depends highly on interactions with the colloidal organic matter fraction. Depending on the nature of the organic matter, these interactions may have opposite effects. Association with insoluble organic compounds tends to reduce metal availability, whereas organic chelators tend to mobilize a metal from insoluble minerals and may also increase its availability (Tate 1987).

Due to the dominant role of pH for the soil chemical milieu and in particular metal chemical speciation, it is very difficult and often impossible to distinguish experimentally properly between metal toxicity sensu stricto and other toxic effects related to pH. Investigating effects of soil pH and heavy metal contents, Monteil (1989) and Schindler et al. (1991) found the rates of microbial processes to be reduced by high metal contents and surprisingly, by increased pH. This pH effect could be interpreted in terms of a competitive Langmuir-type metal sorption by microorganisms and soil particles. In relation to the microbially sorbed concentrations, Cd(II) exhibited the highest toxicity followed by Pb(II) and Cu(II), respectively. Welp and Brümmer (1989) indicated that the toxicity of metals for microorganisms generally decreases in the following sequence: $Hg^{2+} > Cd^{2+} > Cu^{2+} > Ni^{2+} > Zn^{2+} > Pb^{2+}$, whereas the chemical boundary conditions have not been precisely defined.

Inorganic anions affect the speciation of metals and thus their toxicity to microorganisms, e.g., the toxicity of Cd decreased at higher $Cl^-$ concentration due to the formation of negatively charged cadmium-chloride complexes. These complexes have lower affinities for net negatively charged cell surfaces than $Cd^{2+}$. Certain ions like $CO_3^{2-}$, $PO_4^{3-}$ and $S^{2-}$ form insoluble salts with most heavy metals that are unavailable for uptake by microorganisms (Collins and Stotzky 1989).

Cations, present in high concentrations such as $Ca^{2+}$ or $Mg^{2+}$, can compete with cationic metal species for negatively charged sites on the microbial cell surface thus influencing the toxicity (Collins and Stotzky 1989). For example; it has been reported that the toxicity of Zn and Hg to *C. vulgaris* decreased with increasing concentrations of Ca (Rai et al. 1981). Metal interactions with soil organisms are further complicated by the fact that different metabolic processes may be affected simultaneously, but with different sensitivity. Further, the action of one metal may tend to cancel or to amplify the effect of another metal, while the combined action may also depend on the metal concentrations. However, the detailed knowledge of the ecological relevance of such interactions seems to be scarce (Stadelmann and Santschi-Fuhrmann 1987). Figure 3.5 shows a concept of the various ways in which SOM turnover can be affected by heavy metals. Although this picture is very much simplified in relation to the complexity of real soil systems, it nonetheless illustrates how different effects may depend on the partitioning of the metals into various pools.

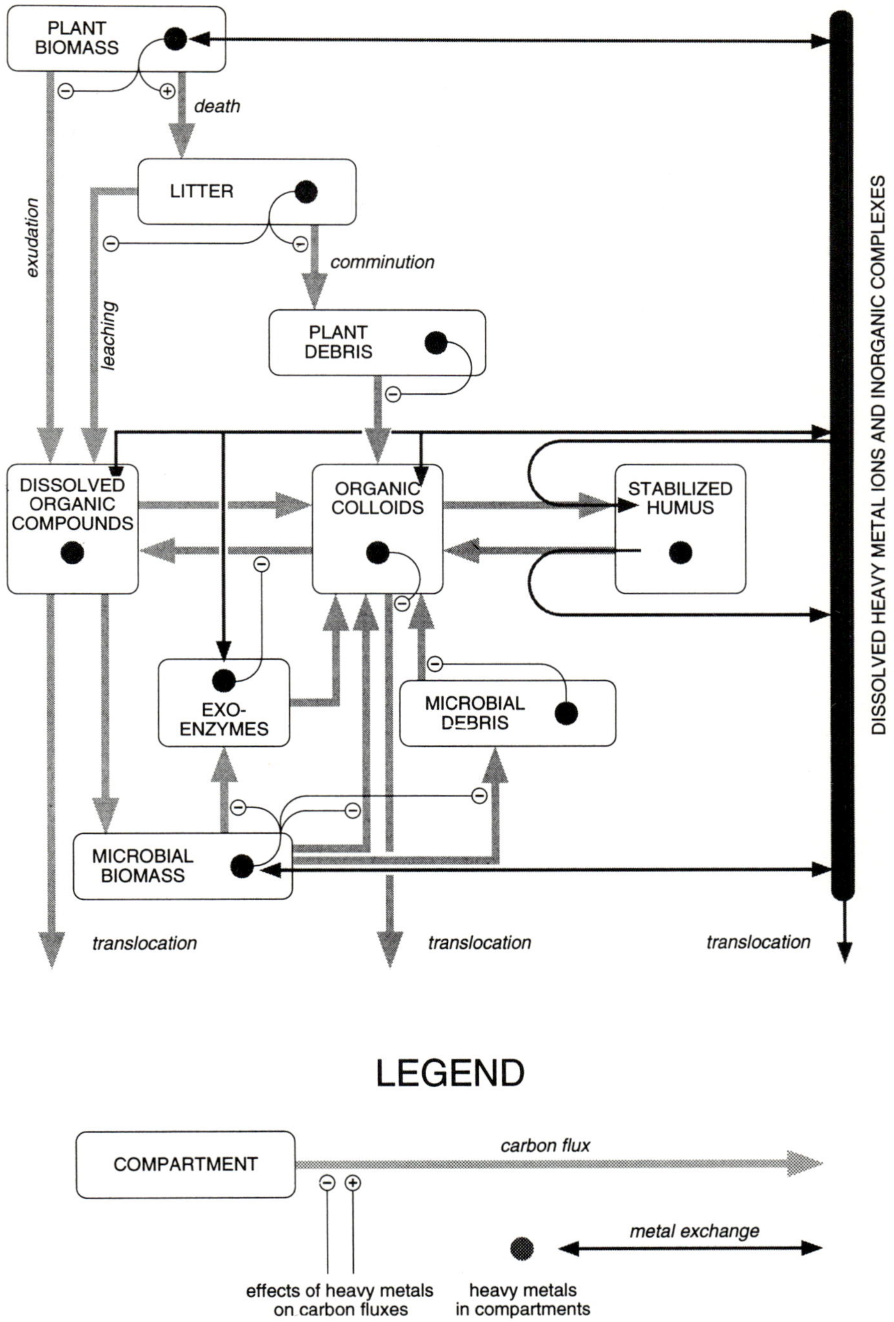


**Fig. 3.5.** Influences of heavy metal pollution on SOM turnover with respect to the compartments of organic matter in which the heavy metals are bound. Some pathways of carbon fluxes (e.g., microbial exudation and plant uptake of metal-organic complexes) and all fluxes of heavy metals between the compartments have been omitted for clarity. Postulated effects of heavy metals in the various compartments on carbon fluxes are represented by the *positive* or *negative signs*

**Microbial Impacts on Metal-SOM Interactions.** Whereas heavy metal stress can have strong effects on soil biological processes, soil organisms may modify in turn heavy metal availability in various ways, in particular through the production and release of acidity, chelators, colloidal polymers, reducing agents, oxygen consumption, and breakdown of organo-metallic complexes (Tate 1987). The effects are manifold:

- The modification of pH and redox potential by microorganisms increases or reduces metal availability.
- Microbially synthesized organic acids may enhance mineral weathering through simple protonation or deprotonation reactions or by chelation.
- Complex-forming ligands excreted by microorganisms support the transport of metals into deeper soil horizons.
- Microbial mineralization of the organic ligands favours the formation of inorganic metal complexes.
- Bioaccumulation of metals in capsular or slime layers and along cell walls can remove metallic cations from the soil solution (Macrae and Edwards 1972).
- Many metal transformations are enzymatically mediated at rates far in excess of those of chemically catalyzed reactions.

On the basis of such interactions, a large variety of feedback mechanisms are conceivable. For example, carbon mineralization may be inhibited by liberated metals from the organic matter being decomposed (Martin et al. 1966). As the storage capacity of organic matter for metals tends to increase with the degree of humification (Stevenson 1982), it may be speculated that such a negative feedback is stronger for the decomposition of stabilized humic matter than for labile organic compounds.

**Ecological Impacts of Heavy Metal Pollution on SOM.** The ecological relevance of heavy metal effects on SOM turnover has been demonstrated or at least strongly suggested by many studies. For example, Watson et al. (1976) found accumulation of undecomposed litter on heavy metal polluted soils near a mining-smelting complex. Strojan (1978) observed considerable forest floor organic matter accumulation on zinc polluted soils in the vicinity of a zinc smelter in Pennsylvania. Plant litter accumulation in heavy metal contaminated woodland soils may also be caused by a decreased activity of soil invertebrates (Hopkin 1989).

Tyler (1981) investigated the relationship between the level of metal pollution and the decomposition rate of various litter fractions. He found that inhibitory effects may be small during the first weeks when easy degradable substances are available for microorganisms, but thereafter may become significant. These results are in accordance with Strojan (1978) who also showed that the inhibitory effect of metals on decomposition increased with time.

The inhibition of microbial activity on metal-polluted sites has been found to be enhanced by sparse vegetation coverage (Shaw 1989), nutrient loss (Strojan

1978), dryness, changes of pH values, diminution of litter input (Jordan and Lechevalier 1975) and protection of degradable SOM. The results of Chander and Brookes (1991) suggest that both, decreased efficiency of litter decomposition and conversion into microbial biomass limit the production of biomass on metal-contaminated soils. The latter mechanism is assumed to be more important since the availability of fresh substrates is a major regulator of the growth of soil microbial biomass (McGill et al. 1981, Merckx et al. 1985; Martens 1990).

Despite the large body of field work indicating toxic effects due to heavy metal pollution, there is little conclusive evidence with respect to the causal mechanisms, because under field conditions it is generally not possible to determine the role of a given factor independently of other factors.

**Long-term Adaptation of Soil Organisms to Heavy Metal Stress.** Long-term sensitivity of soil organisms to heavy metals can be very different from short-term effects. Due to different responses to changes in metal concentrations, including evolutionary processes, the population spectrum in soil will change with time (Stadelmann and Santschi-Fuhrmann 1987). Such adaptations will also find their expression in correlations of the spatial distribution of species with metal concentrations in the soil. For example, Nordgren et al. (1983) observed that the population density of some fungal genera increased, whereas that of others decreased along a gradient of soil contamination by copper and zinc around a brass mill. Compared to the metal effect, variations of SOM and soil moisture content were statistically of little influence. Duxbury and Bicknell (1983) analyzed bacterial population responses in a metal-contaminated soil and an uncontaminated soil. Their data suggest a subdivision of bacterial populations into a group with wide metal tolerance and a group of narrow tolerance. Soil fungi are more resistant than soil bacteria (Beck 1981); gram-negative bacteria more tolerant than gram-positive bacteria. In plants occurring on soils with high natural levels of certain heavy metals, high tolerance to these metals is a commonly observed phenomenon.

Development of metal resistance in a population will usually be very slow. Even in bacteria, where genes for metal tolerance are usually carried on plasmids, it may take several years (Tate 1987). Because of this and because of low heavy metal mobility in soils, biological effects may persist for years. McGrath et al. (1988) found that even 20 years after severe heavy metal contamination of a soil due to sewage sludge application, clover rhizobia was still ineffective and, microbial biomass was still reduced by 50%, whereas growth of grass appeared to be unaffected.

The selection of metal tolerant species may results in a decreased diversity of microorganisms. This can have significant ecological effects: (1) impairment of the steady-state situation between microbial phytophags and saprophytes in favor of the phytophags; (2) impairment of population regulating mechanisms (e.g., predation, grazing, microparasitism); and (3) impairment of the organism succession and thereby the microbial regulated matter flow and matter stabilization in the microbial biomass (Domsch 1985). Strojan (1978) revealed a decrease in diversity of arthopods in areas with high metal pollution. In grasslands near a smelter no earthworms and snails were found anymore (Geiger 1990). The breakdown of single functional groups of the food webs may result in a change of the food webs dynamics and thus strongly influence SOM.

## 3.5 CTB Risks Associated with Heavy Metal Retention in SOM

According to the concept of CTB in soils and sediments by Stigliani et al. (1991), the identification of a CTB risk involves three questions:

1. Which processes lead to an accumulation of the chemical under consideration in soil or sediment?
2. Which mechanisms can result in a rapid release of the chemical with harmful effects to man and environment?
3. Which kinds of gradual environmental change may eventually trigger the release mechanism?

**Accumulation Processes.** The first question refers to possible charging processes leading to a CTB. Accumulation will occur if the mass balance of inputs and sources versus outputs and sinks is positive. With respect to heavy metals, input rates have increased during the last decades due to human activity. Compared to the rates of heavy metal mobilization by mineral weathering and of leaching losses, export by harvesting plant biomass is the only significant output pathway in cultivated soils. In noncultivated soils the deposition of airborne metal pollutants is usually approximately equivalent to net accumulation. Due to its high specific sorption capacity and its concentration at the soil surface, organic matter is a major storage compartment for this input. Even in mineral soils with less than 5% organic matter content, it accounts for up to 50% of heavy metal binding (Kabata-Pendias and Pendias 1992). Heavy metal inputs are enhanced in the vicinity to the emitting sources and on sites where lateral transfer processes converge into accumulation areas, e.g., by soil erosion and sedimentation or by trapping of contaminated plant litter in lee-side positions.

SOM appears to have a very high intrinsic stability under heavy metal stress and thus be suited for long-term storage. Feedback effects of heavy metals on SOM cycling such as the inhibition of litter decomposition tend to increase the storage capacity in response to increasing concentrations. Reduction of soil biological activity will decrease the comminution of detritus, the uptake and mineralization of organic substances, as well as the decomposition of organic matter by extra-cellular enzymes. The inhibition of primary production seems to be less effected since due to the immission pathway the contaminants are concentrated close to the soil surface, and in general do not affect root activity in the mineral soil horizons below to the same degree as in the topsoil. Copper pollution of vineyards provides an illustrative example. Soils on acidic bedrock may behave quite differently though and therefore deserve special attention.

As a direct consequence of increased metal concentrations in soil, stability of organic matter will be increased by enhanced cation bridging between negatively charged organic and clay mineral surfaces and flocculation. Because of their high number of dissociated carboxylic groups, bridging will in particular involve humic and fulvic acids as well as polyuronic acids. As the binding strength between molecules and particles are correlated with the number of bridges formed, low-weight molecules will be less immobilized than high-weight molecules.

Heavy metal retention by SOM is affected in particular by pH, organic matter content and composition, concentrations of other cations, ionic strength, type and concentration of mineral surfaces, water content, nutrient availability, redox potential and temperature. These factors are not independent of each other. Some such as pH and organic matter composition will have a more direct influence than others such as temperature and nutrient availability. The latter will affect storage capacity primarily of biotic processes such as decomposition rate and chelator production.

**Rapid Release of Heavy Metals.** Some heavy metals like mercury and chromium take different oxidation states in the variation range of soil redox conditions. Changing redox conditions will directly affect their binding properties and reactivity. Moreover, redox conditions affect the speciation of metals through their influence on the composition of the surrounding soil chemical milieu. These changes are mediated through biotic processes and thus are strongly dependent on kinetic effects. In particular, a decrease in redox potential will eventually result in the reduction of sulfur and the immobilization of metals by the precipitation of insoluble metal sulfides. In addition, reduction will increase pH as protons are consumed, enhancing the immobilization effect. On the other hand, anaerobic conditions are known to favor the accumulation of strongly chelating aliphatic acids in the soil solution, e.g., citric, oxalic, and 2-ketogluconic acid (Stevenson and Fitch 1986). This will promote metal mobility. If the organo-metal complexes are rapidly exported by leaching, stabilizing negative feedback mechanisms will tend to be mitigated. Eventually the decomposition process may even be accelerated if preferential sorption of metals to the chelators would result in an increased detoxification. This case then would represent a potential CTB release mechanism. The fate of specific trace elements will, however, strongly depend on the presence of competing elements, in particular metals such as iron and manganese (Lindsay 1981).

Much more severe situations are expected to arise when metals, which have been accumulated under anaerobic conditions as sulfide precipitates, are remobilized by subjecting them to aerobic conditions, e.g., by dredging harbor sludges or drainage of heavy metal contaminated wetlands. Organic matter may have a role in the trigger mechanism serving as the reducing agent. Thus, a stop in the supply of organic matter to the anaerobic microfauna for some reason, may cause a rise of the redox potential even if the oxygen supply remains unchanged. Since in this scenario organic matter is not the relevant storage compartment, however, it will not be considered any further here. There may be, however, also situations of anaerobic accumulation conditions where SOM plays the key role of the storage compartment for heavy metals. If, for example, a peat soil is percolated with metal polluted water, the organic matter will sorb large quantities of the metals which can be released again by aerobic decomposition upon drainage of the peat. If such a change to aerated conditions is connected with a displacement of the peat to intensive human land use, e.g., by a flood, a landslide or peat exploitation, the consequences may be very harmful, qualifying such a scenario as a CTB.

Increasing temperature in general accelerates biotic processes. As decomposition seems to be in general more dependent on temperature than primary production, higher temperature will tend to decrease SOM content and release bound metals. Negative feedback effects through increased toxicity may be less reduced at the same time, as long as temperature shifts towards optimum conditions will increase the metal tolerance of microorganisms (Collins and Stotzky 1989). On the other hand, increased mineralization of low-molecular-weight organics leads to reduced metal mobilization by chelate and mixed complexes. Presumably the effects of temperature change on metal release from organic matter will be buffered by a change in the composition of SOM pool towards less acidic high-molecular-weight humic and nonhumic substances and a larger pool of living biomass. Such effects will probably stabilize the metal content rapidly. Metal mobilization merely in response to a rise in average soil temperature thus appears to be of very short duration, and to depend very sensitively on the dynamics of the transition processes.

The opposite of soil warming is the case of a climate change with a temperature drop. The preferential slow-down of decomposition activity will result in an accumulation of SOM and an amplification of podzolization processes. Metals will be mobilized through increased formation of mobile organo-metallic complexes, in particular complexes involving fulvic acids. Increased generation of organic acids entails proton release, and thus, via a pH drop, will also tend to increase free metal concentrations. This might enhance toxicity and amplify the inhibition of decomposition processes. Again, other mechanisms may be expected to counteract these changes. The accumulation of immobile organic matter provides more binding sites and will thus reduce free metal availability by increased sorption. In the pH range below 3.5 humic matter aggregation is enhanced by hydrogen bonding and possibly van der Waals forces (Schnitzer 1986).

Other impacts such as a change in the nutrient regime or the ionic strength and composition of the infiltrating water will also affect the retention capacity of heavy metals by SOM, either directly through cation exchange, complexation with oxyanions (phosphate, sulfate), and colloid stability or more indirectly through biotic and abiotic transformation of the organic matter complex. At the present time, it is hard to see a mechanism by which such impacts by themselves would result in a substantial and rapid release of heavy metals.

Besides the redox potential, acidification is the single most important soil factor which is able to undergo rapid change in response to environmental alterations and to trigger a CTB with respect to heavy metals bound to SOM. Soil acidification due to acidic atmospheric depositions is a process which will presumably lead to substantial soil degradation of large areas in Europe over the next decades (Stigliani et al. 1991). A decrease in pH will shift chemical equilibria toward increasing concentrations of free metals. Basic nutrient cations will be increasingly displaced from the exchanger complex and leached, and mineral weathering will accelerate. Biological activity will be negatively affected in response to the pH drop itself as well as by increased metal toxicity, although some evidence indicates that below some pH toxicity may even decrease metal sorption by organic matter as it becomes less competitive relative to clay mineral surfaces (Monteil 1989; Schindler et al. 1991).

Decreased decomposition activity in acidic metal-polluted topsoils will result in accumulation of humic substances and also in a change of the ratio between soluble and insoluble organic matter (Peterson 1990). Sorption of humic and fulvic acids by clays increases significantly when pH is dropped from 7 to 3.5 (Hayes and Himes 1986). Organic acids are also increasingly stabilized by amorphous oxyhydroxides at low pH values when most surface functional groups become positively charged. Besides anion exchange, this sorption mechanism possibly also involves ligand exchange processes (Hayes and Himes 1986). Continuous leaching of the divalent base cations $Ca^{2+}$ and $Mg^{2+}$ will, however, weaken the aggregation of negatively charged soil colloids. With time, this will lead to a decrease in nutrient quality of decaying plant litter which in turn will further promote humus accumulation and soil acidification.

**Gradual Environmental Changes.** On the basis of the preceding discussion, acidification, transition from aerobic to anaerobic conditions, and eventually cooling, can be considered as primary candidate factors for triggering rapid heavy metal release from SOM. Independently of SOM, the transition from anaerobic to aerobic conditions is generally regarded to be an important factor in the release of heavy metals to the environment (e.g., dredged harbor sludge). Of these changes, acidification appears to be the most severe, as it has a direct influence on metal speciation which may override negative biotic feedback effects. In the case of the other two factors (decrease of temperature and change of redox potential) the outcome of counteracting mobilization and immobilizing effects is less predictable. Their effects are highly dependent on changes in the chemical composition of the humus. Assuming that both anaerobic conditions and low temperatures tend to enhance the inhibition of microbial activity through heavy metal stress, the mobilizing effects tend to be small for most metals. Mercury, which can be subject to methylation, is a noteworthy exception.

Looking at current trends of globally important environmental changes, acidification and, to a lesser degree, decrease of soil redox potentials but not temperature decrease, are found to be relevant. Transition to lower redox potentials is a rather local phenomenon. It may be due to increased waterlogging of soils or to the deposition of translocated soil material in wet areas or waters, e.g., by soil erosion, landslides, or by mining operations. Waterlogging of soils may be caused by soil compaction, rising groundwater tables, increased precipitation, or decreased evapotranspiration. Many human activities lead to such effects, e.g., intensification of agriculture, restoration of wetlands, deforestation, or building of reservoir dams.

**Potential Non-CTB Long-term Effects.** Mobilization of heavy metals in the topsoil may not necessarily lead to catastrophic consequences. In the best case, metals are merely transferred into a less acidified deeper horizon, where they are insolubilized again by sorption, colloid coagulation and precipitation. On acidic bedrocks, however, serious problems with groundwater contamination and plant toxicity might arise.

Plant growth reduction and mortality by heavy metal poisoning has the benefit of limiting the metal flux into herbivores and thus to serve as a certain barrier of heavy metal transfer in the food chain. If metal concentrations increase slowly enough to allow the evolution of metal resistance in plants and soil organisms, the consequences will be much more dramatic as higher trophic levels will become increasingly exposed as targets of metal poisoning. Thus, the CTB effect will be buffered in the soil but only to be delayed and shifted toward animals and humans. If anything is certain in the speculations of this chapter, it is that the latter alternative is definitely the least desirable.

## 3.6 Conclusions and Final Remarks

The foregoing discussion shows that conditions which would result in a CTB-type release of heavy metals accumulated in SOM fortunately would appear to occur only under very extreme circumstances. The most probable CTB scenarios are those involving oxidation and mineralization of organic soils upon aeration and those driven by accelerated soil acidification. It seems that SOM is quite resilient against the impact of heavy metal pollution alone, in the sense that it does not seem to be much prone to self-amplifying breakdown processes triggered by the pollution process alone. On the contrary, it appears that catastrophic developments as envisaged by the CTB concept in general require quite drastic environmental changes as triggers. Such changes may occur in response to natural catastrophes such as floods or landslides. However, more often they have to be expected as the consequence of human impacts on the environment, in particular soil acidification by atmospheric immissions and changes in land management such as land drainage, deforestation, or taking land out of cultivation.

It must be emphasized once more that due to the complexity of the subject and the uncertainty of our knowledge with respect to the dominant factors and mechanisms, such considerations are extremely speculative. They are necessary though to demonstrate the need of comprehensive models allowing to simulate consistent scenarios on the basis of clearly formulated concepts and assumptions. Only formally rigorous models can provide a tool to assess the consequences of even more complex situations. In the foregoing discussion, possible implications of the spatial and temporal variability of soils related to potential CTB scenarios have been completely neglected. Such effects may not be unimportant since parameters such as pH and redox potential may vary substantially even on the scale of a few millimeters and that water content and temperature may exhibit extreme short-term fluctuations. Compared to homogeneity, such conditions allow many more interactions and can speed up decomposition processes considerably.

The predictability of a soil CTB is limited by the available knowledge on soil functioning. SOM is one of the oldest research subjects in soil science. Despite the huge bulk of scientific knowledge that has accumulated in the meantime, the concepts of SOM are still quite fuzzy. A major reason is the intricate structure of feedback effects which makes it extremely difficult to clearly separate different factors and mechanisms. Important knowledge gaps exist, also with respect to the control mechanisms for the kinetics of processes in SOM turnover, in particular their dependence on soil structure. However, also many equilibrium relationships are still not very well known such as equilibrium constants of organo-metallic complexes. Knowledge in general becomes more and more fuzzy the more it concerns biotic processes and the higher the involved trophic levels are. On one hand, this situation is closely related to the complexity of the subject; on the other hand, it is also a consequence of the poor support and appreciation that has been given by our society to ecological research.

While many of the identified scientific gaps will probably be filled in the foreseeable future, a principal problem of risk scenarios remains to be our lack of imagination about what humans are able to do. This also holds for the assessment of environmental risks. The human factor is the major unknown. One example is the ever increasing displacement of materials including soils leading to a dispersion of pollutants which – as history continuously shows – rapidly gets out of control. There is, however, probably still a long way to go from a strategy of *a posteriori* detection and remediation of CTBs in soils and sediments to an environmental management designed for the a priori avoidance of such situations.

## References


Alexander M, Scow KM (1989) Kinetics of biodegradation in soil. In: Sawhney BL, Brown K (eds) Reactions and movement of organic chemicals in soils, Soil Sci Soc Am, Spec Publ No 22, Madison WI, USA, pp 243-269

Anderson DW, Paul EA (1984) Organo-mineral complexes and their study by radiocarbon dating. Soil Sci Soc Am J 48:298-301

Bååth E (1989) Effects of heavy metals in soil on microbial processes and populations (a review). Water, Air, Soil Pollut 47:335-379

Babich H, Stotzky G (1977) Sensitivity of various bacteria, including actinomycetes, and fungi to cadmium and the influence of pH sensitivity. Appl Environ Microbiol 33:681-695

Beck T (1981) Untersuchungen über die toxische Wirkung der in Siedlungsabfällen häufigen Schwermetalle auf die Bodenmikroflora. Z Pflanzenern Bodenkd 144:613-627

Birch HF (1958) The effect of soil drying on humus decomposition and nitrogen availability. Plant Soil 10:9-31

Bloomfield C, Sanders JR (1977) The complexing of copper by humified organic matter from laboratory preparations, soil and peat. J Soil Sci 28:435-444

Blundell MR, Wild DG (1969) Inhibition of bacterial growth by metal salts: a survey of the effects on the synthesis of ribonuclei acid and protein. Biochem J 115:207-212

Bosatta E, Ågren GI (1985) Theoretical analysis of decomposition of heterogeneous substrates. Soil Biol Biochem 17:601-610

Bosatta E, Ågren GI (1991) Theoretical analysis of carbon and nutrient interactions in soils under energy-limited conditions. Soil Sci Soc Am J 55:728-733

Boudot JP (1992) Relative efficiency of complexed aluminium, noncrystalline Al hydroxide, allophane and imogolite in retarding the biodegradation of citric acid. Geoderma 52:29-39

Boudot JP, Bel Hadj Brahim A (1989) Biodegradation of synthetic organo-metallic complexes of iron and aluminium with selected metal to carbon ratios. Soil Biol Biochem 21(7):961-966

Bouwman, AF (ed)(1990) Soils and the greenhouse effect. Wiley, Chichester, UK, 575 pp

Browder JA, Volk BC (1978) Systems model of carbon transformations in soil subsidence. Ecol Modelling 5:269-292

Brussaard L, Bouwman LA, Geurs M, Hassink J, Zwart KB (1990) Biomass, composition and temporal dynamics of soil organisms of a silt loam soil under conventional and integrated management. Neth J Agric Sci 38:283-302

Buffle J (1988) Complexation reactions in aquatic systems – an analytical approach. Ellis Horwood Limited, Chichester, West Sussex, UK, 692 pp

Bunnell FL, Tait DEN, Flanagan PW (1976) Microbial respiration and substrate weight loss, II. A model of the influences of chemical composition. Soil Biol 9:41-47

Bunzl K, Schmidt W, Sansoni B (1976a) Kinetics of ion exchange in soil organic matter IV. Adsorption and desorption of $Pb^{2+}$, $Cu^{2+}$, $Cd^{2+}$, $Zn^{2+}$ and $Ca^{2+}$ by peat. J Soil Sci 27:32-41

Bunzl K, Wolf A, Sansoni B (1976b) Kinetics of ion exchange in soil organic matter V. Differential ion exchange reactions of $Cu^{2+}$, $Cd^{2+}$, $Zn^{2+}$ and $Ca^{2+}$-ions in humic acid. Z Pfanzenern Bodenk 139:475-485

Chander J, Brookes PC (1991) Plant inputs of carbon to metal-contaminated and non-contaminated soil and effects on the synthesis of soil microbial biomass. Soil Biol Biochem 23:917-925

Clark MD, Gilmour JT (1983) The effect of temperature on decomposition at optimum and saturated soil water contents. Soil Sci Soc Am J 47:927-929

Cole MA (1977) Lead inhibition of enzyme synthesis in soil. Appl Environ Microbiol 33:262-268

Coleman DC (1985) The role of microfloral and faunal interactions in affecting soil processes. In: Mitchell MJ, Nakas J (eds) Microflora and faunal interactions in natural and agro-ecosystems. Nijhoff, Dordrecht, pp 317-348

Coleman DC, Anderson RV, Cole CV, Elliott ET, Woods L, Campion MK (1978) Trophic interactions in soils as they affect energy and nutrient dynamics, IV. Flows of metabolic and biomass carbon. Microb Ecol 4:373-380

Collins YE, Stotzky G (1989) Factors affecting the toxicity of heavy metals to microbes. In: Beveridge TJ, Doyle RJ (eds) Metal ions and bacteria. Wiley, New York, 461 pp

Davies RI, Chesire MV, Graham-Bryce IJ (1969) Retention of low levels of copper by humic acid. J Soil Sci 20:65-71

Doelman P, Haanstra L (1979a) Effect of lead on soil respiration and dehydrogenase activity. Soil Biol Biochem 11:475-479

Doelman P, Haanstra L (1979b) Effects of lead on the decomposition of organic matter. Soil Biol Biochem 11:481-485

Domsch KH (1985) Funktion und Belastbarkeit des Bodens aus Sicht der Mikrobiologie. Kohlhammer, Stuttgart, 67 pp

Doyle JJ, Marschall RT, Pfander WH (1975) Effects of cadmium on the growth and uptake of cadmium by microorganisms. Appl Microbiol 29:562-564

Duxburry T, Bicknell B (1983) Metal-tolerant bacterial populations from natural and metal-polluted soils. Soil Biol Biochem 15:243-250

Eijsackers H, Zehnder AJB (1990) Litter decomposition: a Russian matriochka doll. Biogeochem 11:153-174

Elliott ET, Anderson RV, Coleman DC, Cole DC (1980) Habitable pore space and microbial trophic interactions. Oikos 35:327-335

Fogel R, Cromack K (1977) Effect of habitat and substrate quality on Douglas fir litter decomposition in western Oregon. Can J Bot 55:1632-1640

Follett RF, Stewart JWB, Cole CV (eds)(1987) Soil fertility and organic matter as critical components of production systems. Soil Sci Soc Am Madison WI, USA, Spec Publ No 19, 166 pp

Freckman DW (1988) Bacterivorous nematodes and organic-matter decomposition. Agriculture, Ecosystems and Environment 24:195-217

Frimmel FH, Christman RF (eds)(1988) Humic substances and their role in the environment. Wiley, New York, 271 pp

Frissel MJ, Veen JA van (eds)(1981) Simulation of nitrogen behavior of soil-plant systems. Pudoc, Wageningen, 277 pp

Gabteni N, Gallali T (1988) Etude expérimentale des interactions entre éléments métalliques et la minéralisation de la matiére organique d'une boue résiduaire ajoutée à un sol. Cah ORSTOM Ser Pedol 24:255-261

Gamble DS, Schnitzer M, Hoffman I (1970) $Cu^{2+}$-fulvic acid chelation equilibrium in 0.1 M KCl at 25°C. Can J Chem 48:3197-3204

Geiger G (1990) Keimverhalten und Schwermetallaufnahme von Pflanzen nach verschiedenen Sanierungsmethoden an einem Boden in Dornach. Diplomarbeit, Institut für terrestrische Ökologie, Labor für Bodenchemie, ETH Zürich, 64 pp

Gisi U, Schenker R, Schulin R, Stadelmann FX, Sticher H (1990) Bodenökologie. Thieme, Stuttgart, 304 pp

Greenland DJ, Hayes MHB (eds)(1981) The chemistry of soil processes. Wiley, Chichester, 714 pp

Hassett JJ, Banwart WL (1989) The sorption of nonpolar organics by soils and sediments. In: Sawhney BL, Brown K (eds) Reactions and movement of organic chemicals in soils. Soil Sci Soc Am, Madison WI, USA, Spec Publ No 22, pp 31-44
Hatcher PG, Spiker EC (1988) Selective degradation of plant biomolecules. In: Frimmel FH, Christman RF (eds) Humic substances and their role in the environment. Wiley, New York, pp 59-74
Hayes MHB, Himes FL (1986) Nature and properties of humus-mineral complexes. In: Huang PM, Schnitzer M (eds) Interactions of soil minerals with natural organics and microbes. Soil Sci Soc Am, Madison WI, USA, Spec Publ No 17, pp 103-158
Hering JG, Morel FMM (1990) Kinetics of trace metal complexation: ligand exchange reactions. Environ Sci Technol 24:242-252
Hopkin SP (1989) Ecophysiology of metals in terrestrial invertebrates. Elsevier, London, 366 pp
Huang PM, Schnitzer M (eds)(1986) Interactions of soil minerals with natural organics and microbes. Soil Sci Soc Am, Madison WI, USA, Spec Publ No 17, 606 pp
Hunt HW, Coleman DC, Cole CV, Ingham RE, Elliott ET, Woods LE (1984) Simulation model of a food web with bacteria, amoebae and nematodes in soil. In: Klug MJ, Reddy CA (eds) Current perspectives in microbial ecology. Am Soc Microbiol, Washington DC, USA, pp 346-352
Hunt HW, Coleman DC, Ingham ER, Ingham RE, Elliott ET, Moore JC, Rose SL, Reid CPP, Morley CR (1987) The detrital food web in a shortgrass prairie. Biol Fert Soils 3:57-68
Ingham RE, Trofymow JA, Coleman DC (1985) Interactions of bacteria, fungi, and their nematode grazers: effects on nutrients cycling and plant growth. Ecol Monogr 55(1):119-140
Ingham ER, Trofymow JA, Ames RN, Hunt HW, Morley CR, Moore JC, Coleman DC (1986a) Tropic interactions and nitrogen cycling in a semi-arid grassland soil, I. Seasonal dynamics of the natural populations, their interactions and effects on nitrogen cycling. J Appl Ecol 23:597-614
Ingham ER, Trofymow JA, Ames RN, Hunt HW, Morley CR, Moore JC, Coleman DC (1986b) Tropic interactions and nitrogen cycling in a semi-arid grassland soil, II. System responses to removal of different groups of soil microbes or fauna. J Appl Ecol 23:615-630
Jenkinson DS (1977) Studies on the decomposition on plant material in soil IV. The effect on rate addition. J Soil Sci 28:417-423
Jenkinson DS, Rayner JH (1977) The turnover of soil organic matter in some of the Rothamsted classical experiments. Soil Sci 123:298-305
Jenny H (1941) Factors of soil formation. McGraw-Hill, New York, 218 pp
Jordan MJ, Lechevalier MP (1975) Effects of zinc-smelter emissions on forest soil microflora. Can J Microbiol 21:1855-1865
Kabata-Pendias A, Pendias H (1992) Trace elements in soils and plants, 2nd edn. CRC Press, Boca Raton, 365 pp
Knapp EG, Elliott LF, Campbell GS (1983) Carbon, nitrogen, and microbial biomass interrelationship during the decomposition of wheat straw: a mechanistic simulation model. Soil Biol Biochem 15:455-461
Kögel-Knabner I (1992) Forest soil organic matter: structure and formation. Bayreuther Bodenk Ber No 24, 103 pp
Küster E, Grün I (1984) Cadmium und Bodenorganismen. Angew Botanik 58:31-38
Ladd JN, Oades JM, Amato M (1981) Microbial biomass formed from $^{14}C$, $^{15}N$-labelled plant material decomposing in soils in the field. Soil Biol Biochem 13:119-126
Liang CN, Tabatabai MA (1977) Effects of trace elements on nitrogen mineralisation in sols. Environ Pollut 12:141-147

Lindsay WL (1981) Solid phase-solution equilibria in soils. In: Stelly M, Kral DM, Cousin MK (eds) Chemistry in the soil environment. Spec Publ No 40, Am Soc Agr and Soil Sci Soc Am, Madison WI, USA, pp 183-202

MacCarthy P, Clapp CE, Malcolm RL, Bloom PR (eds)(1990) Humic substances in soil and crop sciences: selected readings. Am Soc Agr and Soil Sci Soc Am, Madison WI, USA, 281 pp

Macrae IE, Edwards JF (1972) Adsorption of colloidal iron by bacteria. Appl Microbiol 24:819-823

Marinsky JA, Gupta S, Schindler P (1982) The interaction of the Cu(II) ion with humic acid. J Colloid Interface Sci 89:401-411

Martens R (1990) Contribution of rhizodeposits to the maintenance and growth of soil microbial biomass. Soil Biol and Biochem 22:141-147

Martin JP (1971) Decomposition and binding actions of polysaccharides in soil. Soil Biol Biochem 3:33-41

Martin JP, Ervin JO, Shepherd RA (1966) Decomposition of the iron, aluminium, zinc, and copper salts of complexes of some microbial and plant polysaccharides in soil. Soil Sci Soc Am J 30:196-200

McGill WB, Hunt HW, Woodmansee RG, Reuss JO, Paustian KH (1981) Formulation, process controls, parameters and performance of PHOENIX: a model of carbon and nitrogen dynamics in grassland soils. In: Frissel MJ, Veen JA van (eds) Simulation of nitrogen behaviour of soil-plant systems. Pudoc, Wageningen, pp 171-191

McGrath SP, Brookes PC, Giller KE (1988) Effects of potentially toxic metals in soil derived from past applications of sewage sludge on nitrogen fixation by *Trifolium repens L.* Soil Biol Biochem 20:415-424

Meentemeyer V (1978) Macroclimate and lignin control of litter decomposition rates. Ecology 59:465-472

Melillo JM, Aber JD, Muratore JF (1982) Nitrogen and lignin control of hardwood leaf litter decomposition dynamics. Ecology 63:621-626

Merckx R, Hartog A den, Veen JA van (1985) Turnover of root-derived material and related microbial biomass formation in soils of different texture. Soil Biol Biochem 17:565-569

Monteil MB (1989) Der Einfluss von Cd, Cu und Pb auf die $CO_2$-Produktion von Waldböden. Inaugural dissertation, Universität Bern, 122 pp

Moore JC, Zwetsloot HJC, Ruiter PC (1990) Statistical analysis and simulation modelling of the below-ground food webs of two winter wheat management practices. Neth J Agric Sci 38:303-316

Morrissey RF (1975) The influence of heavy metals on microbially mediated nitrogen transformations in soil: nitrification and ammonification. Diss Abstr Int B Sci Eng 36:3237

Nordgren A, Bååth E, Söderström B (1983) Microfungi and microbial activity along a heavy metal gradient. Appl Environ Microbiol 45:1829-1837

Oades JM (1988) The retention of organic matter of soils. Biogeochemistry 5:35-70

Oades JM (1989) An introduction to organic matter in mineral soils. In: Dixon JB, Weed SB (eds) Minerals in soil environments. Soil Sci Soc Am, Madison WI, USA, SSSA Book Series, No 1, pp 89-159

Parnas H (1975) Model for decomposition of organic materials by microorganisms. Soil Biol Biochem 7:161-169

Parton WJ, Schimel DS, Cole CV, Ojima DS (1987) Analysis of factors controlling soil organic matter levels in Great Plains grasslands. Soil Sci Soc Am J 51:1173-1179

Parton WJ, Stewart JWB, Cole CV (1988) Dynamics of C, N, P, and S in grassland soils: a model. Biogeochemistry 5:109-131

Paul EA, FE Clark (1989) Soil microbiology and biochemistry. Academic Press, San Diego, Calif, 273 pp
Paustian K (1985) Influence of fungal growth pattern on decomposition and nitrogen mineralization in a model system. In: Fitter AH (ed) Ecological interactions in soil. Brit Ecol Soc, Blackwell Sci Publ, Oxford, Spec Publ No 4, pp 159-174
Pearce F (1994) Peat bogs hold bulk of Britain's carbon. New Scientist vol 144, no 1952, 6
Peterson RC Jr (1990) Effects of ecosystem changes (eg, acid status) on formation and biotransformation of organic acids. In: Perdue EM, Gjessing ET (eds) Organic acids in aquatic ecosystems. Dahlem Konferenzen 1990. Wiley, New York, pp 151-166
Piscator M (1986) The dependence of toxic reactions on the chemical species of elements. In: Bernhard M, Brinkman RE, Sadler PJ (eds) The importance of chemical 'speciation' in environmental processes. Dahlem Konferenzen 1986. Springer, Berlin Heidelberg New York pp 59-70
Rai LC, Gaur JP, Kumar HD (1981) Protective effects of certain environmental factors on the toxicity of zinc, mercury, and methylmercury to *Chlorella vulgaris*. Environ Res 25:250-259
Rich S, Horsfall JG (1963) Fungicides as metabolic inhibitors. In: Hochster RM, Quastel JH (eds) Metabolic inhibitors. Academic, New York, pp 263-284
Schindler PW (1990) Co-adsorption of metal ions and organic ligands: formation of ternary complexes. In: Hochella MF Jr, White AF (eds) Mineral-water interface geochemistry. Mineralogical Soc of America Washington DC, pp 281-307
Schindler PW, Galicia H, Purghart BG, Marquis L, Monteil M, Nyffeler UP (1991) Schwermetallbelastung und -toxizität von Waldböden. In: Pankow W (ed) Ergebnisse aus dem Nationalen Forschungsprogramm 14: Belastung von Waldböden Bd 6 Verlag der Fachvereine, Zürich, pp 45-100
Schlesinger WH (1986) Changes in soil carbon storage and associated properties with disturbance and recovery. In: Trabalka JR, Reichle DE (eds) The changing carbon cycle: a global analysis. Springer, Berlin Heidelberg New York pp 195-220
Schmitt HW, Sticher H (1991) Heavy metal compounds in the soil. In: Merian E (ed) Metals and their compounds in the environment – occurrence analysis and biological relevance.VCH Verlagsgesellschaft, Weinheim, Germany, pp 311-331
Schnitzer M (1991) Soil organic matter – the next 75 years. Soil Sci 151:41-58
Schnitzer M (1986) Binding of humic substances by soil mineral colloids. In: Huang PM, Schnitzer M (eds) Interactions of soil minerals with natural organics and microbes.Spec Publ No 17 Soil Sci Soc Am Madison, WI, USA, pp 77-101
Schnitzer M, Hanson EH (1970) Organo-metallic interactions in soils: 8. An evaluation of methods for the determination of stability constants of metal-fulvic acid complexes Soil Sci 109:333-340.
Schnitzer M, Kahn SU (1972) Humic substances in the environment. Marcel Dekker Inc, New York, 327 pp
Schwarzenbach RP, Westall C (1981) Transport of nonpolar organic compounds from surface water to groundwater. Laboratory sorption studies. Environ Sci Technol 15:1360-1367
Shaw AJ (1989) Heavy metal tolerance in plants: evolutionary aspects. CRC Press, Boca Raton, 355 pp
Smith DD, Wischmeier WH (1957) Factors affecting sheet and rill erosion. Transact Am Geophys Union 38:889-896
Smith OL (1979) An analytical model of the decomposition of soil organic matter. Soil Biol Biochem 11:585-606
Sørensen LH (1974) Rate of decomposition of oranic matter in soil as influenced by repeated air drying-rewetting and repeated additions of organic material. Soil Biol Biochem 6:287-292

Sposito G (1984) The surface chemistry of soils. Oxford Univ Press, New York, 234 pp
Sposito G (1986) Sorption of trace metals by humic materials in soils and natural waters. CRC Critical Rev Environ Control 16(2):193-229
Sposito G (1989) The chemistry of soils. Oxford Univ Press, New York, 277 pp
Stadelmann FX, Santschi-Fuhrmann E (1987) Beitrag zur Abstützung von Schwermetall-Richtwerten im Boden mit Hilfe von Bodenatmungsmessungen. Bericht der Forschungsanstalt für Agrikulturchemie und Umwelthygiene (FAC) Bern-Liebefeld Schweiz 105 pp
Stevenson FJ (1976) Stability constants of $Cu^{2+}$ $Pb^{2+}$ and $Cd^{2+}$ complexes with humic acids. Soil Sci Soc Am J 40:665-672
Stevenson FJ (1982) Humus Chemistry. Wiley, New York, 443 pp
Stevenson FJ, Fitch A (1986) Chemistry of complexation of metal ions with soil solution organics. In: Huang PM, Schnitzer M (eds) Interactions of soil minerals with natural organics and microbes.Soil Sci Soc Am Madison, WI, USA Spec Publ No 17 pp 29-58
Stigliani WM, Doelman P, Salomons W, Schulin R, Meulen-Smidt GRB ter, Zee SEATM van der (1991) Chemical time bombs – predicting the unpredictable. Environment 33(4):4-9 and 26-30
Strojan CJ (1978) Forest leaf litter decomposition in the vicinity of a zinc smelter. Oecologia 32:201-212
Stumm W, Furrer G (1987) The dissolution of oxides and aluminum silicates; examples of surface-coordination controlled kinetics. In: Stumm W (ed) Aquatic surface chemistry. Wiley, New York, pp 197-217
Stumm W, Morgan JJ (1981) Aquatic chemistry – an introduction emphasizing chemical equilibria in natural waters, 2nd edn. Wiley, New York, 780 pp
Swift MJ, Heal OW, Anderson JM (1979) Decomposition in terrestrial ecosystems. Studies in ecology, vol 5. Blackwell Scientific Publications, Oxford, 372 pp
Tate RL (1987) Soil organic matter. Wiley, New York, 291 pp
Tiessen H, Stewart JWB, Hunt HW (1984) Concept of organic matter transformations in relation to organo-mineral particle size fractions. Plant Soil 76:287-295
Tyler G (1981) Heavy metals in soil biology and biochemistry. In: Paul EA, Ladd JN (eds) Soil biochemistry, vol 5. Dekker, New York, pp 371-414
Veen JA van (1987) The use of simulation models of the turnover of soil organic matter: an intermediate report, vol 6. Trans XIII Congr Int Soc Soil Sci. Hamburg, pp 626-635
Veen JA van, Paul EA (1981) Organic carbon dynamics in grassland soils, I. Background information and computer simulation. Can J Soil Sci 61:185-201
Veen JA van, Ladd JN, Frissel MJ (1984) Modelling C and N turnover through the microbial biomass in soil. Plant Soil 76:257-274
Veen JA van, Ladd JN, Amato M (1985) Turnover of carbon and nitrogen through the microbial biomass in a sandy loam and clay soil incubated with ($^{14}C(u)$) Glucose and ($^{15}N$) $(NH_4)_2SO_4$ under different moisture regimes. Soil Biol Biochem 17(6):747-756
Verberne ELJ, Hassink J, Willigen P de, Groot JJR, Veen JA van (1990) Modelling organic matter dynamics in different soils. Neth J Agric Sci 38:221-238
Waller PA, Pickering WF (1990) The lability of copper ions sorbed on humic acid. Chem Spec Bioavail 2(4):127-138
Watson AP, Hook RI van, Jackson DR, Reichle DE (1976) Impact of a lead mining-smelting complex on the forest floor litter arthropod fauna in the new lead belt region of southeastern Missouri. ORNL/NSF/EATC-30 Oak Ridge National Laboratory, 180 pp
Welp G, Brümmer GW (1989) Wirkung von Schwermetallen auf Boden-Mikro-organismen in Beurteilung von Schwermetallkontaminationen im Boden. Dechema-Fachgespräche Umweltschutz. 21-22 Jan 1988 in Oberursel/Ts, pp 253-269

# 4 Mobilization of Heavy Metals as Affected by pH and Redox Conditions

*A. C. M. Bourg and J. P. G. Loch*

## 4.1 Introduction

Water is a major vector of transport of heavy metals in the lithosphere. The solids present in soils, aquifers and surface water bodies (i.e., suspended and deposited sediments in rivers, lakes, oceans) can trap significant quantities of toxic heavy metals and act as reservoirs in the various hydrocycles taking place at the earth's surface. The geochemical phenomena controlling the retention of heavy metals are adsorption and precipitation, while dissolved complexation influences advective and dispersive transport. For all of these, pH and redox conditions are master variables controlling the potential release of stored pollutants to the aqueous phase and therefore their dispersion in the environment and their availability to biota. While the redox status of natural environments is usually expressed as Eh (potential measured at an electrode), by analogy with pH, it is sometimes useful to express redox potentials in terms of pe (-log of electron activity). In this way electrons can be treated like classical reactants and products so that both chemical and electrochemical reactions can be expressed by a single equilibrium constant. Various solids control the fixation of heavy metals: clay minerals, organic matter, oxides and hydroxides of Fe, Mn and Al for adsorption, and low solubility sulfide, carbonate and phosphate minerals for precipitation. Complexing by organic matter, also very important, is addressed in Chapter 3 (Geiger et al.).

The permanence of storage is a subject of concern. Changes in pH and redox status of soils and sediments are dependent on three major factors: soil vulnerability, chemical inputs, and land use policies and practices (Stigliani 1993). The vulnerability of soils and sediments is defined here as their ability to resist changes in pH and pe (redox potential). Chemical inputs correspond to the total source of not only pollutants but also protons and electron donors and acceptors (reductants and oxidants, respectively). Practices and policies lead to man-made changes which, though not necessarily viewed as detrimental, can indirectly induce significant changes in the solubility of the chemicals stored in the lithosphere solids.

## 4.2 The Chemistry of Natural Waters

As water circulates in the hydrosphere (sediments, soils and aquifers), it acquires its intrinsic chemical composition through heterogeneous geochemical reactions. This chemical signature is the result of biological activity, water–solid interactions, water–air exchanges, and advective and dispersive transport. Since these natural processes depend strongly on climate, lithology and relief, the chemical composition of water varies with space. Anthropic activities (industry, mining, agriculture, transportation, urbanization, etc.) as well as the natural evolution of the earth's landscape (pedogenesis and diagenesis) can lead to significant variations with time in the water composition.

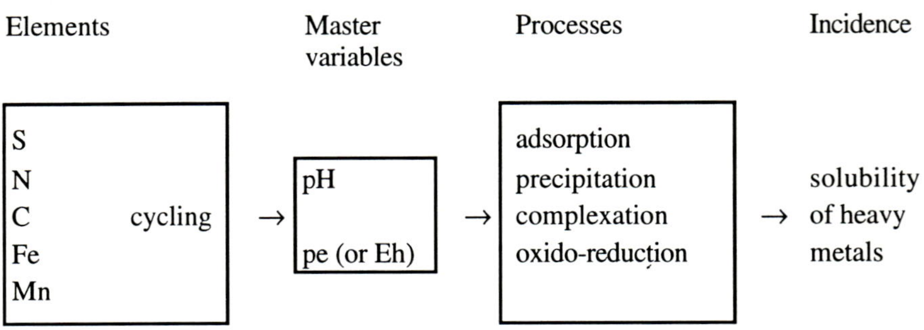


**Fig. 4.1.** Major element cycles and solubility of heavy metals (modified from Salomons 1993)

The natural cycling of major elements (C, N, S, Fe, Mn) and water and the changes induced in these cycles by human activities can provoke variations in pH and pe, two master variables for the geochemical processes controlling the solubility of heavy metals (Fig. 4.1.). These in turn can significantly influence the mobilization of heavy metals in the hydrocycle.

## 4.3 pH and Redox Related Processes Controlling the Solubility of Heavy Metals

### 4.3.1 pH

In aquatic systems, heavy metal cations tend to desorb or dissolve from solids as pH decreases. As pH increases, they adsorb or precipitate, but only up to a certain limit after which they are again solubilized (Fig. 4.2). The high pH range at which metal solubility increases is, however, not common in natural or mildly polluted environments.

A number of heavy metals (transition metals and metalloids) occur as oxyanions. Their solubility pattern is the opposite of that of cations. They tend to adsorb and/or precipitate more as pH decreases (Fig. 4.2).

The pH dependence of the solubility of heavy metal cations strongly complexed to natural dissolved organic matter (e.g., Cu, Pb) is anion-like when the metal solubility is controlled by the solubility of the organic matter (Bourg and Schindler 1985). Depending on the nature of the metal and on the relative concentration of metal, organic matter and adsorbing solid, heavy metal cation solubility can vary from being entirely controlled by its direct interaction with an inorganic surface (cation-like) to being completely subjected to the sorption and solubility pattern of the organic matter (anion-like).

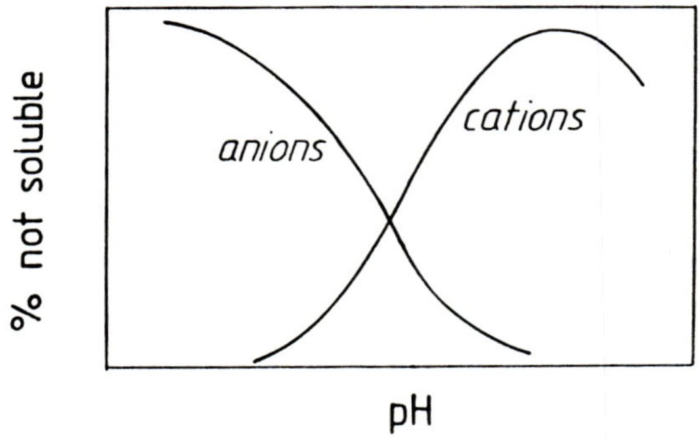


**Fig. 4.2.** Typical adsorptive and precipitative behavior of heavy metal cations and anions as a function of pH

### 4.3.2 Redox

Redox conditions have a direct and significant influence on the solubility of heavy metals like iron, manganese or chromium. Depending on their valency the metals occur under more or less soluble forms (reducing environments favor the solubility of Fe and Mn, but decrease the solubility of Cr). The mobility of other heavy metals whose redox state is not affected by the range of redox conditions encountered in natural waters is nevertheless indirectly affected by the availability of complexing agents, adsorbing Fe and Mn (hydr)oxides (electron acceptors) and redox-sensitive anions which can induce precipitation (e.g., reduced sulfur forms).

Fe and Mn oxides and hydroxides, and particulate organic matter usually occur as coatings of other solids. Their reactivity is therefore usually much greater than what could be expected solely on the basis of their weight fraction. While this implies that they are extremely powerful adsorbents, it also means that they are very subject to dissolution through reductive processes.

### 4.3.3 Redox and pH

Many, if not most, redox couples present in aquatic systems involve multielectron transfers, often with a phase change in at least one of the reduced or oxidized forms (see Table 4.2). These reactions are not necessarily rapid and redox potentials measured with a Pt electrode or calculated using various redox couples are, therefore, usually of little value for a precise interpretation of the observed water chemistry. They can, however, be used in pe-pH diagrams to give a general picture of the potential mobility of heavy metals, assuming equilibrium conditions and the absence of significant metal-organic interactions, as a function of the two master variables pH and pe.

A series of pe-pH diagrams is given to illustrate the typical solubility pattern of heavy metals and metalloids (Fig. 4.3).

Under oxidizing and slightly reducing conditions, the solubility of heavy metal cations decreases due to the precipitation of oxides, hydroxides or carbonates and, as mentioned above, only for neutral to mildly alkaline pH values (for these pH and pe conditions, amorphous or crystalline e.g., reduced sulfur forms).

Fe and Mn solids can scavenge heavy metals, either as strong adsorbents or as coprecipitating matrices. When the environment is only slightly reducing or slightly oxidizing and for medium to acidic pH range, the surface reactive Fe and Mn oxyhydroxides are solubilized, when Fe(III) and Mn(IV) are reduced to soluble species. If, under these conditions, the solution becomes slightly alkaline, Fe and Mn may precipitate as carbonates, but these minerals are certainly much less adsorbant than the corresponding oxides and hydroxides.

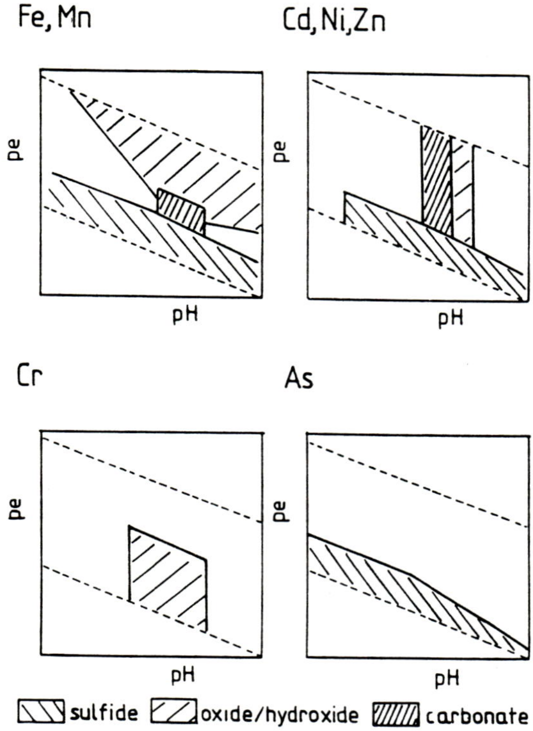


**Fig. 4.3.** Typical patterns for pe-pH diagrams. The assumed total dissolved activities are $10^{-3}$ for both total inorganic C and total S. Adapted from Richard and Bourg (1991) for Cr; and from Brookins (1988) for the others

Under reducing conditions, heavy metal cations and metalloids can be removed from solution as sulfide minerals if sufficient sulfur is available. If not (i.e., in some soil environments such as gley-type soils), reduction implies the loss of binding capacity by Fe and Mn (hydr)oxides for cations and consequently induces contaminant mobility and bioavailability (Fig. 4.3). Chromium is also insoluble under reducing conditions, as Cr(III) hydroxide, but only for neutral to mildly alkaline pH values.

The solubility of heavy metals as a function of pH and pe is summarized by the diagrams in Figs. 4.4 and 4.5. In systems containing a sufficient supply of sulfur (i.e., bottom sediments in river, estuarine or marine environments), low pH and moderate to high redox conditions are the most favorable for the solubilization of many heavy metals. For systems with a low S content (i.e., gleyic and gley soils) anaerobic conditions tend to induce a high solubility, and therefore a high mobility and bioavailability of toxic heavy metal cations.

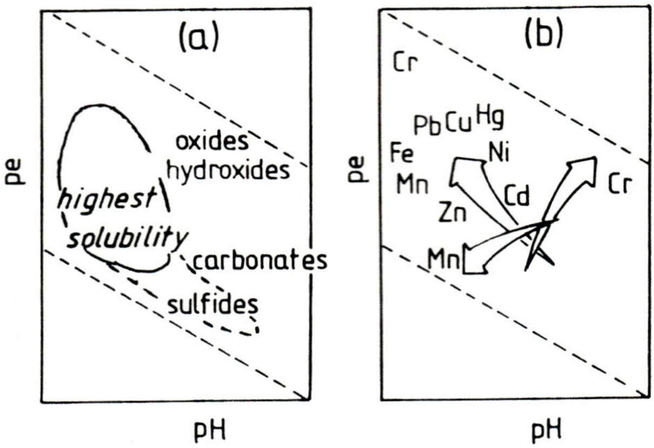


**Fig. 4.4 a–b.** Trends in solubility of heavy metals in relation to pH and pe (in the absence of dissolved and solid organic matter) **a** Main minerals controlling the solubility of heavy metals. **b** Trends of increasing solubility (*dotted lines* indicate increased solubility and mobility when the S supply is limited.)

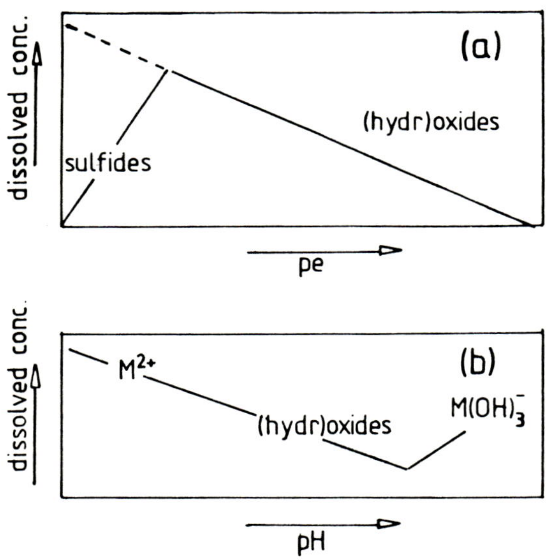


**Fig. 4.5 a–b.** Effect of pe (**a**), and pH (**b**) on the solubility of heavy metals. *Dotted lines* indicate increasing solubility and mobility when the S supply is limited. (Bourg 1994)

## 4.4 Environmental Changes Causing pH and Redox Variations

A number of pe and pH measurements in a variety of natural environments have been reported by Baas Becking et al. (1960) (Fig. 4.6). The pH and pe of most soils range from 3.5 to 9 and - 6 to +12, respectively.

Redox processes can induce strong acidification or alkalinization of soils and other aquatic systems (van Breemen 1987). Oxidized components are more acidic ($SO_3$, $N_2O_5$) or less basic ($Fe_2O_3$) than their reduced counterparts ($H_2S$, $N_2$, $NH_3$, FeO). As a result, alkalinity and pH tend to increase with reduction and decrease with oxidation. In closed systems, these processes are totally reversible. However, the dissolved nitrogen gas produced during denitrification (Table 4.2) is lost to the atmosphere and, depending on whether the acidic or the basic component of an acid-base pair is more mobile, removal of one of the pairs of components leads to a system that will become increasingly acidified or alkalinized (van Breemen 1987).

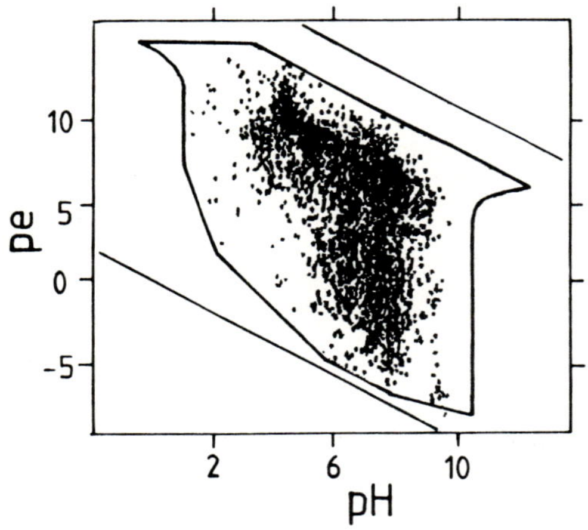


**Fig. 4.6.** Eh and pH measurements of natural aqueous environments (from Baas Becking et al. 1960)

### 4.4.1 Acidification

Several processes can modify the proton balance in a natural terrestrial or aquatic system (Table 4.1). Continued acidification (e.g., emissions from industry, traffic, mining and afforestation activities) would increase the solubility of most cationic heavy metals. This is a major concern in soil and sediment systems which are poorly buffered with respect to pH. Oxidation reactions (e.g., nitrification and

**Table 4.1.** Processes which can modify the proton balance (Zobrist et al. 1987)

| Processes which supply protons |
|---|
| *Atmospheric deposition* (wet and dry) |
| *Oxidation reactions* (e.g., nitrification, sulfide oxidation, ferrous iron oxidation) |
| **Processes which consume protons** |
| *Weathering of carbonate minerals* |
| $MCO_3 + 2H^+ \rightarrow M^{2+} + H_2CO_3$ (with M = Ca and/or Mg) |
| *Weathering of silicate minerals* |
| primary rock mineral $+2H^+ + H_2O \rightarrow$ cations$+H_4SiO_4+$secondary (weathered) minerals |
| *Ion exchange* |
| $2ROH + SO_4^{2-} \rightarrow R_2SO_4 + 2OH^-$ |
| $NaR + H^+ \rightarrow HR + Na^+$ |
| *Reduction reactions* (e.g. denitrification, sulfate reduction, iron oxide reduction) |
| **Amphoteric processes** |
| Production of biomass/respiration of biomass (Redfield equation) |

sulfide oxidation) also supply protons. Several heterogeneous reactions can consume protons and efficiently buffer the pH of natural water systems.

Carbonate solids (mostly Ca, Mg or mixed Ca/Mg minerals) are efficient buffers in the neutral to basic pH range (pH > 6.2). When protons are added, an initial decrease in pH will be followed by dissolution of the solid carbonate and therefore by the establishment of a new state of equilibrium, reducing the pH shift that would occur in the absence of dissolved calcium. At pH > 6.2, as long as carbonate solids are present in sufficiently fine particles or as grain coatings they will provide pH buffering. When they are gone the aluminum silicates will be attacked by protons.

Aluminum silicate minerals contained in the soil or sediment solids can provide a considerable resistance to pH changes (Stumm and Morgan 1981). At pH 8, for example, the anorthite-kaolinite equilibrium system has a buffer capacity three orders of magnitude higher than that of a millimolar carbonate solution.

$$\text{Ca-feldspar (anorthite)} + 2H^+ + H_2O \;=\; \text{kaolinite} + Ca^{2+}$$

Equilibrium systems with a sufficient number of coexisting phases have, in theory, an infinite pH buffering capacity (Stumm and Morgan 1981). However this does not apply to some natural systems (e.g., sandy soils). Significant pH variations are observed regularly in surface waters and in soils due to the alternating phenomena of photosynthetic activity and microorganism respiration (e.g., Carlton and Wetzel 1987; Fuller and Davis 1989). The buffering capacity of solids in a given system is exhausted when the solids are depleted (i.e., not available for dissolution) in one of the components of an acid-base pair. It can also be limited by reaction kinetics and by faster transport of one of the components of the acid-base pair away from the reaction zone.

### 4.4.2 Reduction of Soils

Many things can cause changes in redox conditions. Flooding, rising water table, excess organic matter, sludge spreading, increased biological activity due to increased nutrient supply or temperature, deterioration of soil structure (compaction), and deforestation can increase reduction, while resuspension of sediments, dredging, or disposal of mine tailings can induce oxidation.

The ability of an aquatic environment to resist redox changes (towards more reducing or oxidizing conditions) depends on the availability of oxidized or reduced species. Redox buffering can be estimated by measuring oxidation and reduction capacities (Barcelona and Holm 1991; Heron et al. 1993). The main types of redox reactions involve both aqueous species and solids.

**Table 4.2.** Typical electron-accepting reactions in aquatic systems (Heron et al. 1993)

| | |
|---|---|
| $O_2 + 4H^+ + 4e^- \rightarrow 2H_2O$ | (aq) |
| $NO_3^- + 6H^+ + 5e^- \rightarrow 0.5N_2 + 3H_2O$ | (aq) |
| $NO_3^- + 10H^+ + 8e^- \rightarrow NH_4^+ + 3H_2O$ | (aq) |
| $MnO_2 + 4H^+ + 2e^- \rightarrow Mn^{2+} + 2H_2O$ | (solid) |
| $Fe(OH)_3 + 3H^+ + e^- \rightarrow Fe^{2+} + 3H_2O$ | (solid) |
| $SO_4^{2-} + 9H^+ + 8e^- \rightarrow HS^- + 4H_2O$ | (aq + solid) |
| $CH_2O + 4H^+ + 4e^- \rightarrow CH_4 + H_2O$ | (aq + solid) |

The total oxidation (OXC) and reduction (REC) capacities are given by:

$$OXC = 4[O_2] + 5[NO_3^-] + 2[Mn(IV)] + [Fe(III)] + 8[SO_4^{2-}] + 4[\text{oxidized C}]$$
$$REC = 8[NH_4^+] + [Fe(II)] + 2[Mn(II)] + 8[S(-II)] + 4[\text{reduced C}]$$

All concentrations are given in moles per cubic meter of the aquatic system.

For soils, aquifers and bottom sediments (i.e., high solid/liquid ratio), iron oxy-hydroxides and organic matter (if reducible) and, to a lesser extent, manganese oxides are the major contributors to the oxidation capacity. In surface aquatic systems, nitrogen, sulfur, organic carbon and oxygen are the predominant electron acceptors. The redox buffering of soils and aquifers is much greater than that of surface aquatic systems (rivers, lakes,...).

**Table 4.3.** Concentrations of oxidized and reduced species and their estimated potential contribution to the oxidation (OXC) and reduction (REC) capacities of aquatic systems [a]

| | Aquifer: | | | River: | | |
|---|---|---|---|---|---|---|
| | Conc. | OXC (equiv/m³) | REC (equiv/m³) | Conc. | OXC (equiv/m³) | REC (equiv/m³) |
| $O_2$ | 10 mg $l^{-1}$ | 0.44 | | 10 mg $l^{-1}$ | 1.25 | |
| $NO_3^-$ | 20 mg $l^{-1}$ | 0.6 | | 5 mg $l^{-1}$ | 0.4 | |
| $NH_4^+$ | <1 mg $l^{-1}$ | | <0.16 | - | | - |
| Mn(s) | 0.3 mg $g^{1}$ | 18 | | 1 mg $g^{-1}$ | 0.002 | |
| Mn(aq) | 0.01 mg $l^{-1}$ | | 0.0001 | 0.01 mg $l^{-1}$ | | 0.0004 |
| Fe(s) | 6 mg $g^{-1}$ | 175 | | 50 mg $g^{-1}$ | 0.09 | |
| Fe(aq) | 0.1 mg $l^{-1}$ | | 0.001 | 0.05 mg $l^{-1}$ | | 0.002 |
| S(-II)(s) | 0.03 mg $g^{-1}$ | | 12 | - | | - |
| $SO_4^{2-}$ | 40 mg $l^{-1}$ | 1.2 | | 10 mg $l^{-1}$ | 0.83 | |
| $C_{org}$(s)[b] | 0.3 mg $g^{-1}$ | 165 | | 30 mg $g^{-1}$ | 0.50 | |
| $C_{org}$(aq) | 1 mg $l^{-1}$ | 0.12 | | 2 mg $l^{-1}$ | 0.67 | |
| Total | | 360 | 12 | | 3.74 | 0.002 |

[a]Assuming a porosity of 0.35 and a bulk density of 1.6 kg $l^{-1}$ for the aquifer and a suspended solid concentration of 50 mg $l^{-1}$ for the river.

[b]In the case of soils, the particulate organic content is usually much higher (>10 mg $g^{-1}$ and a large part of it is oxidizable and therefore contributes to the REC).

### 4.4.3 Oxidation of Soils and Sediments Previously Exposed to Sulfur-Reducing Conditions

Weathering under oxidizing conditions of solids containing reduced sulfur will cause a release of soluble metals and acidity (e.g., Stumm and Morgan 1981). A typical solubility pattern for divalent cationic heavy metals is shown in Fig. 4.7. As reduced sulfur becomes depleted (*sulfide solid* curve), the heavy metal concentration increases (line *A*) until its solubility becomes controlled by adsorption (*adsorption* curve). If pH decreases (because of the simultaneous production of acidity and an absence of pH buffering), the metal solubility is also limited by adsorption but at a higher value (line *B*).

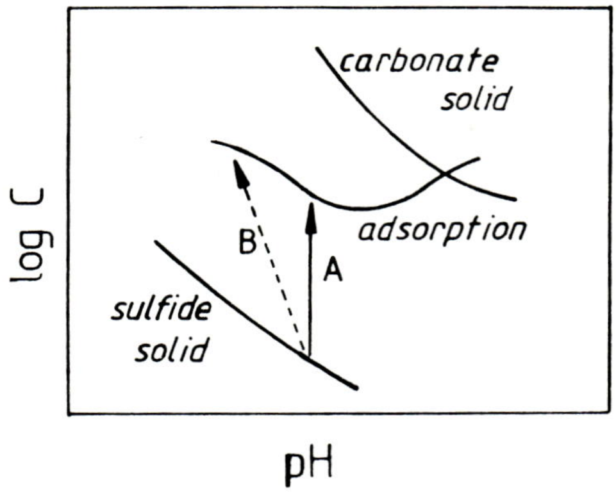


**Fig. 4.7.** Theoretical pathway for increase in solubility of a heavy metal cation as reduced sulfur is oxidized to sulfate and the corresponding precipitated metal sulfide is dissolved. *Solid line A* represents the case of strong pH buffering; *dotted line B:* the case of simultaneous acidification; the *curved lines:* the various solubility controlling processes; and the *straight lines:* the two possible solubilization pathways

## 4.5 Sensitive Environments

### 4.5.1 Moderately to Highly Polluted Local Sources

Landfills and other localized areas where waste materials are disposed of represent real and major chemical time bombs.

Mine tailings and dredged materials both have a high heavy metal load and a significant reduced sulfur content (Salomons and Förstner 1988). Moreover, the biogeochemical processes responsible for the accumulation in both solid types take place under anaerobic conditions. When left at the soil surface, the oxidation of sulfides generates acid and if the pH buffer capacity of the neighboring, receiving environment is insufficient, acidic conditions may develop which, for most metals, provoke an increasing solubility (e.g., Chapman et al. 1983).

Leachate produced during waste degradation in landfills can be rich in heavy metals. The leachate entering soil and aquifer environments can be highly reduced, leading to production of methane and reduction of sulfate (e.g., Lyngkilde and Christensen 1992). The typical redox reactions taking place in aquifers are listed in Table 4.2. Although the reduced sulfur redox zone is favorable to the trapping of heavy metals (precipitated as sulfide minerals; Christensen et al. 1994), in some cases (e.g., Ni), a significant heavy metal load can be transported over large distances (> 200 m in 8 years) due to competitive complexation with organic molecules and carbonate ions in the dissolved phase. (Bourg et al. 1994).

Moreover, anionic contaminants (such as Cr) were observed, in long-term field tracer experiments, to migrate even under suboxic conditions (Kent et al. 1994).

When the reducing capacity of organic matter in the leachate coming out of a landfill diminishes and the system becomes oxidizing, the metal sulfide trapped near the landfill should undergo oxidative dissolution resulting in a release of heavy metal cations. This process is accompanied by the production of protons and therefore may induce acidification of the system when its pH buffer capacity is exceeded.

### 4.5.2 Weak to Moderate Pollution Risk at the Regional Scale (from Loch et al. 1993)

In the absence of excess reduced sulfur, or under conditions that are only moderately reducing (e.g., dissolved sulfate is present and Fe and Mn (hydr)oxides are reduced and solubilized), the soils most vulnerable to changing heavy metal fixation capacities are:

- Aerobic, low pH soils (pH < 5), implying the absence of available carbonate solids, low base saturation, good drainage and a moderate organic matter content (< 5%). The oxide content is usually low but its capacity to adsorb oxyanions and heavy metal cations is high. When acidified, the Al oxides and ultimately the Fe oxides will be depleted because of the low pH buffering capacity, resulting in a release to the aqueous phase of previously bound metals. At low pH, the moderate organic matter content provides insufficient substitute binding capacity.
- Poorly drained soils with a high water table (i.e., gleysols). Such soils tend to have a low Fe oxide content, which is usually distributed as mottles in the soil profile, surrounded by reduced oxide-depleted soil. As above, the oxides, though low in content, provide a high adsorbing capacity. A rise in piezometric level or increased humidity will lead to the redistribution or complete depletion of Fe oxides and the release of contaminants. The role of a moderate organic matter content is the same as above.
- Organic soils. Most organic matter soils have a low pH and are prone to reduction when humidity increases. Oxides present will be easily depleted by reductive dissolution enhanced by the presence of the high dissolved organic carbon concentration.

This assessment of vulnerability assumes that there are few if any soils on earth that do not contain some oxides and hydrous oxides (Marshall 1964) and these are, together with solid organic matter, major contributors to the adsorption capacity of natural solids (e.g., Jenne 1968, Loch et al. 1993).

To identify the most vulnerable European soils, the following procedure was adopted (Loch et al. 1993). For each European country, 2 to 7 major soil units were selected from the FAO/UNESCO World Soil Map, scale 1:5 000 000

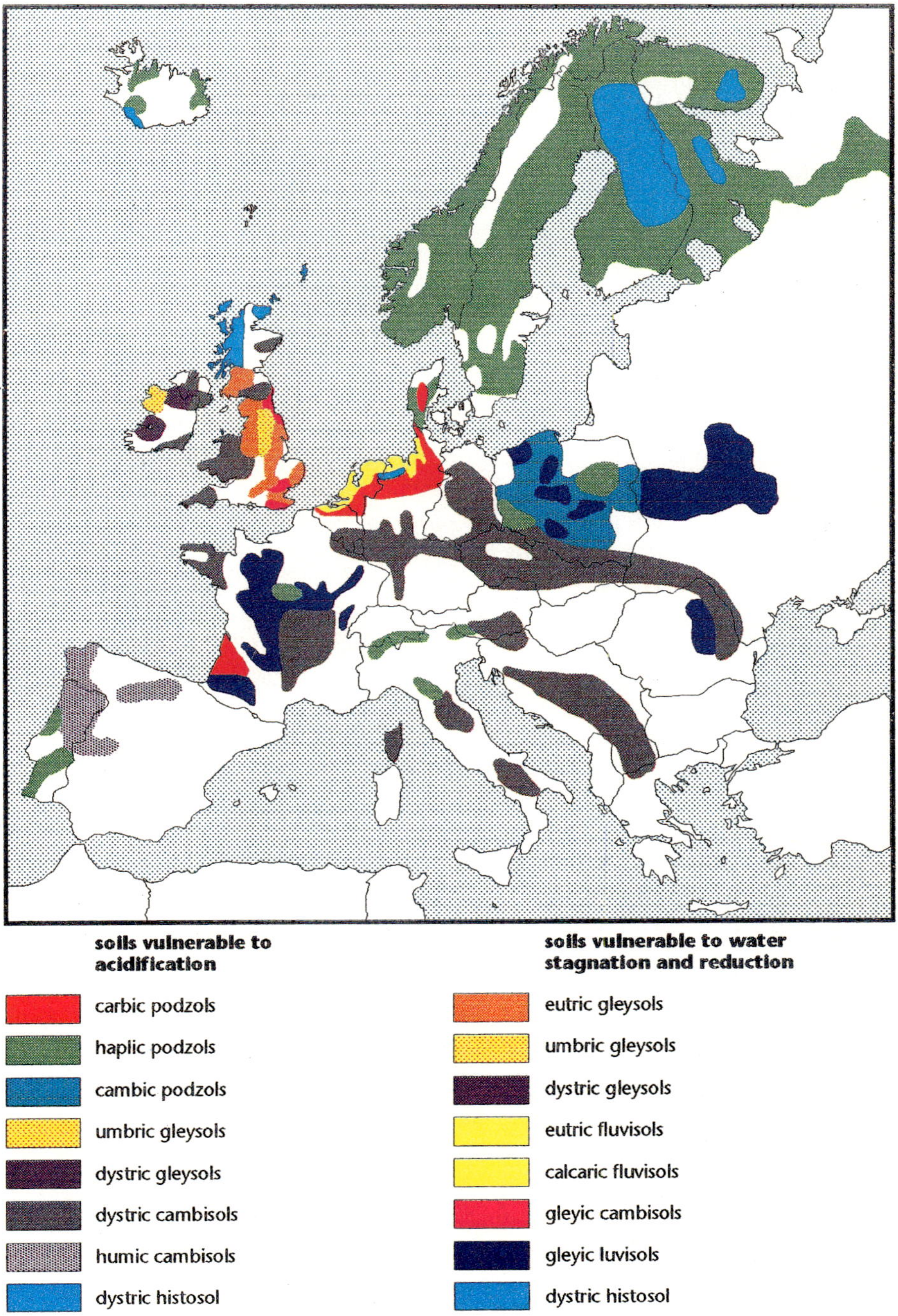


**Fig. 4.8.** Major European soils vulnerable to enhanced solubility of heavy metals (Loch et al. 1993)

(1978). This resulted in a list of 27 units. For each of these units, the following properties were considered: gley or water stagnation within the upper 50 or 100 cm, % base saturation of the adsorption complex, calcareous character and presence of argic, calcic, mollic or histic horizons (FAO/UNESCO 1988). The resistance of these soils to changing redox potentials and pH was inferred from this information.

## 4.6 Summary and Conclusion

In the presence of excess sulfur, highly reduced conditions (favorable to the occurrence of reduced S) generally protect against the mobility and dissolution of heavy metals. When conditions become oxidizing, the heavy metal cations are solubilized and released to the aqueous (mobile) phase.

For other situations, in the presence of low to moderate organic matter content in the solid phase, increased acidity and moderate to strong reducing conditions can remobilize previously trapped heavy metal cations.

For metal anionic forms (e.g., oxyanions) and for cationic metals forming strong complexes with dissolved organic compounds (if the organic carbon content is high), moderately acidic conditions may maintain pollutants in the solid phase.

A better understanding of the solubility behavior of heavy metals in the presence of natural solids is necessary for both the evaluation of the delayed potential remobilization and for chemical decontamination.

**Acknowledgements**. Many thanks to Anna Kay Bourg for editing the English and to various collaborators (R. Scott Altmann, Catherine Crouzet and Monika Kedziorek) for fruitful discussions.

### References

Baas Becking LGM, Kaplan IR, Moore D (1960) Limits of the natural environment in terms of pH and oxidation-reduction potentials. J Geol 68:243-284

Barcelona M, Holm TH (1991) Oxidation-reduction capacities of aquifer solids. Environ Sci Technol 25:1565-1572

Bourg ACM (1994) Speciation of heavy metals in soils and groundwater and implications for their natural and provoked mobility. In: Salomons W, Förstner U, Mader P (eds) Heavy metals: problems and solutions, Springer, Berlin Heidelberg New York, in press

Bourg ACM, Schindler PW (1985) Control of trace metals in natural aquatic systems by the adsorptive properties of organic matter. Proc Internat Conf Heavy Metals in the Environment, Sept 1985, Athens, 1:97-99

Bourg ACM, Williams GM, Bannon MP (1994) The role of complexants in enhancing the transport of nickel in a sandy aquifer down gradient from a liquid waste disposal. Submitted for publication

Brookins DG (1988) Eh-pH diagrams for geochemistry, Springer, Berlin Heidelberg New York, 176 pp

Carlton RG, Wetzel (1987) Distributions and fates of oxygen in peryphyton communities. Can J Bot 65:1031-1037

Chapman BM, Jones DR, Jung RF (1983) Processes controlling metal attenuation in acid mine drainage streams. Geochim Cosmochim Acta 47:1957-1973

Christensen TH, Kjeldsen P, Albrechtsen H-J, Heron G, Nielsen PH, Bjerg PL, Holm PE (1994) Attenuation of pollutants in landfill leachate polluted aquifers. Crit Rev Environ Sci Technol, in press

FAO/UNESCO (1988) Soil map of the world, revised legend. World soil resources. Report 60. FAO/UNESCO/ISRIC. Rome.

Fuller CC, Davis JA (1989) Influence of coupling sorption and photosynthetic activity processes on trace element cycles in natural waters. Nature 340:52-54

Heron G, Christensen TH, Tjell JCh (1993) Oxidation capacity of aquifer sediments. Environ Sci Technol 28:153-158

Jenne EA (1968) Controls on Mn, Fe, Co, Ni, Cu and Zn concentrations in soils and water: the significant role of hydrous Mn and Fe oxides. Am Chem Soc Symp Ser 93:337-387

Kent DB, Davis JA, Anderson LCD, Rae BA, Waite TD (1994) Transport of chromium and selenium in the suboxic zone of a shallow aquifer: influence of redox and adsorption reactions. Water Resour Res 30:1099-1114

Lindsay WL, Sadiq M (1980) cited in Lindsay WL (1979) Chemical equilibria in soils, John Wiley and Sons, New York, 449 pp

Loch JPG, Berg GA van den, Weijden ÇH van der (1993) The stability of Fe-, Mn- and Al-(hydr)oxides and their binding capacity for contaminants in changing soil and sediment environments. In: Meulen GRB ter, Stigliani WM, Salomons W, Bridges EM, Imeson AC (eds) Chemical time bomb. Proc European State of the Art Conf on Delayed Effects in Soils and Sediments, Foundation for Ecodevelopment "Stiching Mondiaal Alternatief", Hoofdorp, The Netherlands, pp 48-51

Lyngkilde J, Christensen TH (1992) Redox zones of a landfill pollution plume (Vejen, Denmark). J Contam Hydrol 10:273-289

Marshall CE (1964) The physical chemistry and mineralogy of soils. Vol 1: Soil materals, John Wiley and Sons, New York

Richard FC, Bourg ACM (1991) Aqueous geochemistry of chromium: a review. Water Res 25:807-816

Salomons W (1993) Non-linear response of toxic chemicals in the environment: a challenge for substainable management. In: Meulen GRB ter, Stigliani WM, Salomons W, Bridges EM, Imeson AC (eds) Chemical time bomb. Proc European State of the Art Conf on Delayed Effects in Soils and Sediments, Foundation for Ecodevelopment "Stiching Mondiaal Alternatief", Hoofdorp, The Netherlands, pp 31-43

Salomons W, Förstner U (1988) Chemistry and biology of solid waste. Dredged materials and mine tailings. Springer, Berlin Heidelberg New York, 301 pp

Stigliani WM (1993) Overview of the chemical time bomb problem in Europe. In: Meulen GRB ter, Stigliani WM, Salomons W, Bridges EM, Imeson AC (eds) Chemical time bomb. Proc European State of the Art Conf on Delayed Effects in Soils and Sediments, Foundation for Ecodevelopment "Stiching Mondiaal Alternatief", Hoofdorp, The Netherlands, pp 13-29

Stumm W, Morgan J (1981) Aquatic chemistry, 2nd edn. Wiley Interscience, New York, 780 pp van Breemen N (1987) Effects of redox processes on soil acidity. Neth J Agric Res 35:271-279.

Zobrist J, Sigg L, Schnoor JL, Stumm W (1987) Buffering mechanisms in acidified alpine lakes. In: Barth H (ed) Reversibility of acidification. Elsevier Applied Sci, London, pp 95-103

# 5 Erosion and Transport of Pollutants from the Terrestrial to the Aquatic Environment

*A. C. Imeson*

## 5.1 Introduction

The pathways followed by eroded and transported soil or mine spoil on its way to an aquatic environment involve an interrupted journey that is sometime completed in minutes or hours but which usually takes hundreds of years. The fate of most eroded sediment is to be periodically transported very short distances and to spend long periods at rest on hill slopes. The sites on slopes where transported soil accumulates, only temporarily retain soils and sediments (colluvium) so that should these be contaminated, they inevitably form a potential hazard. The main objective of this chapter is to explain the processes behind this risk. It aims to explain how long-term changes in some basic soil properties of accumulating and eroding soil affect rates of soil erosion, accumulation and transport of contaminants and nutrients in terrestrial environments.

Erosion leads to the concentration of toxic substances because of the selective sorting and deposition of different size fractions. Also, in time, the transformation of accumulating sediments on hill slopes can lead to renewed risk of erosion. Consequently, the occurrence or erosion is highly non-linear, related not just to the direct effect of erosive forces that drive erosion but also to the gradual changes that take place over tens of years in the properties of soils that govern their erodibility. Consequently, erosion results from interacting processes that operate over many time-scales. This and the complexity resulting from the non-linear character of most of the individual processes makes the forecasting of erosion risk difficult. Furthermore, the exact amount of pesticide or nutrient eroded is very sensitive to the amount of rainfall or other weather conditions (Ahuja and Lehman 1983, Edwards et al. 1993a 1993b and Rudra et al. 1993).

To illustrate the relevance of erosion in the biogeodynamics of contaminants, the case of mine-spoil waste containing heavy metals is an obvious one (e.g. Sidle et al. 1993) Eroded material is often transported over long distances, selectively extending the effects of contamination in soils and in water courses. This erosion is undesirable because transport causes the sorting and selective deposition of colloidal clay and organic matter, concentrating adsorbed, toxic, substances. Erosion can, therefore, cause the build-up of concentrations of normally non-toxic substances to toxic levels at locations where transported material is deposited. Many well known examples of this concern agro-chemicals adsorbed to transported soil colloids. Thus pesticide concentrations can be increased in colluvium by at least 10 to 20 times, creating hazardous sinks unforeseen at the time of application (Harrodd 1992).

As well as threatening and creating sinks of sediments that may emerge as delayed risks, erosion is in fact increased by soil contamination. An increase in erosion could be an advanced and sensitive indicator of a contamination problem. A low soil erodibility is often a result of an imbalance in the relative rates of degradation and stabilization processes in the soil. Microbiological stabilization processes in the soil are sensitive to soil contamination. Easily surveyed soil characteristic indicative of erodibility (e.g. aggregation or bulk density) could, therefore, be used for mapping and assessing potential delayed effects of contaminants.

Chapter 5 is divided into four parts. Section 5.1 provides background information on soil erosion in the context of the transport of contaminants. Section 5.2 considers the changes of some soil properties affecting soil erosion and capacity controlling properties and Section 5.3 describes a few examples. Finally Section 5.4 is a summary of the main problems and conclusions. For more detailed information concerning the relevance of soil erosion to the CTB problem, reference is made to Imeson et. al. (1992).

## 5.2 Soil Erosion and Delayed Non-linear Responses

### 5.2.1 Introduction and Context

Soil erosion occurs when the processes of, wind erosion, rain-wash, rill and gully erosion detach and transport soil particles, at or near the surface of the soil. This erosion depends upon the erosive force or energy of these processes (erosivity) and on the ability of the soil to resist them (erodibility). These two aspects of erosion (erosivity and erodibility) involve many interacting chemical, biological and mechanical processes, that operate at different hierarchically-nested scales. In the context of this chapter, emphasis is placed on erodibility; in particular to soil properties that are dynamic over tens or several hundreds of years (Mahboubi et al. 1993) and that can lead to the sudden emergence of soil erosion problems.

Soil erosion does not threaten all soils equally. It affects largely, the upper 20 to 30 cm of soils on agricultural land or soils that are frequently burnt or heavily grazed and trampled. The upper soil layer contains most organic matter so that it has a high cation exchange capacity (CEC), and can accumulate potentially dangerous chemical substances. For example, in most parts of the world, unless erosion has occurred, this layer still contains most of the $Cs_{137}$ deposited during the period of atmospheric nuclear testing. Studies of heavy minerals and metals in soils show that these are mixed by ploughing and by bioturbation throughout this 20 to 30 cm or more thick layer. Over longer time periods, during the Holocene, Jungerius and Mücher (1972) have shown that mixing occurs throughout a much thicker layer, 80 cm or more in depth. Consequently, the upper soil layer does not always contain the same soil. Contaminants found throughout the 80-cm-thick layer of mixing, can be transferred upwards to the surface layer of potential erosion risk (20–30 cm).

Under certain conditions, the movement of water through macropores in the soil can lead to the subsurface erosion of soil material. This erosion (tunnel and pipe erosion) occurs under specific sets of conditions associated with ploughing and the creation of plough pans and with soils having dispersive B or BC soil horizons. It is associated with new pathways of water movement that bring contaminated water to more distant and deeper locations much more rapidly than might otherwise be expected. Furthermore, the arrival of concentrated subsurface flow downslope can trigger the erosion of sediments in more low lying colluvium.

Changes in material properties can lead to the evolution of conditions that allow rill and gully erosion to occur and threaten deeper soil layers. Rills are the small shallow channels that develop when surface or subsurface run-off becomes concentrated on the surface. Gulleys are larger channels, that can not be removed by ploughing and which can be eroded in recent sediments and weathering materials. With respect to non-linear response of pollutants this erosion is a potential hazard for landfill sites and mine spoil, where material properties favour these processes. Another group of threatened soils are those found on alluvial lowlands and wetlands. Not only are these the sinks for chemicals transported to them by overbank flow from rivers, they also accumulate chemicals from industrial and agricultural sources via groundwater and surface drainage from roads and ditches. River bank erosion, that usually supplies at least 40% of the sediment carried by most rivers can be a threat to these deposits. In fact as meander bends migrate downstream along flood plains, they repeatedly erode and then redeposit the active flood plain sediments and soils. Archaeological remains on flood plains are seldom old suggesting that deposits never remain at one place for longer than 600 years in humid regions to as little as 60 years in semi-arid environments (Schumm 1977).

Specific soils directly threatened by erosion and erodibility problems also include lowland agricultural soils, particularly when these are either silty, highly calcareous or have an impermeable horizon near the surface. Paradoxically, soils in mountainous areas are often gravelly or stony, have high infiltration rates and a low soil erodibility. Severe misuse is needed before they constitute a risk, as for example is the case with soils degraded by frequent wild fires, footpaths or alpine sporting activities.

Hydrological changes brought about by various activities in one part of a drainage basin may lead to the erosion of soils elsewhere. The problem of threatened soils, therefore, needs to be examined in the context of the changes that are taking place in entire drainage basins. For example, soil conservation on slopes, by reducing sediment loads, but not the run-off reaching the river channels, increases the transport capacity of the channel flow and leads to the erosion of channel sediments.

The most persistent erosion problems are on soils that only slowly recover from a disturbance (soils having a low inertia and/or resilience; Westmann 1987). Mediterranean, alpine and tundra soils are good examples.

### 5.2.2 The Data of Soil Erosion

Although a vast amount of information and data on soil erosion is available, it is difficult to apply much of this, other than in a very general way, to the problem of non-linear response, that have to address processes of change. Nearly all data on soil erosion are highly site specific and only of value within the context for which the data were collected. Measurements of soil erosion rates have proved to be highly sensitive to initial conditions describing the state of the soil. This has up until now hindered the validation of detailed deterministic erosion models. Furthermore, the complex interactions and feedbacks linking the component biological, chemical and mechanical processes affecting erosion rates are only recently being studied (Brussard and Kooistra 1993). The future course of soil erosion can unfortunately not be predicted with statistically derived parameters that can only just about describe soil erosion today. One of the greatest needs in soil erosion research is to develop a process-oriented methodology that can be applied to study the impacts of environmental change.

### 5.2.3 Soil Erodibility

Soil erodibility can be considered as a secondary behavioural property of the soil reflecting amongst other things, soil structure, texture, soil mineralogy, soil moisture content and (micro) biological activity. It is a dynamic property that responds to changes which may be intrinsic or extrinsic to the soil system. In practise it closely reflects past and present land use. Traditionally, soil erodibility has been approached from within the conceptual framework of the USLE (Wischmeier and Smith 1978). This treats soil erodibility as a more or less static long-term soil property evaluated using statistical models derived from soil analysis and plot experiments.

Soil erodibility is also traditionally evaluated with respect to (a) the resistance of the soil to mechanical detachment and breakdown and (b) the ability of the soil to maintain high rates of infiltration and water penetration. Research has demonstrated that amongst the most important soil characteristics that can be used to describe erodibility are those that are reflected in the organization, water-stability and dynamics of the soil structure. Under many circumstances, soil structural attributes can be used to provide an indication of the ability of the biological

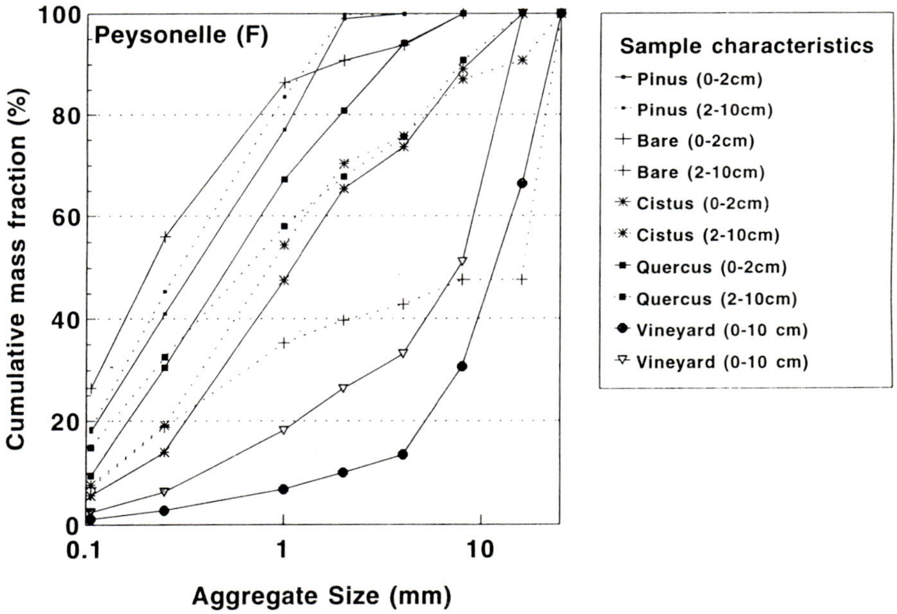


**Fig. 5.1.** Aggregate size distributions of adjacent soils vary according to land use. Note that the bare soil characteristic refers to bare patches of forest soil. The degraded soils are in the vineyards where there is little aggregated fine material

activity in the soil to generate and maintain a medium with a high water stability and water retention potential. Various (micro) biological processes produce water-stable aggregates of different sizes that can resist compaction and detachment by raindrop impact. This is important because water-stable aggregates prevent surface seals or crusts from forming. A non-erodible soil generally has less soil that is not aggregated than an erodible soil and also coarser micro-aggregates.

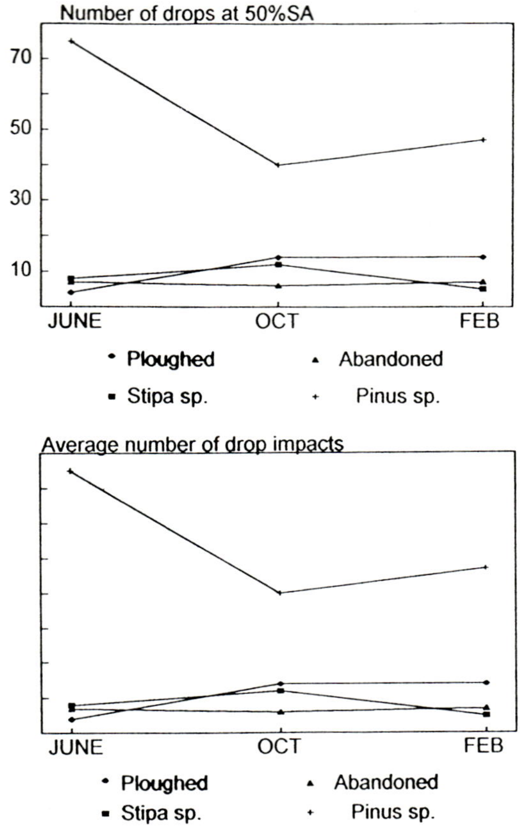


**Fig. 5.2.** The aggregate stability of soils (determined by the single drop test) for sites on tilled soil, of Esparto grass, under pine trees *Pinus sp.*, and land abandoned 12 years ago. Only under the pine trees is there a significantly higher aggregate stability which is the highest during the summer. The figure shows the number of water drops required to break down 50% of the aggregates tested. Redrawn from Cerda et al. (1994)

Some data that illustrate how both aggregate stability and size distributions are related to the state of soil degradation, are shown in Figs. 5.1 and 5.2. These data show how at a site in the south of France, soil aggregation clearly relates to land use and is affected by land abandonment. Data such as these provide examples of how soil structure changes with land use or with increased environmental stress. It can also be used to establish rates of change.

The effects of soil organic matter and (micro) biological processes on soil structure are summarized in the hierarchical model of Tisdale and Oades 1982. This states that different organic substances/components stabilize aggregates of different size. The importance of fungal-hyphae and very small roots on stabilizing the larger aggregates in this model is stressed. Various polysaccharides play an important part in stabilizing smaller micro-aggregates. Although the amount of soil organic matter has usually been found to correlate with soil erodibility, the correlations are the result of many different processes and effects. For example, stabilization depends on the food source and population dynamics of the soil micro-fauna and flora, on the digestion of soil by earthworms, on plant root exhudates and on microbial effects. A recent discussion of the processes involved at different hierarchies of scale is given by Oades and Waters (1991) and Oades et al. (1990). The nature of these processes mean that both spatially and temporarily, soil erodibility is highly variable. Annual trends in erodibility reflect the effects of soil micro-fauna and flora on soil stabilization but also both the seasonal cycle of climate and shorter term variations in temperature and soil moisture.

A simple way of distinguishing between erodible and non-erodible soils, is by means of the ratio (R) between the rates of soil structure development and decline. This ratio can be evaluated from field and laboratory measurements and from modelling. Values of $R < 1$ will ultimately lead to an erodible state of the soil if they persist for several years.

### 5.2.4 Medium and Long-term Trends in Soil Erodibility

Long and medium term are used here to refer to changes which occur respectively over several hundreds or several tens of years. With respect to the medium term there are important differences in the magnitude and the character of the changes that occur under different land-use types.

During the last century organic farming that formerly maintained soil fertility, has almost disappeared as chemical fertilizers have replaced the use of manure. This has caused a general decline in organic matter levels and in the physical properties (including erodibility) of most cultivated soils. Together with the negative effects on soil structure produced by soil compaction under heavy farm vehicles, and the decline in soil biological activity under monoculture, this has resulted in severe and widespread soil degradation. Superimposed on this general decline are seasonal or yearly variations in erodibility levels that reflect variations in climate, tillage methods and crop rotations. One consequence of this trend is that the combination of a low erodibility of the topsoil and a subsurface ploughpan results in situations in which subsurface bypass flow can develop. This leads to a very much more rapid movement of water through the soil than would otherwise occur. Another consequence is that many soils are so badly degraded that the base-level of erodibility may have been reached. Any change that improves soil erodibility would have an impact on the hydrology and sedimentation processes that are now prevalent.

For semi-natural environments very little is known about long-term trends in soil erodibility. Patterns of soil erodibility at different scales in the landscape are specific to particular vegetation types. These patterns buffer the effects of local changes so that the situation is rather stable. In natural areas most erosion, and the highest soil erodibilities occur along pathways and animal tracks. Animal population dynamics and various human activities in the forest are the causes of change. These changes can have a great impact on storm run-off production and can mobilize sediments stored in river channel systems.

Data from Spain indicating that rates of change are shown, suggest that on abandoned land a good aggregate stability exists for the larger macro-aggregates after about 5 years (Cammeraat et al. 1994). Stabilization proceeds very rapidly as soon as a site is shaded by litter or vegetation. As far as micro-aggregates are concerned, between 20–60 microns, it may take as long as 15–20 years for these to recover following a disturbance. Recent work on the residence time of organic matter in different sized soil aggregates (Buyanovsky et al.) suggests the same trend. Long and sometimes medium-term changes relevant for soil erodibility occur as soil properties evolve, minerals weather and salts accumulate. This can be illustrated by a few examples. Dietrich (1985) has shown how in the past in California, colluvium has repeatedly accumulated in hill-slope hollows from which it is periodically eroded every few hundred years. This is because the gradual thickening and compaction of the continually accumulating colluvium together with weathering decreases the permeability and shear strength of the B and C horizons. The surface slope of deposited colluvium also steadily increases so that with waterlogging and positive pore pressures, mass movements inevitably erode the soil in the hill-slope hollow. Colluvium in the hollow gradually accumulating over many hundreds of years, is suddenly removed by gully erosion or mass movements during one critical storm, allowing accumulation to begin once more. From this it can be concluded, that all contaminants stored in the colluvium in hill-slope hollows presents a high risk.

Analogous changes in soil properties occur in the soils that accumulate behind soil or water conservation terraces and in earth dams. In this context the accumulation of water-soluble salts may be important in dry areas promoting dispersion and tunnel erosion. There are very many changes that occur in soils accumulating behind terraces. Inevitably, however, in very many situations both types of change lead to changes in the soil and hill-slope hydrology that will in the long term bring about the collapse or erosion of the terrace or necessitate its reconstruction.

### 5.2.4 Sediment Sinks

Information is required concerning the residence time of sediments in the various sinks that exist different terrestrial environments. Much indirect information exists on this subject obtained from studies of the maximum age of colluvium accumulating behind terraces or in hill-slope hollows. Studies on extremely steep slopes in Oregon and California reported by Dietrich (1978) suggest that the maximum age of colluvium is about 2000 years. In small steep watersheds, the sediment delivery ratio (ratio or erosion in a catchment to the sediment transport from it is high – often about 0.3 to 0.6). This is because there are few opportunities to storage. As catchments get larger and slopes decrease, the ratio of the volume of the annual sediment yield to the amount of sediment in storage, increases very rapidly so that the residence time of sediment in storage as alluvium also increases to tens or hundreds of thousands of years. However, sediment budget data can be misleading because gross outputs are considered and as mentioned meander bend migration regularly erodes material and redeposits it maybe a few hundred metres down valley. This effect is not reflected in sediment budgets. Counter intuitively, (but on the basis of little data), the maximum residence time of about 600 years described for sediment in active valley alluvium is, therefore, about half of that of hillside colluvium.

Thus once individual grains are deposited in alluvial sediments, the risk of erosion, except near channels water courses, is in the short and medium term negligible.

One of the biggest problems is that land-use changes in drainage basins can bring forward the moment in time when sediment in the colluvial sinks is eroded. Almost all gulleys on hillsides (excluding shales) occur in colluvial sediments.

## 5.3 Long-term Changes in Soil Properties

### 5.3.1 Introduction

Any changes in soil erodibility, either positive or negative increases the risk of contaminant mobilization. As already mentioned, a decrease in soil erodibility can convert sediment, stored as colluvium or alluvium from a sediment sink into a sediment source. On the other hand, an increase in erodibility can have the opposite effect but it can lead to run-off following alternative pathways transferring creating a risk at another location; especially when erosion stops and less sediment reaches river channels, river channel erosion is triggered. Schumm (1978) describes in his theory of complex response, how a period of sediment supply to a river caused by headwater erosion results in a non-linear output of sediment in the form of a series of sediment output waves generated by the one event.

The effect of erodibility changes on the partitioning of rainfall between overland flow, infiltration and percolation deserves special consideration. This is relevant for the storage and transport of chemicals in the soil and for the way it effects the erosivity of run-off in rivers and river channels. It is also important for the integrity of landfill cover systems.

In practise the most important causes of erodibility change are related to changes in land use or farming activities that influence organic matter composition and content, soil compaction and soil (micro)aggregate formation and stabilization.

With respect to erosion, the soils on a hill slope can be can be considered as (1) areas that are unaffected by erosion; (2) areas of net soil loss; or (3) areas of net soil gain. The areas of net gain represent locations of temporary storage where adsorbed chemicals can accumulate. The physical capacity of a store is often limited by topography so that there is a limit to any potential increase in volume. Colluvial stores are in fact subject to periodic erosion. Deitrich (1985), as described above has demonstrated how in California, colluvium accumulates over several hundreds of years in hill-slope hollows until it reaches a critical condition that permits erosion. The accumulation is by means of what Deitrich (1985) terms diffusive processes, largely of biological origin. The sudden erosion of this material occurs every few thousand years because of changes in mechanical and hydraulic properties brought about by an increase in weight and the accumulation of clay.

Mainly due to changes in erodibility, the rates of soil loss on a slope can be increased from almost zero to several cm per year. Over decades this results in the truncation of soil profiles and in the accumulation of valley bottom colluvium, up to several metres thick. Almost every soil on every cultivated slope shows a varying degree of soil loss on steep slope segments, and of colluviation at slope foot positions.

One method of establishing the recent magnitude of erosion and colluviation in relation to storage capacity is by means of the distribution of $Cs_{137}$ in the landscape. It is assumed that $Cs_{137}$ was uniformly deposited during the period of atmospheric nuclear testing and retained in the upper soil. Cesium isotope studies, although carried out to study erosion patterns are of great relevance to the CTB problem because they indicate the magnitude of depletion and accumulation of soil particles transporting adsorbed pollutants. These studies also indicate the depth of mixing.

An analogous method of studying long-term rates of erosion and deposition is to examine the distribution in soil profiles of heavy minerals derived from volcanic eruptions. This has been done by Jungerius and Mucher (1978) in Luxembourg who studied the heavy minerals from the eruptions in the Eiffel mountains during the Allerod period. The data provide information on transport (erosion) rates and soil mixing depths on different types of parent material for a major part of the Holocene.

### 5.3.2 Changes of Erodibility in the Erosion Zone

Within a soil profile soil erodibility usually is low in the upper A horizon and increases with depth until stony (rocks, concretions) or sometimes clayey layers are reached.

Occasionally, (e.g. in loess) the depth of the potentially highly erodible layer is large but in many upland soils it is low and these soon become stony.

It is a paradox that soils that still have a low erodibility because of a high amount of organic matter present perhaps the biggest risk in terms of the delayed release of contaminants. This is because these soils have high CEC values and a potentially high retention capacity for chemicals. The loss of organic matter could trigger at the same time erosion and the release of bound substances.

Furthermore, an organic rich stable surface horizon layer may be protecting a potentially erodible subsoil so that a slow decline in the erodibility of the topsoil could lead to the sudden triggering of high rates of erosion of the deeper soil layer sometimes by processes of rill and gully erosion. This often occurs when such soils are converted from forest/pasture to cultivated land. Soils that have a good structure and do not generate overland flow still supply material downslope when cultivated due to the down slope displacement of soil material by biogenic transport processes and ploughing.

The erodibility of the surface soil can be increased as a result of various degradation processes that reduce the inputs of soil stabilizing substances. An effect of this degradation (accompanied by a lower organic matter content and CEC) is usually to decrease (macro) porosity, to increase bulk density and to reduce soil aggregate stabilization. In addition, the potential soil moisture storage and fertility are lowered and erosion processes are able to transport soil aggregates. The development of an impermeable subsurface horizon can, in soils that are not inherently erodible, lead to extensive (rill and tunnel) erosion and sediment transport over long distances. When occasionally a highly erodible E or BC or C horizon is subject to erosion, as much as 15 cm of soil can be lost in a day and more than 50 cm of colluvium deposited at slope foot positions.

In most cases, erosion leads eventually to the residual accumulation of course rock fragments and stones that protect the soil from further rapid soil loss and eventually encourage a regeneration of the structure in the soil between the stones. Run-off may remain high, however, and lead to the erosion of colluvium. Once a surface is protected by stones, any further loss of soil requires some external disturbance (e.g. ploughing, overgrazing and fire).

### 5.3.3 Changes of Erodibility in the Accumulation Zone

Colluvium is usually composed of soil material similar to that in the supply zone except that certain fractions are preferentially excluded (large stones and sometimes fine silt and clay) or concentrated (coarse silt and sand).Accumulation of water-soluble salts and lack of stones sometimes makes colluvium more erodible than the original soil. Although this is frequently the case and colluvial soils are often the first to generate surface run-off and to be eroded sometimes the opposite is true. An example of this is on badland slopes where colluvium no longer contains the high concentrations of dispersive clay and salt found in the original regolith.

In spite of a high initial erodibility colluvium is not eroded frequently by overland flow. The greatest risk is the evolution of soil properties favouring rill and gully erosion, as mentioned above. Where colluvium is derived from organic rich soil, it may also have a very low erodibility.

In time colluvium may develop a favourable non-erodible structure and have a high value as a sink. Hydrologic conditions tend to favour water and nutrient transport to colluvial zones which is positive in areas where evaporation does not exceed the precipitation. If this occurs, salt accumulation can lead clay dispersion and trigger gulley erosion and piping, as frequently occurs in dry climates.

In the context of soil contamination, it is significant that there may be both rapid and gradual changes in the erodibility of colluvium that influence its potential role as a sink.

### 5.3.4 Changes in Capacity Controlling Properties

The relations of soil erosion to the delayed mobilization of contaminants, can be considered with respect to the *capacity controlling properties* of the soil. The *capacity controlling properties* (CCP) of a soil are the properties that control the loading, retention and release of chemicals (Stigliani et al. 1990, 1991). These include the size of the soil store; its pH, CEC, redox potential, water-soluble salt and organic matter content; and also physical properties that influence water storage and flow pathways. Some of the CCP could also be considered as erosion controlling properties (ECPs). In this section the effects of erodibility on other CCP will be considered briefly with respect to the soil store loading, retention and release.

### 5.3.5 Erodibility and the Loading, Retention and Release of Chemicals

The capacity controlling properties of soils control the loading storage and release of chemicals like heavy metals and organic pollutants. In this section the more important CCP will be discussed briefly with respect to heavy metals. Although soil material is the ultimate and most important sink of these elements in the terrestrial environment it has a finite capacity to bind them. Exceeding this capacity may lead to increased mobility in the soil system resulting in severe environmental damage. Especially significant is the enrichment of heavy metals to potentially dangerous levels in the accumulation zone by the processes of colluviation (deposition of eroded sediment on slope). The potential danger depends on a number of environmental and pedological conditions.

Geochemically, a heavy metal entering into the soil or sediment system may accumulate in the following ways: (1) dissolved in the soil solution as a simple organic solute or complexed organic ligand; (2) specifically or electrostatically adsorbed on exchanged sites of organic or inorganic constituents like humic material, clay minerals and or pedogenic oxides; (3) occluded or fixed into soil minerals (clays and/or pedogenic oxides); (4) precipitated with other compounds (secondary minerals); and (5) possibly incorporated into biological material. The first two forms are the mobile components and are directly bio-available within the time frame of this study; the last three forms are immobile and are not directly bio-available. With respect to erodibility, the last four forms of accumulation seem to be particularly important because they are all highly vulnerable to erosion, transport and accumulation. The potential danger is in fact mainly governed by specific soil chemical conditions like hydraulic conductivity, organic matter, clay (mineral) and pedogenic oxide contents, pH and redox conditions and the soluble organic completion capacity. Although the regulation mechanisms are not so well understood and prediction of the fate and behaviour of heavy metals seems to be difficult, especially at the scale of the landscape, heavy metal activity in the soil (solution) system is generally considered to be a result of metal equilibrium with the above mentioned chemical factors. In the following paragraph a short review will be given of the most important ones.

*Organic Matter.* Because soils that are high in organic matter have a high binding capacity and low erodibility, it is these soils that have the potential to generate surprises should the erodibility decrease. A possible situation that could generate such a surprise might occur when the litter input into the soil is reduced or when the soil with a high aggregate stability becomes unshaded. The following explanation and example illustrate this.

In relation to erosion, soil organic matter has the following important functions. In the form of accumulating litter it protects the soil from raindrop impact and it helps maintain a cool moist soil micro-environment favourable for the growth of roots the soil (micro) fauna and flora that stabilize the different hierarchies of soil aggregate (described in Sect. 5.3.5). It also functions as a source of energy. Although correlations between organic matter and soil erodibility are high when soils are compared, for specific locations they are often poor because the dynamics of the processes involved are missed when data are treated statistically. For example in the data from Spain shown in Fig. 5.3, two situations exist, one in a shaded position beneath a tree, where there is a good micro-environment and sufficient organic matter for the different stabilization process. The other is an open environment with a high rate of mineralization due to high temperatures and a lack of litter and root growth. The low erodibility at some unshaded sites is the result of carbonate rich surface seals. This interpretation indicates that trigger mechanisms and erosion thresholds might be related to the number of days of soil biological activity and to the supply of organic matter. The potential risk of a time bomb effect would be analogous to the switching between the shaded

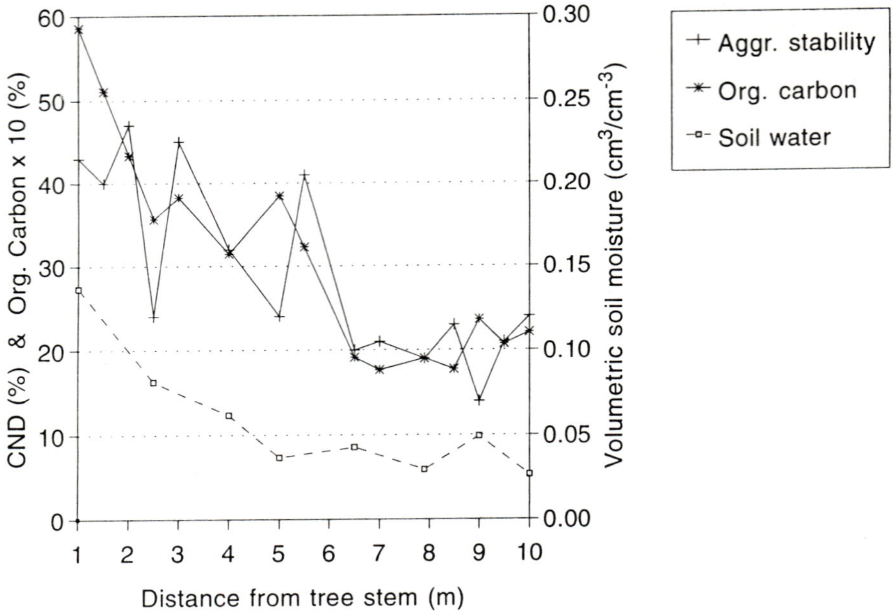


**Fig. 5.3.** Variances in the aggregate stability (CND), the organic carbon content and soil moisture content in semi-natural conditions, from beneath a pine tree, where the soil is completely shaded to an open patch. Both aggregate stability and organic carbon tend to be either low (in the open) or high (in the shade)

and unshaded state, shown in Fig. 5.3. In a sense all cultivated soils are in the degraded state and this is why they erode. For natural soils the critical ratio between the rates of aggregate regeneration and degradation has been mentioned above. Whether a triggering mechanism is reached will depend upon the resilience of the soil-vegetation (structure and patterns) and the recovery time as influence by grazing fire and drought.

*Clay (minerals).* Clay and clay minerals are very important for erodibility. For a full discussion reference is made to Emerson (1988). Triggering mechanisms that influence erodibility are related to the charge and to the way in which the different clay minerals are affected by swelling and dispersion.

There are many examples where the accumulation of small accumulations of water-soluble salts have converted non-erodible colluvium into a highly erodible state slowly over a number of years (see Imeson and Emmer 1992). When dispersive conditions occur, the processes of erosion can become different (rill and gully erosion) so that deeper soil horizons become effected.

Very large areas of southern Europe have soils that are affected by dispersive conditions for a few critical periods each year, allowing severe erosion to occur. These areas should be identified as potential areas of risk.

## 5.4 Some Examples of Potential Non-linear Delayed Responses

### 5.4.1 Abandoned Land

Since the turn of the century, marginally economic agricultural land has been abandoned in southern Europe so that, for example vast areas of abandoned terraced land are found in Spain. Land abandonment has had two contrasting effects, according to geological and pedological conditions. These effects depend on whether abandonment leads to a decrease or increase in hill-slope erosion.

In drier areas on marls abandoned terraces, after an initial period of stability, have suddenly started to erode. This is due either to lack of maintenance or to the accumulation of water-soluble salts that lowered soil erodibility over periods of from say 5 to 50 years. Sediments stored for the last centuries in these terraces are now being transferred to river systems where they are accumulating. This accumulation can lead to shifts in the position of river channels and the mobilization of sediments. Depending on the materials accumulating with the eroded sediments a potential time bomb effect exists on this abandoned land.

Where terraces are not eroded, for example on limestone or sandstone, as well as on unterraced land, soil erosion is reduced after abandonment. The gradual reduction in erosion as soil erodibility decreases leads to a reduction in sediment supply to river channels. This increases the sediment transport capacity of flow and can result in the mobilization of alluvial sediments. When these contain heavy metals that have accumulated through mining activities in the past centuries, a potential hazard may be created. For example, on the abandoned terraced sandstone hills in the north of Valencia Province, Spain, mercury mines abandoned early this century left large areas of contaminated sediments potentially vulnerable to altered hydrologic regimes.

### 5.4.2 Conservation Tillage

Another delayed response can be predicted in areas where minimum or zero-tillage practises have been introduced to control erosion. Examples of such implementations have been described in southern Portugal and southern Spain (ESSC 1990). As in other areas they rely on heavy applications of herbicides for weed control. Minimum tillage in Belgium, Spain and Portugal (ESSC 1990; Giraldez 1990) produces a more compact surface soil with a low erodibility. Run-off remains high and again since this contains low amounts of sediment, erosion (rill, gully and channel erosion) occurs in downslope colluvial and alluvial deposits. In the United States (Logan et al. 1987) there is concern about the way in which minimum tillage leads to the development of preferential flow paths in the soil and the contamination of groundwater.

### 5.4.3 The Loess Belt of Northern Europe

A third type of example, where soil erodibility increases with time, is illustrated by the loess belt in northern Europe. In this region, soil degradation caused by compaction, the excessive use of liquid manures and a lowering in the amount of organic matter and soil biologic activity is extensive. When degradation has proceeded to a certain stage, almost any moderately extreme rainfall event will produce run-off and erosion, leading to the loss of topsoil that may be contaminated by agro-chemicals and lead to the concentration of these at depositional sites. The upper 50 cm of loess colluvium studied by Mücher (pers. comm.) in Belgium are characterized by a very low number of earthworms, presumably as a result of land-use practises during the last twenty years or so. The poisoning of the soil mesofauna by ammonia is sometimes observed directly after liquid manure applications.

Another example, mentioned in the introduction, is the case of mine spoil. Studies have shown that the rapid weathering and compaction of spoil from coal mines in England and eastern Europe change the hydraulic and soil mechanical properties of the material. In particular impermeable zones can develop that lead to preferential flow paths, tunnel erosion and mass wasting. This can lead to catastrophic movements of spoil during wet weather. The same processes of weathering that lead to erosion also may result in the release of toxic material. A particularly dramatic example is provided by the extensive Santa Domingo copper mines in Portugal which were worked until a few decades ago. The mine spoil is highly toxic and acidic. Waste water leaking from drainage channels has devastated the vegetation leaving the soil bare and polluted along a 200-m-wide 40-km-strip of land. Also, drainage from mine spoil used as ballast for the railway line from the mines to the Guadiana River has drained into and polluted the small rivers in the drainage basins it crosses leaving a succession of polluted, unvegetated slopes and channels.

## 5.5 Conclusions

Knowledge of soil erosion and soil erodibility is basic to understanding how potential contaminant sinks can be lost or created.

This chapter emphasizes soil erodibility – a highly dynamic soil characteristic that is continually evolving in response to land-use and other changes. The erodibility can suddenly rise or decrease and in so doing influence the hydrology of entire drainage basins. Some aspects of erodibility reflect the dynamics of aggregation and soil solution chemistry. The balance between the rates of aggregate evolution and aggregate decomposition is an important as this could provide an advanced indicator of a potential shift from a non-erodible to an erodible condition.

Soil erosion is a non-linear, largely unpredictable phenomenon. This must be acknowledged when addressed in soil contamination. Much of the available data on soil erosion are not directly measured but based on extrapolations from statistically-based linear models. Although this data is a useful starting point it can not be used to seriously answer questions about future change.

Imeson et al. (1993) describe how a hierarchical, evolutionary modelling framework could be developed, analogous to that being developed for desertification studies in southern Europe. This would be based on delineating in a landscape what might be called *delayed erosion response units*. This approach would enable studies of chemicals and processes operating at different rates and over different spatial and temporal scales to be integrated. Implementing such a study would involve establishing the long-term dynamics of sediment sinks in drainage basins. The risks at the sinks, formed by colluvial and alluvial areas, depends on both the evolution of the material in the sink itself and on erosion processes that couple source and sink areas. Consequently, risk assessment must consider sinks and sources together.

In terms of policy, it is already possible to make general conclusions about the extent and nature of the problem. Attention should now be given to establishing the rates of key processes affecting non-linear response on potentially threatened soils as a basis for model development. The effects of land-use change and policy on these processes should be made visible by means of decision support models and information systems that link soil degradation processes and the delayed response problem to decisions that influence land-use and other policy that could trigger changes.

## References

Ahuja LR, Lehman OR (1983) The extent and nature of rainfall-soil interactions in the release of soluble chemicals to runoff. J Environ Qual 12:34-40

Allen PM (1988) Evolution why the whole is not greater than the sum of the parts. Ecodynamics, contributions to theoretical ecology. pp 2-30

Baker JL (1992) Effects of tillage and crop residue on field losses of soil-applied pesticides. pp 175-187 In: Schnoor JL (ed) Fate of pesticides and chemicals in the environment, Wiley, New York, 436 pp

Brussard L, Kooistra MJ (1993) Soil structure/soil biota interrelationships. Elsevier, Amsterdam

Buyanovsky GA, Aslam M, Wagner GH (1994) Carbon turnover in soil physical fractions. Soil Soc Am J 58:1167-1173

Cammeraat LH, Hin J, Imeson AC The evolution of soil structure on abandoned vineyards. Manuscript

Cerda A (1993) La infiltracion en los suelos del pais valencio factores y variaciones espacio-temporales. Academic Thesi, University of Valencia, 357 pp

Cerda A, Garcia Alvarez A, Cammeraat LH, Imeson AC Agregacion del suelo en una catena afectada por el abandono del cultivo en la cuenca del Guadalentin (Murcia) 1. Esdtabilidad y distribucion de las agregados del suelo

Drury CF, McKenney DJ, Findlay WI, Gaynor JD (1993) Soil Sci Coc Am J 57:797-802

Edwards DR, Daniel TC, Moore PA, Sharpley AN (1994a) Solids transport and erodibility of poultry litter surface-applied to Fescue. Trans ASAE, 37(3):771-776

Edwards DR, Daniel TC, Moore PA, Vendrell PF (1994b) Drying interval effects on quality of runoff from Fescue plots treated with poultry litter. Trans ASAE, 37(3):837-843

ESSC (1990) Proceedings seminar on Interaction between Agricultural Systems and soil conservation in the Mediterranean Belt. Lisbon 4-8 th September 1990

Giraldez JV, Carrasco C, Otten A, Ietswaart H, Laguna A, Pastor M (1990) The control of soil erosion in olive orchards under reduced tillage (9 pp in ESSC 1990)

Harald (1992) Fate of pesticides used in flower cultivation, presented at the ESSC meeting, Silsoe, April 1992

Imeson AC, Kwaad FJPM (1990) The response of tilled soils to wetting by rainfall and its implication for soil erosion. In: Boardman J, Foster DL, Dearing A (eds) Soil erosion on agricultural land. Wiley, Chichester, pp 3-14

Imeson AC, Perez-Trejo F (1992) Desertification response units, a dynamic systems framework for understanding and combatting land degradation. (Manuscript)

Imeson AC, Verstraten JM (1989) The micro-aggregation of erodible and non-erodible soils. Catena Supplement 14:1-14

Jungerius PD, Mucher HJ (1972) Holocene slope development in the Lias cuesta area, Luxembourg, as shown by the distribution of heavy minerals. Zeitschrift für Geomorphol 14, 2:127-136

Logan TJ, Davidson JM, Baker JL, Overcash MR (eds)(1987) Effects of conservation tillage on groundwaterquality-nitrates and pesticides. Lewis Publ, Chelsea, Michigan, 292 pp

Mahboubi AA,Lal R, Faussey NR (1993) Twenty-eight years of tillage effects on two soils in Ohio. Soil Sci Soc Am J 57:506-512

Naveh Z (1987) Biocybernetic perspectives of landscape functions and land use patterns. J Landscape Ecology and Management 1:75-85

Nietfeld H, Priesack, Beese F (1992) A model of solute transport and microbial growth in aggregates. Modelling of Geo-Biosphere Processes 1:1-12

Oades JM, Waters AG (1991) Aggregate hierarchy in soils. Aust J Soil Res 29:815-828

Perez-Trejo F (1990) Landscape response units: process base self-organizing systems. Manuscript for Landscape Ecology and Geographical Information Systems, 22 pp

Perez-Trejo F (1992) A model of the dynamics of aggregation in structured soils. Manuscript submitted to Geo-Biosphere Processes.

Plimmer JR (1992) Dissipation of pesticides in the environment. pp 79-82 In: Schnoor JL (ed) Fate of pesticides and chemicals in the environment, Wiley, New York, 436 pp

Rudra RP, Dickinson WT, Euw EL von (1993) The importance of precise rainfall inputs in nonpoint source pollution modeling. Transactions of the ASAE, 36:445-450

Salomons W, Stigliani WM (1994) Biogeodynamics of pollutants in soils and sediments. Risk assessment of delayed and non linear responses. Springer, Berlin Heidelberg New York

Schumm SA (1977) The fluvial system. Wiley, New York

Sidle RC, Brown RW, Williams BD (1993) Erosion processes on arid minespoil slopes. Soil Sci Soc Am J 57:1341-1347

Szabolcs (1990) Effects of predicted climatic changes on European soils, with particular regard to salinization. In: Boer MM, Groot RS de (eds) Landscape ecological impact of climatic change. Proceedings of Lunteren Conference 3-7 December 1989, pp 177-193

Tisdale JM, Oades JM (1982) Organic matter and water stable aggregates in soils. J Soil Science 33:141-163

Westmann WE Resilience concepts and measures. In: Dell B, Hopkins AJM, Lamont B (eds) Resilience in Mediterranean type ecosystems. Junk, Dordrecht, 168 pp
Wischmeier NH, Smith Dd (1978) Predicting rainfall erosion losses. A guide to conservation planning. US Dept Agriculture Handbook No 357, 58 pp

# 6 Long-term Fate of Pesticides in Soil

*G. Kuhnt*

## 6.1 Introduction

The history of pesticide production and application covers more than one century. Before the Second World War, most of the pesticides were based upon the extremely toxic effects of some heavy metals such as lead, arsenic, mercury, or natural substances like pyrethrum or nicotine. Due to their well-known toxicity, however, they were only used in comparably low quantities for special purposes (cf. Ruhnau et al. 1991).

During the Second World War, the development and industrial production of halogenated, especially chlorinated, organic compounds such as 2,4-D; MCPA; 2,4,5-T and of course, DDT, marked the starting point of intensive pesticide application in agriculture, industry and household, resulting in a ubiquitous distribution of those products (Egler 1964).

For about one decade, the substances were regarded as readily degradable and totally harmless to man and animal. But then it became obvious that the accumulation and magnification of some lipophillic organochlorine compounds in fatty tissues could cause serious health problems (Klimmer 1957). Moreover, traces of persistent organochlorine pesticides were found in nearly every environmental compartment, even in places where they have never been used (Dustman 1966, Kenaga 1968).

Today, pesticide residues have to be regarded as a ubiquitous component of the environment and although in many cases the amounts are comparably low they must not be neglected (Schmidt 1986).

Numerous investigations have been carried out on soil bound residues of pesticides (cf. Edwards 1973) but only few are known on their long-term remobilization potentials. It is assumed that in the course of climate, land use or soil–water balance changes harmful effects may derive from the mobilization of soil bound residues of pesticides. Therefore, in the following some aspects related to pesticide residues in soil will be discussed in the light of the chemical time bombs concept.

## 6.2 Behaviour and Fate of Pesticides in Soil

In 1985 about 2.5 million tons of pesticides (active ingredients) have been used worldwide (Knirsch 1991). Most of them are applied directly to the soil where they are subject to evaporation, volatilization, uptake by plants or animals, adsorption, leaching or degradation. The fate and distribution of pesticides in soil are highly dependent on the properties of the substance as well as the characteristics of the soil environment. While under advantageous conditions the substances may be transformed into lesser or non-toxic compounds, other situations could lead to accumulation, higher persistence or leaching into groundwater. Consequently, serious groundwater contaminations have been reported for some areas under intensive agricultural use (cf. Giessl and Hurle 1984, Stock et al. 1987, Werner 1987, RIVM 1991). Some soils, however, show high sorption potentials for pesticides. At first glance this circumstance appears positive for the protection of groundwater and food-chain. But it has to be regarded in this context that strong sorption in most cases increases the persistence of potentially degradable compounds and leads to an accumulation of the respective substance.

In order to predict the potential behaviour of a certain chemical in soils, basic data like molecular weight, water-solubility, vapour pressure or environmentally relevant data like partition coefficients, sorption constants or potential degradability have consequently to be matched to additional data on the properties of the soil (Kuhnt 1986). In most cases, adsorption and desorption which both and in conjunction determine the buffering capacity of a soil, play a major role with respect to pesticide behaviour. Soil constituents with high specific surface and net charge, i.e. organic matter, clay minerals, metal oxides and hydroxides, largely determine adsorption and desorption. The relevant boundary conditions are concentration and dissociation or polarity of the chemical on the one hand, soil moisture, temperature, pH value and oxidation and reduction potentials on the other. These inherent filtering and buffering capacities enable soils to reduce the impact of potentially toxic substances on the biosphere, making them important temporary sinks for environmental chemicals.

Nevertheless, pesticides bound to soil particles are neither degraded nor finally immobile. On the contrary, a strong adsorption of chemicals often increases their persistence and results in accumulation effects. Sorbed pesticides can also be subject to translocation by e.g. run-off, percolating water or soil cultivation. Therefore, only a rapid and complete degradation of pesticides leads to a detoxication and only in those cases soils can be called effective sinks for this group of environmental chemicals. Regarding the complex internal structures of the various soil types and the manifold interrelationships with chemicals showing different properties it becomes obvious that the behaviour and fate of environmental chemicals in natural soils are always extremely difficult to predict. As a consequence, even low amounts of bound residues of persistent pesticides in soil have to be regarded as potentially hazardous, because remobilization could eventually lead to unexpected toxic effects.

## 6.3 Pesticide Residues in Soil

A closer look at soil bound residues of pesticides reveals that even those substances which are no more applied due to administrative regulations can still be found in some soils. Especially chlorinated hydrocarbon insecticides like DDT (DDD, DDE), aldrin, endrin, dieldrin, chlordane, heptachlor, lindane or toxaphene show ubiquitous distribution and at least traces can be detected in nearly every soil. Based on the available data, the average persistence of some of the above insecticides were calculated as shown in Table 6.1.

The given ranges are mainly a result of the different soil properties and therefore under certain conditions more rapid degradation is as likely as much longer persistence. In 1972, DDT and its metabolites were detected in soils from five apple orchards although at that time the pesticide was not directly applied for 13 years (Kuhr et al. 1974). Lichtenstein et al. (1971) found that 15 years after DDT application 18 to 24% of the applied dosage were still present in the upper layer of an agricultural loam soil. The authors also report that traces of lindane

**Table 6.1.** Persistence of some chlorinated hydrocarbon insecticides in soil (after Edwards 1966 and Schmidt 1986)

| Pesticide | Calculated half-life time (months) | Time for 95% disappearance (years) |
|---|---|---|
| Aldrin | 3 - 8 | 1 - 6 |
| Chlordane | 10 - 12 | 3 - 5 |
| DDT | ~ 30 | 4 - 30 |
| Dieldrin | ~ 27 | 5 - 25 |
| Heptachlor | 8 - 10 | 3 - 5 |
| Lindane | 12 - 20 | 3 - 10 |

were detected 11 years after application and that about 5% of aldrin were recovered in the form of dieldrin. More recent results of Bossi and Larsen (1992) show that at least traces of organochlorine pesticides can be detected in some European soils which have never been exposed to direct pesticide application. A comparative evaluation of the amounts of DDT and its metabolites DDD and DDE in five different soils reveals that degradation is still in process which seems to indicate active desorption and degradation mechanisms.

Organophosphorus pesticides which are commonly regarded less stable than chlorinated hydrocarbon insecticides also show relatively low degradation rates. Residues of parathion for example have been detected many years after application, especially in dry soils (Streit 1991). Fryer et al. (1975) report that the herbicide paraquat may be broken down only slowly if at all in the soil under field conditions.

For atrazine the average half-life time is about 75 days but in the literature, values ranging from 29 up to 151 days or, under disadvantageous conditions, up to 1 year are also given (cf. Kuhnt and Fränzle in print). Even pesticides which are regarded to be readily degradable may remain unaltered in the soil system much longer than expected. Fränzle (1982) detected considerable amounts of the herbicide 2,4-D in a sandy podzol soil 9 months after application although after Audus (1964) the substance should be totally degraded within 4 weeks.

The few examples given illustrate that pesticides and their metabolites may remain in soils over long periods. The specific properties of a certain substance in combination with the characteristics of the soil mainly determine the persistence and the accumulation potential of the respective chemical. As long as potentially hazardous pesticides are strongly bound to soil colloids, toxic effects resulting from plant uptake or groundwater pollution are unlikely. However, changes in soil boundary conditions caused by erosion, alterations of the global climate or land-use changes could reactivate the toxic potentials of soil bound residues of pesticides.

## 6.4 Remobilization Potentials of Pesticide Residues in Soils

In the soil environment, pesticides are metabolized by various chemical and microbial degradation processes. In some cases, a total breakdown results in the formation of non-toxic components like $CO_2$ and $H_2O$. If the degradation process is not that complete, metabolites may be formed which are sometimes even more stable and toxic than the initial substance. A very important degradation mechanism for pesticides is the biotic degradation by aerobic micro-organisms. Although for some substances, anaerobic breakdown is reported to be more effective than aerobic (Hill and Mccarty 1967), the greater part of pesticides is degraded by aerobic micro-organisms. Table 6.2 displays some boundary conditions in soils under which effective pesticide degradation favourably occurs.

**Table 6.2.** Best prerequisites for the degradation of organic chemicals by aerobic micro-organisms (after Kuhnt and Knief 1991)

| Micro-organisms | High population diversity; adapted species |
|---|---|
| Soil moisture | 50 - 80% of field capacity |
| Soil temperature | 25°–35 °C |
| Soil aeration | Permanent aeration, no water-logging effects |
| Organic matter | 2–3% |
| C/N ratio | Narrow (< 10–15) |
| pH value | Around neutrality |
| Soil mineral composition | Balanced grain size distribution |
| Soil texture | Fine crumb to very fine subangular blocky |

Regarding the fact that soils are highly heterogeneous entities which are extremely variable in space and time, the optimum conditions mentioned in Table 6.2 cannot be expected for the majority of soils treated with pesticides. Therefore, it has to be considered that a certain amount of pesticides – especially persistent ones – is subject to either leaching or accumulation. Table 6.3 tries to summarize possible effects and expected results of pesticide behaviour in soils under varying boundary conditions.

It derives from the highly generalized table that due to the specific combination of soil properties either accumulation and increased persistence or higher evaporation, plant uptake and leaching may occur. Consequently, changes in soil boundary conditions may lead to an alteration of physical and chemical properties in soil. In this case, immobile residues of pesticides could be remobilized and neither groundwater pollution nor intake into the food-chain can be excluded.

In the following some situations will be discussed in which remobilization of soil-bound residues of pesticides may occur. Since, as mentioned above, only few investigations have been performed on the reactivation of toxic potentials of stored pesticides, the scenarios have to be discussed on a more or less hypothetical basis.

**Table 6.3.** Effects of various boundary conditions on pesticides' behaviour in soil

| SITUATION | POSSIBLE EFFECTS | POTENTIAL RESULTS |
|---|---|---|
| Properties of Pesticide:<br>- High molecular weight/diameter<br>- Low water solubility<br>- High persistence | High sorption tendency<br>Low amounts in soil solution<br>Low degradation rates | High accumulation tendency<br>Bound residues over longer periods<br>Increased persistence |
| Intensive pesticide use<br>(e.g. viniculture, fruit and vegetable growing, house and allotment gardens) | High input rates<br>Reduction of biotic activity/population diversity<br>High initial sorption<br>Blocking of exchange surfaces<br>Higher amounts in liquid phase | Long-term accumulation<br>Reduced degradation<br>Continous increase of leaching/plant uptake<br>Reduction of soil buffering capacity |
| Soil buffering capacity, mainly determined by:<br>- Quantity and quality of organic matter<br>- Clay content<br>- pH value<br>- Content of amorphous oxides | Low buffering capacity:<br>Rapid depletion of buffering capacity<br><br>High buffering capacity:<br>Long-lasting sorption, enhanced persistence | Unpredictable exchange reactions<br>Higher pesticide mobility, leaching<br><br>Accumulation of pesticides<br>Soil bound residues |
| Reduced microbial activity<br>Diminished population diversity | Reduced degradation potential<br>Highly specialized microorganisms | Higher persistence of pesticides<br>Accumulation in soil according to sorption capacity<br>Longer lag phases |
| Intensive use of artificial fertilizers | Reduction of organic matter content<br>Increased ionic strength of soil solution | Reduction of buffering capacity<br>Higher desorption rates of bound residues |
| Low soil moisture content<br><br><br>High soil moisture content | Reduced leaching and plant uptake<br>Reduced microbial activity<br><br>Increased desorption, water percolation, plant uptake | Enhanced sorption of pesticides; accumulation<br>Higher persistence of substances<br><br>Higher intake into plant and groundwater<br>Re-mobilization of bound residues |
| Low soil temperature<br><br><br><br>High soil temperature | Reduced evaporation<br>Reduced microbial activity<br>Higher sorption tendency<br><br>Higher evaporation rates<br>Higher microbial activity<br>Higher desorption tendency | Enhanced sorption of pesticides<br>Higher amounts of substances in soil<br>Higher persistence<br><br>Pesticide loss via evaporation/vaporization<br>Higher degradation rates<br>Higher amounts of pesticides in soil solution |

### 6.4.1 Effect of Soil Erosion on Pesticide Residues

Soil erosion by wind, water or inadequate cultivation methods is a well-known problem of areas under intensive agricultural use. The aerial removal of topsoil material results in accumulation and the formation of colluvial soils on valley floors, terraces or lower slopes. Pesticide residues strongly bound to soil particles are translocated together with the eroded soil material. Since most of the persistent chemicals are stored in the topsoil layer which is mainly subject to erosion, a certain enrichment should be expected in colluvial soils. Dependent on the thickness of the accumulated material larger parts of the newly formed soil profile may be contaminated by pesticide residues. Regarding the fact that in some cases even low pesticide concentrations show negative effects on soil flora and fauna as well as plant roots (cf. Schmidt 1986), enrichment processes by soil erosion could increase the impact of pesticides on the environment at those locations where accumulation mainly takes place.

A considerable amount of eroded soil material is directly washed into rivers or lakes. Under totally different environmental conditions even strongly sorbed substances may be then subject to desorption. For the aquatic environment, intensive biomagnification has been reported especially with respect to those persistent pesticides showing low or moderate water-solubility (Petrocelli et al. 1973).

A future intensification of land use as well as climate changes resulting in occasional but heavy rainfall or stronger storms may significantly increase the degree of soil erosion. Regarding the fact that particularly dry and firm soils are exposed to erosion by natural processes and that especially those soils have higher accumulation potentials for pesticides (cf. Table 6.1), soil-bound residues of pesticides in combination with soil erosion can be regarded as chemical time bombs under disadvantageous conditions.

### 6.4.2 Pesticide Residues in Changing Soil Environments

From Table 6.3 it becomes obvious that the tendency of a pesticide to be mobile or bound in the soil environment is a function of the chemicals' properties on the one hand and the boundary conditions of the soil system on the other. Consequently, accumulated pesticides can be remobilized either by slight or drastic changes of the main parameters controlling the behaviour of chemical substances in soil. Since soils are highly complex systems, an extremely sensitive reaction to alterations of natural conditions or anthropogenic influences can be stated. Climate or land-use changes, irrigation or drainage, changing of cultivation methods (e.g. deep ploughing) or a rising groundwater table usually lead to completely different conditions in the soil environment. The alteration of important parameters determining pesticide behaviour in soil (e.g. buffering capacity, soil–water balance, microbial activity, pH value) may result in a remobilization of bound residues, facilitating the entry of pesticides into the food-chain via groundwater contamination or plant uptake.

In this context another important parameter concerning the microbial degradation of pesticides in soil has to be mentioned. As long as certain pesticides are continuously applied to soils under more or less stable boundary conditions, specifically adapted micro-organisms are able to effectively degrade those chemicals in relatively short time. Numerous investigations have shown that frequent applications of potentially degradable pesticides result in shorter lag phases and accelerated breakdown of the initial substance (Edwards 1966, Burns 1975, Fränzle 1981). For those pesticides, however, being no more used today due to their high persistence and toxic nature but still being present in soils as bound residues, no specifically adapted micro-organism population exists to rapidly metabolize those chemicals. Therefore, it has to be assumed that some pesticides, if remobilized, will today be degraded less effectively than expected from former lab experiments or field research.

It has to be stated that the quantitative aspect of a potential remobilization of pesticide residues in soil cannot be extrapolated from the existing data. To what extent changes in the soil system may reactivate the initial toxicity of bound residues is unclear. Nevertheless, also in this case some situations appear conceivable where soil-bound residues can be regarded as chemical time bombs.

### 6.4.3 Pesticide Residues and Land-use Changes

With respect to land-use changes, two different situations will be discussed in the following.

Hesterberg et al. (1992) developed a system model showing the potential consequences of a situation where farmland is converted to forest. This is in close compliance with current EEC policy and therefore it has to be expected that large areas of arable lands will be converted to forest in future. According to Hesterberg et al. (1992) this will result in considerable changes of the capacity-controlling properties of the soil. The decrease of pH values, changes in microbial population, the alteration of quantity and quality of organic matter, reduced salinity and a modified structure of the topsoil will lead to a decrease of pesticide residue degradation in combination with altered pesticide solubility, reduced drainage and enhanced erosion. Their findings match perfectly with the above statements on the effects of soil erosion or changes in the soil environment and therefore no further arguments have to be presented with respect to this particular situation.

But also the conversion of intensively used arable land to grassland involves certain risks. Usually arable land is treated with higher dosages and a greater variety of different pesticides than grassland. Under specific circumstances, accumulation of pesticides may result in high amounts of bound residues. After conversion to grassland the newly implemented land use of the respective areas determines the fate of the accumulated residues. If the grassland is merely used for hay-harvesting the entry of large amounts of pesticide residues into the food-chain is unlikely due to relatively low transfer rates. However, a completely different situation can be given if the grassland is used as pasture. Recent findings show a considerable uptake of soil material by grazing cattle (Schulz 1992). It is assumed that grazing animals incorporate 700–1 400 grams of soil daily. In this case especially lipophillic pesticides magnify in milk or fatty tissue and even if acute toxicity cannot be observed, a certain reactivation of a pesticides' toxicity has to be considered especially with respect to the exposure of man.

### 6.4.4 Observations and Control of Pesticide Residues in Soil

The last point to be discussed here is not primarily related to the scientific but to a policy-related problem of soil-bound residues of pesticides. As mentioned above, the use of some pesticides, especially chlorinated hydrocarbon insecticides like DDT, aldrin, dieldrin, heptachlor or toxaphene is meanwhile prohibited or restricted. In Germany, the application of atrazine, a herbicide which has been used intensively over three decades, has been prohibited quite recently due to the detection of up to 17.52 mg $l^{-1}$ in groundwater samples (Stock et al. 1987). With respect to other types of pesticides or commercial formulations restrictions are planned for the near future.

The ban of highly toxic and/or persistent pesticides and a substitution by newly developed products of better degradability and less polluting properties or a complete abandonment of chemical plant protection in general has to be regarded as a positive development. Decreasing levels of organochlorine pesticide concentration in various environmental compartments underline this statement. Nevertheless, caution still appears advisable. A closer look at the scientific literature dealing with pesticide behaviour in soil reveals that much more attention is paid to existing or new products being applied today. Also in environmental policy the tendency to focus on actual problems is obvious. Decreasing concentrations of formerly used pesticides in toxicologically sensitive environmental compartments seem to be directly correlated to a decreasing interest in these chemicals. In the light of the above arguments such a view appears dangerous. As long as soil-bound residues of persistent chemicals can be detected and as long as hazardous effects cannot safely be excluded there is no reason to forget about pesticides which are no longer in use. Well-known hazards are under control by careful observation, but if measurements are terminated due to the assumption that certain pesticides are past problems, the explosion of such a chemical time bomb strikes harder than necessary.

## 6.5 Conclusions

On the basis of well-known facts concerning the behaviour and fate of pesticides in soil and on the background of measured pesticide residues, some situations have been presented under which a remobilization of soil-bound residues of pesticides may lead to harmful effects for man and the environment. Due to a nearly complete lack of information on the long-term fate of stored chemicals in soil, the scenarios developed are mainly of hypothetical nature. Apart from the fact that intensified research on this item appears necessary it seems advisable to keep a close look on soil-bound residues of pesticides and their toxic potentials. Moreover, in the course of development and notification of new pesticides special emphasis should be laid on the prevention of accumulation and the formation of soil-bound residues. Soils do not have unlimited buffering capacities and degradation potentials and therefore the existence of chemical time bombs has to be kept in mind not only for heavy metals but also for those substances which are potentially degradable.

**Acknowledgements.** The author wishes to thank Mrs. Katja Bertram for her valuable help in pre-selecting the relevant literature and Mr. Thomas Hertling for the fruitful discussions on this complex item and some very important hints.

## References

Audus LJ (1964) The Physiology and biochemistry of herbicides. London, New York

Bossi R, Larsen BR (1992) Determination of organochlorine compounds. In: Kuhnt G, Murphy P, Muntau H (eds) Euro-soils: identification, collection, treatment, characterization. EEC Joint Research Centre Ispra, pp 71-77

Burns R (1975) Factors affecting pesticide loss from soil. Soil Biochem 4:103-144

Dustman EH (1966) Scientific aspects of pest control. Nat Acad Sci Publication N° 1402 Washington

Edwards CA (1966) Insecticide residues in soils. Residue Rev 13:83-132

Edwards CA (1973) Persistent pesticides in the environment, 2nd edn. Cleveland

Egler FE (1964) Pesticides in our ecosystem. Amer Scient 52:110-136

Fränzle O (1981) Erfassung von Ökosystemparametern zur Vorhersage der Verteilung von neuen Chemikalien in der Umwelt. Forschungsbericht 106 02 015 im Umweltforschungsplan des Bundesministers des Innern, Kiel

Fränzle O (1982) Modellversuche über die Passage von Umweltchemikalien und ihrer Metaboliten durch die ungesättigte Zone natürlicher Bodenprofile sowie durch Bodenschlämme in Laborlysimetern und im Freiland. Forschungsbericht 106 02 005/02 im Umweltforschungsplan des Bundesministers des Innern, Kiel

Fryer JD, Hance RJ, Ludwig JW (1975) Long-term persistence of paraquat in a sandy loam soil. Weed Res 15:189-194

Giessl H, Hurle K (1984) Pflanzenschutzmittel und Grundwasser. Agrar- und Umweltforschung in Baden-Württemberg, Bd 8, Stuttgart

Hesterberg D, Stigliani WM, Imeson AC (eds)(1992) Chemical time bombs: linkages to scenarios of socioeconomic development. Executive Report 20, CTB Basic Document 2, IIASA, Laxenburg

Hill DW, McCarty PL (1967) Anaerobic degradation of selected chlorinated hydrocarbon pesticides. J Water Poll Controll Federation 39, 8:1259-1277

Kenaga EE (1968) Guidelines for evaluating the properties of pesticides for safe use in the wildlife environment. Down Earth 23, 4:11-18

Klimmer OR (1957) Toxikologie der Schädlingsbekämpfungsmittel. Arch Exp Path Pharm 232:31-47

Knirsch J (1991) Lieferant für den Weltmarkt – Stellenwert, rechtliche Regelungen und Bewertung der bundesdeutschen Pestizidausfuhren. In: Ruhnau M, Altenburger R, Bödeker W (Hrsg) Pestizid-Report -Geschichte, Bedeutung und Folgen einer Pestizid-Wirtschaft in Deutschland, Göttingen

Kuhnt G (1986) Die ökosystemare Erfassung der Verteilung von Umweltchemikalien.- Kieler Geogr Schr 64:109-122

Kuhnt G, Fränzle O (in print) Estimating groundwater pollution by atrazine.

Kuhnt G, Knief K (1991) Ökologisches Verhalten von Tensiden in Böden. Literaturstudie im Auftrage des Umweltbundesamtes, Kiel

Kuhr RJ, Davis AC, Bourke JB (1974) DDT residues in soil water and fauna from New York apple orchards. Pest Monit J 7:200-204

Lichtenstein EP, Fuhremann TW, Schulz KR (1971) Persistence and vertical distribution of DDT, lindane, and aldrin residues, 10 and 15 years after a single soil application. J Agr Food Chem 19, 4:718-721

Petrocelli SR, Hanks AR, Anderson J (1973) Uptake and accumulation of an organochlorine insecticide (dieldrin) by an estuarine mollusc. Bull Env Contamin Toxicol 10:315-320

RIVM (1991) Sustainable use of groundwater – problems and threats in the European communitiues. Report no 600025001, Bilthoven

Ruhnau M, Altenburger R, Bödeker W (Hrsg)(1991) Pestizid-Report-Geschichte, Bedeutung und Folgen einer Pestizid-Wirtschaft in Deutschland, Göttingen

Schmidt GH (1986) Pestizide und Umweltschutz. Braunschweig, Wiesbaden

Schulz D (1992) Dioxine in Böden. In: Rosenkranz D, Einsele G, Harress H-M (Hrsg) Bodenschutz. Ergänzbares Handbuch der Maßnahmen und Empfehlungen für Schutz, Pflege und Sanierung von Böden, Landschaft und Grundwasser, Berlin

Stock R, Friesel P, Milde G (1987) Grundwasserkontaminationen durch Pflanzenbehandlungsmittel in der Niederen Geest Schleswig-Holsteins und im Emsland. In: Milde G, Friesel P (Hrsg) Grundwasserbeeinflussung durch Pflanzenschutzmittel. Stuttgart, pp 209-224

Streit B (1991) Lexikon Ökotoxikologie. Verlag Chemie, Weinheim

Werner G (1987) Strategien und Ergebnisse der Überwachung der Rohwasserqualität von Grundwasserförderungsanlagen auf Kontaminationen durch Pflanzenbehandlungsmittel. In: Friesel P, Milde G (Hrsg) Untersuchung auf Grundwasserkontamination durch Pflanzenbehandlungsmittel – eine Bestandsaufnahme an ausgewählten Grundwasserlagerstätten. Berlin, pp 149-167

# 7 Regional Differences in Potentials for Delayed Mobilization of Chemicals in Europe

*G. R. B. ter Meulen-Smidt*

## 7.1 Introduction

When comparing examples of non-linear releases of chemicals in different regions in Europe, it appears that these may vary considerably, both with respect to major chemicals concerned, soil storage capacities, and processes changing the storage capacities. Although mechanisms follow basic physical, chemical and biochemical laws, values and rates may vary with conditions (Ter Meulen et al. 1993). As major processes and conditions may differ on regional scales, a regional approach of the subject seems warranted for, in spite of the fact that even on a field scale, heterogeneities in soils provide different storage capacities for contaminants.

In a project on human-induced soil degradation (Oldeman et al. 1991), it appeared that Europe is by far the most polluted continent (19 x $10^6$ out of a total of 22 x $10^6$ ha). Industry and bioagricultural activities are the main agents. A comparison of other soil degrading factors showed that erosion is a major factor (83.6% of the total degraded area) all over the world. Relatively high levels of compaction and subsidence of organic soils are found in Europe (15.1% for compaction and 0.9% for subsidence of organic soils), of salinization in Asia and of loss of nutrients in Africa (Oldeman 1993). All these factors in themselves can degrade soil, but they can also serve as a trigger for changes in behaviour of contaminants in soils and sediments.

This chapter will first describe the regional inputs and distribution of contaminants in Europe. The regional differences in the storage capacity of soils and sediments to retain contaminants is treated in Section 7.3. Finally, some examples are presented of how changes in the storage capacity of soils and sediments effect the release of pollutants.

## 7.2 Loads and Loadings

The total number of point sources in Europe is very large, especially in urban and industrialized areas. Major point sources are industrial sites, waste disposal sites, fuel storage and distribution site, line sources (motorways, railways, sewerage systems, etc.) and military sites (RIVM et al. 1992). Major diffuse sources are urban areas, atmospheric deposition, agriculture and tourism.

Pollution trends can be characterized in three phases: in the first and most polluted phase, industrial activities account for the major share of emissions. When industrial emissions are reduced owing to measures enacted to control environmental pollution, pollution from consumer-related wastes grows in relative importance. Finally, pollution and industrial productivity are decoupled, and consumer-related pollution dominates. The wealthy industrial nations have already entered the third phase as a result of extensive decreases in pollution from point sources (Stigliani 1994), mostly reducing air and water pollution. In central and eastern Europe (CEE) the first phase is still more prominent. Most types of emissions in CEE are foreseen to be reduced considerably in the next decades, at first due to economic decline, next to the introduction of western European standards and practices on new investments, leading to cleaner and more efficient production (Bollen et al. 1993).

### 7.2.1 Mining

Mining and its related industry are major pollution sources in Europe. Pollution due to mining and refinery activities has occurred since very early times: analyses of lake sediments from Sweden show that atmospheric lead deposition increased above background levels more than 2600 years ago, with a small, but marked lead deposition peak about 2000 years ago during the Roman period. A more significant increase began 1000 years ago and accelerated during the nineteenth and particularly the twentieth centuries, with a maximum deposition in about 1970 A.D. The cumulative deposition from anthropogenic sources in pre-industrial times (ca. 600 B.C. to 1800 A.D.) was at least as large as the cumulative deposition during the industrial period (1800 A.D. to the present; Renberg et al. 1994). Long-term pollution from old mines even resulted in the formation of a new plant species: the "zinc violet" (*Viola calaminaria*) is found in the Netherlands only in soils with high zinc content. Mining practices comprise three major categories:

coal (hard-coal and lignite); metal and salt; and further oil, gas and sulphur mining. Mining activities and spoils are present in several large regions (Fig. 7.1)

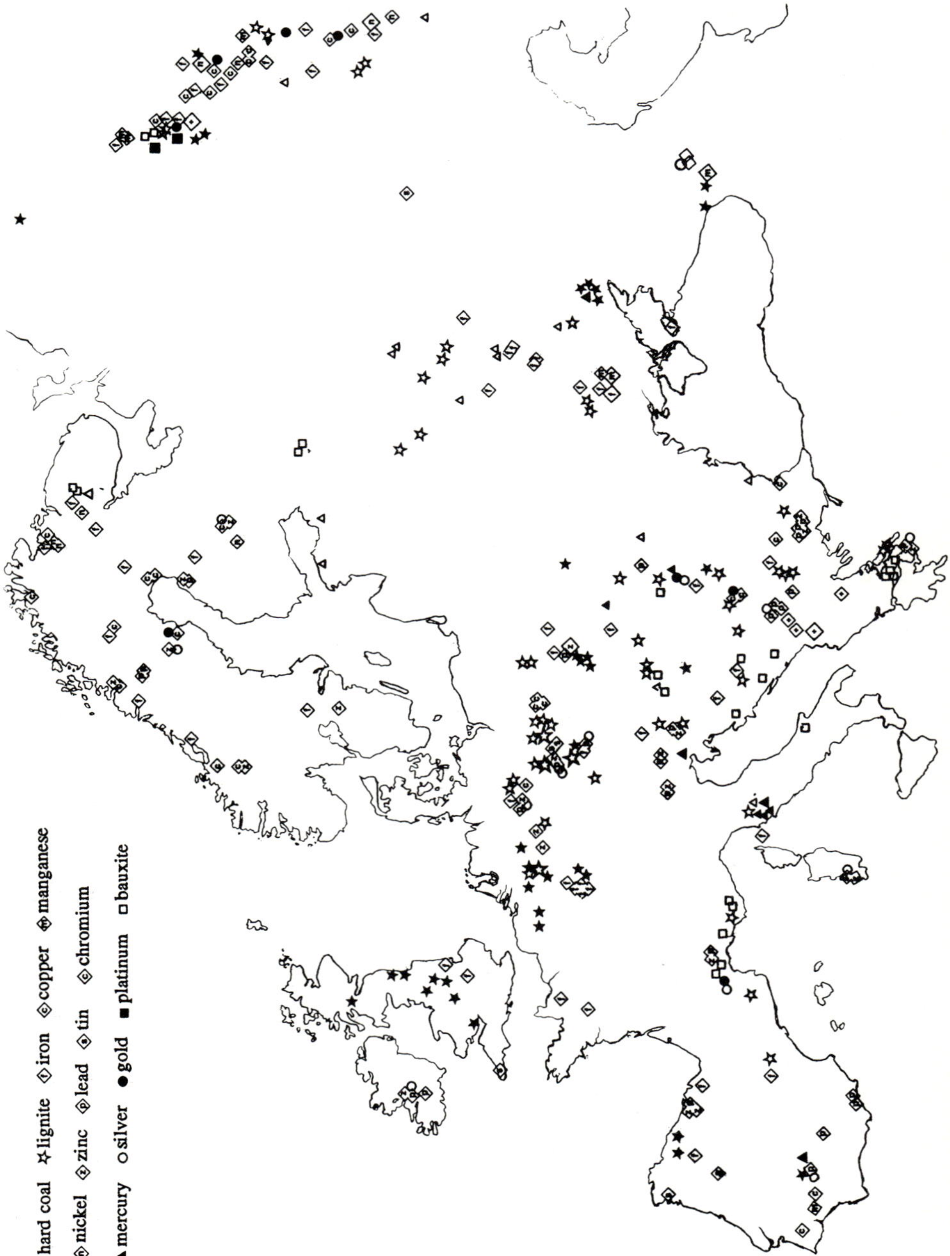


**Fig. 7.1.** Mining activities in Europe

all over Europe. Large hard-coal mining areas are the Ruhr (Germany), Yorkshire-Humberside-East Midlands (UK), Upper Silesia (Poland) and Donbass (Donetsk, Ukraine). Large lignite mining areas are Niederlausitz, Halle-Leipzig, Rhineland (Germany), Upper Bohemia (former Czechoslovakia) and Moscow (Russia). Large salt mining areas are Alsace-Lorraine (France) and Upper Silesia (Poland; RIVM et al. 1992). Metal mining activities are mostly found in central Europe and in the Ural region (Fig. 7.1).

Metal ore mines are often accompanied by processing plants and smelters which contribute significantly to the overall pollution. Mining activities, wastes and industry (refinery) cause contamination of soils, water reservoirs and the atmosphere. In Poland, soils in the neighbourhood of metal mines and smelters show metal concentrations exceeding up to 100 times the background level (Rybicka 1994). Hydrological systems around mining areas can be influenced by mineralized and polluted waters originating from dewatering of mines. They can be rich in $Cl^-$, $SO_4^-$ and can contain high amounts of heavy metals, phenols and natural radioactive isotopes like radium. The acid leachate from coal mining is characterized by high concentrations of total dissolved solids and may be saline. Metal mining waste contains high heavy metal concentrations with sulphides and sulphates. Spoils of salt mining contains several salts, such as chloride and potassium, in high concentrations. If transported into river systems, river and floodplain sediments may be polluted by contaminants derived from mine waters as well as released by ore processing and smelting plants.

Mining activities yield huge volumes of wastes, spoils and cleaning wastes. It has been estimated that 1.0 ton of hard coal may be accompanied by 0.4 ton various wastes, containing variable amounts of heavy metal sulphides. Waste dumps can contaminate adjacent areas due to long-term leaching, but also due to pollution of the atmosphere, vegetation and soils with gases (spontaneous combustion) and dusts (aeolian erosion). Ashes and slags from hard-coal power stations constitute relatively small percentages of wastes (2–3%).

One of the most dangerous hazards is the pollution of surface and groundwater by soluble solids leached from the wastes (products of sulphide weathering/decomposition). Formation of highly acid waters (pH<4) appears to be a common process. Combustion of brown coal results in mass emission of gases ($CO_2$, $SO_2$, nitrogen oxides) and dusts (composed mostly of $SiO_2$, CaO, $Al_2O_3$ and $Fe_2O_3$) with admixtures of heavy metals (Zn, Pb, Cd). Brown coal mines in Poland show sulphur levels between 0.5 and 1.1%. Huge amounts of $SO_2$ are thus released to the atmosphere giving rise to "acid rains" and consequently to the degradation of soils by acidification.

### 7.2.2 Industry

More than 50 large industrial regions can be identified in Europe, most of them situated in central and western Europe. The major ones are in the United Kingdom, the Ruhr area with the Rhine Delta, the area of Saxony and Bohemia, the Po Valley, around Moscow, Donbass (Ukraine) and the Volga area. Smaller ones are around Warsaw, Cracow and Budapest. The industrial sites are often combined with or related to mining activities. Many industrial sites are contaminated by (liquid and solid) waste disposal, spoiled fuel, leaking pipe systems and accidental spills (RIVM et al. 1992). In some Western countries the most polluted industrial sites are being remediated.

A vast amount of pollution, derived from industry, energy plants and transport, is transported through the air and arrives on soils as atmospheric deposition. The compounds emitted by deposition comprise acidic compounds ($SO_2$, $NO_x$ and $NH_4$), heavy metals (e.g. through aerosols) and organic compounds like PAHs. Deposition on forests is generally higher than on, for instance, agricultural lands due to the filtering effect of forest vegetation (Mayer 1993). Even the type of forest makes a difference: deposition loads are higher in coniferous than in deciduous forest (De Vries et al. 1993).

Figure 7.2 shows calculated total deposition for Cd as an example of heavy metal distribution over Europe. Although there are differences, zinc, copper and lead show similar high deposition in the major industrial and urban areas in

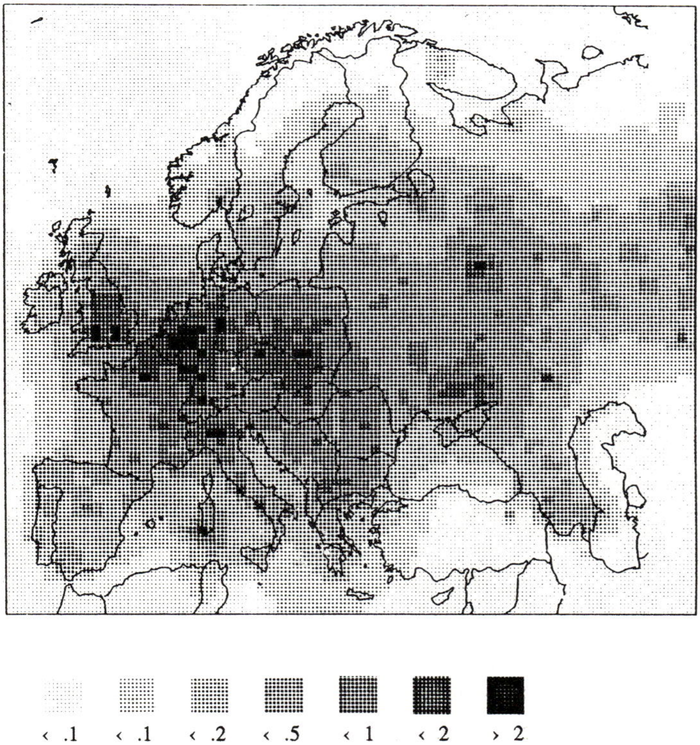


**Fig. 7.2.** Model-estimated total deposition of cadmium (g $ha^{-1}$ $a^{-1}$) for 1989 (Van Jaarsveld, in press)

Europe. Main deposition areas are the Rhine Basin (Ruhr area) and the Black Triangle (the region comprising Upper Silesia, Bohemia and the north Czech Republic). In western Europe, atmospheric deposition has decreased since the 1980s. Decrease in atmospheric deposition in CEE is still less obvious, but according to a recent World Bank report, is already expected for the period 1990–1995 (Bollen et al. in press).

Energy use in both the domestic and the industrial sector is much higher in CEE than in western Europe (3 times in 1990). It is predicted to fall drastically between 1990 and 2000 due to economic decline and price adaptations. High emissions of particulate matter due to intensive use of coal in power stations are found locally in the Black Triangle and in the Donetsk region. The switch from coal to gas and electrostatic precipitators could mean extensive reductions (60% for a scenario of introduction of current Western emission standards and energy efficiency). However, it is still very hard to predict future deposition trends in these countries; these will depend on economies and policies (Bollen et al. in press).

The proportions of the three acidifying components, $SO_2$, $NO_x$ and $NH_3$, varies over Europe: in CEE $SO_2$ emissions, mainly from lignite-fired installations, contribute most, whereas in western and southern Europe $NO_x$ emission from traffic and power generation is relatively large. In countries with intensive cattle breeding (e.g. Denmark, the Netherlands and parts of the UK) $NH_3$ contributes significantly

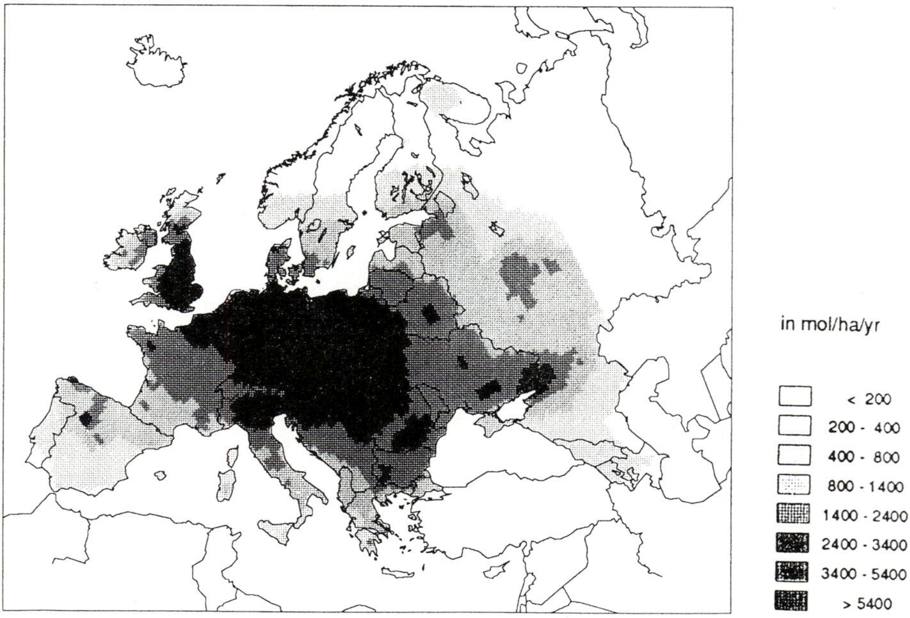


**Fig. 7.3.** Present deposition of potential acidity on coniferous forests in Europe (RIVM et al. 1992)

to acid loads. Power plants generally contribute most to total acid deposition, from about 30% in Finland to ca. 55% in Poland. The contribution from traffic emissions ranges from 7% in Poland to ca. 30% in Finland and Norway (Van Lynden 1993).

Figure 7.3 shows the acidic deposition for coniferous forests. High deposition (>3000 $mol_c\ ha^{-1}\ a^{-1}$) is observed for the region spreading from the Netherlands towards Russia. Deposition in deciduous forests is less, but still ranges over 2000 $mol_c\ ha^{-1}\ y^{-1}$ for this area (De Vries et al. 1993). Acid deposition is less than 400 $mol_c\ ha^{-1}\ y^{-1}$. Relatively low acidic deposition (<1000 $mol_c\ ha^{-1}\ y^{-1}$) is observed in Scandinavia north of 60° latitude (except for the Kola peninsula) and in the Mediterranean region; deposition in Greece and the south of Italy is almost negligible.

### 7.2.3 Urban and Densely Populated Areas

The highest population density in Europe (more than 200 inhabitants per km) is found in north-western Europe (the Netherlands, Belgium, the UK, Germany and the Po-valley). Except for the former USSR, the most densely populated areas are correlated with the major industry/mining areas. Touristic centres like the Mediterranean coast have large increases in population density, and consequently in contaminant loads, during holiday times. Urban areas often contain a large number of small industries (RIVM 1994) In some areas in Europe, like in Katowice in Poland, also heavy industry is found in urban areas. Urban areas exert a permanent load on the soil from traffic emissions and combustion processes, fuel storage, disposal of waste water, leaking of sewage, etc. Particular contaminants are chloride, sulphate, nitrate, oil products, volatile halogenated

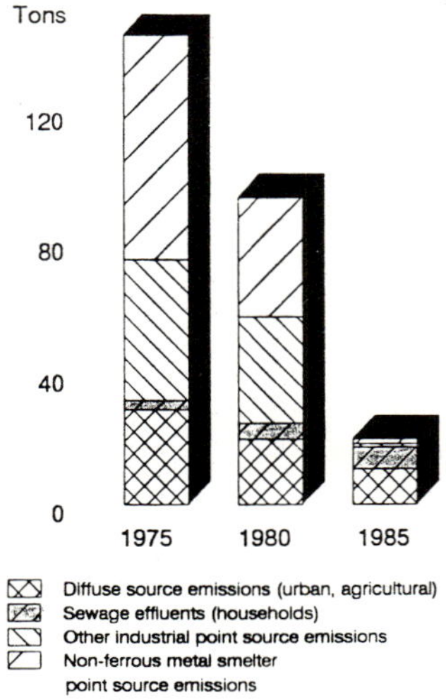


**Fig. 7.4.** Tons of cadmium per year in the Rhine River crossing the German-Dutch border. (Stigliani and Anderberg 1991)

hydrocarbons and heavy metals like lead and zinc. (RIVM 1994). A vast amount from the toxic chemicals is concentrated in street dust (Stigliani and Anderberg 1991). The relative importance of the urban pollution source is increasing (Fig. 7.4). The highly populated areas in west European countries produce the highest deposition of lead; emissions from petrol combustion are still dominating over industrial emissions. The emissions are expected to decrease in western Europe, and to increase in CEE, due to increasing transport.

### 7.2.4 Landfills

In most European countries, especially in CEE, landfilling is still the most common way to remove municipal and industrial waste. In Europe 240 million tons of industrial waste, 104 million tons municipal waste and more than 60 million tons of sediment sludge – 65% have annually been landfilled, 24% incinerated and 11% recycled and composted (Boels and Fleming 1993). The total number of landfills in the EC (municipal and industrial) is estimated at about 120 000–240 000 (one landfill to every 2500–5000 inhabitants). Emission of leachate, consisting of a wide range of heavy metals and organic micropollutants, cannot be avoided, and the area with polluted groundwater increases annually by 1–10%. This means an annual extension of 12–240 $km^2$ of the area of contaminated groundwater in Europe (RIVM and RIZA 1991). Leachate formation depends, next to composition and age of the waste, on the hydrological regime (climate) and on the physical soil conditions, therefore differing considerably throughout Europe (Boels and Fleming 1993). Figure 7.5 shows the estimated groundwater area potentially contaminated by landfills in Europe as a percentage of the total area per region, assuming that in the long term the leachate will reach the groundwater.

The historical approach to landfill was to tip solid waste in a random pattern with limited control on the amount, type or method of compacting, and, once the site was completed, to cover it with a thin layer of topsoil or inert material and abandon it (Boels and Fleming 1993). Only during the last decade and in a few European countries have new landfills been equipped with liners and other facilities to prevent soil and groundwater contamination. In many of the hazardous waste sites the (sealing) constructions or the geohydrological circumstances have proven to be insufficient. Moreover, liners are not designed for the lifetime of the sites operation (Fleming 1992).

The engineering standards applied throughout Europe are highly variable. For newly constructed landfills, for instance those common in the Netherlands, Denmark and Germany, the long-term leachate formation amounts to less than 2% of the annual rainfall on the site; for abandoned sites where precautions are not taken, leachate formation is 20–60% of annual precipitation (Boels and Fleming 1993). The current EEC Directive on landfilling clearly sets out to protect the groundwater regime through very strict engineering standards in important ground-

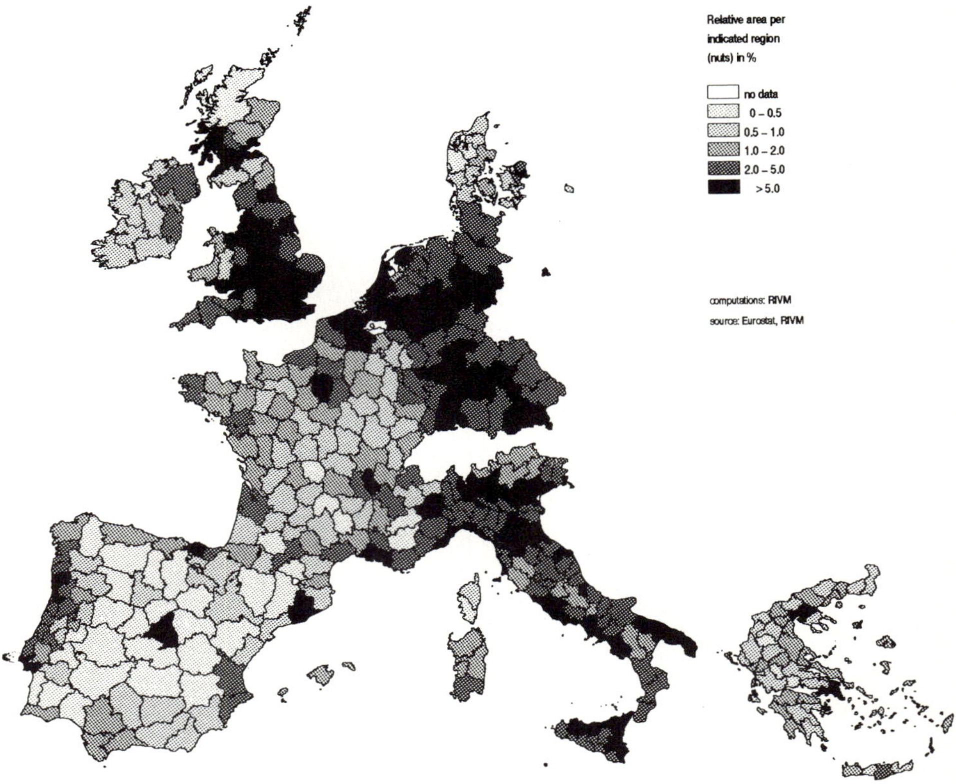


**Fig. 7.5.** Estimated groundwater area contaminated by landfills (RIVM and RIZA 1991)

water areas (RIVM et al. 1992). The most acute regional problems are in Italy, Spain, Greece and CEE where the engineering and management of landfills are presently below standard. This is particularly true for the CEE countries where, due to the present economic difficulties, the economics of landfill technology have not yet caught up. The contamination of soils due to careless waste disposal, especially of chemical waste from mines and industries, is reported to be extensive (Bollen et al., in press).

### 7.2.5 Agriculture

Major agricultural contaminants are nitrate, phosphate and pesticides, and cadmium, copper, lead, arsenic, zinc and fluorine as byproducts of chemical fertilizers, manure and pesticides. In the western part of Europe (EC and EFTA) the trends of the postwar years have been fairly consequent: increased intensified farming and use of chemicals, and heavy increases in yields. Environmental regulations have now started to restrict fertilizer and pesticide use. In the Netherlands, for instance, cadmium loads on agricultural soils decreased by almost 50% in the period from 1980 to 1990 (mostly due to a change to phosphate fertilizer low in cadmium), copper and lead decreased more than 20% (RIVM 1992).

Developments in the "Post-Eastern Block" countries have been more abrupt. Despite considerable differences among the countries, similar trends can be observed since primary goals and strategies were general. Collectivization, which was completed in the 1960s, was prevalent in all these countries, except Poland, where agriculture mostly remained private. Up to 1990, during the period that agriculture was driven by plan economy, subsidized mineral fertilizers and lime were extremely cheap, which contributed to the increase of crop yields (in addition to new, high-yielding varieties, full mechanization and successful pest management). Agricultural output, however, grew less rapidly than in western Europe. After the late 1980s the concept of high input agricultural production changed radically due to the necessity of input rationalization, increasing quality requirements (in line with western Europe), and the hazard of environmental pollution and soil degradation processes (Várallyay 1994).

Fertilizer use increased considerably in western European countries (EC and EFTA members) from 124 kg $ha^{-1}$ in 1964 to 224 kg $ha^{-1}$ in 1984 (RIVM et al. 1992). In eastern Europe and the Balkan States the average application of manure and fertilizer per hectare is half the amount used in western Europe (30.4 kg $ha^{-1}$ in 1964 to 122 kg $ha^{-1}$ in 1984). In central Europe (Poland, Hungary and the Czech and Slovak Republics) the amounts are comparable to western Europe (Bollen et al., in press). The highest fertilizer loads are observed in the Netherlands. In CCE the disappearance of planned economy subsidies combined with the disadvantageous economic circumstances of the last few years have resulted in a decreased application of fertilizers (e.g. a factor of 3 for phosphate in the Danube region) and lime. However, if present policies are going to be continued, the Danube countries will increase phosphate and pesticide use again (Haskoning 1994). In the Mediterranean region chemical inputs are expected to increase. In 1980 inputs of fertilizer in Spain and Italy were 77 and 175 kg $ha^{-1}$ respectively, and are at best expected to be doubled by 2025 (Grenon and Batisse 1989).

Cadmium pollution from phosphate fertilizer use differs across Europe, not only as a result of different rates of application, but also due to the variation in cadmium concentrations (1–170 mg $kg^{-1}$) in the different phosphate ores (Haskoning 1994). In Hungary, for instance, phosphate fertilizer almost totally originated from Kola igneous rock phosphates, which are very low in cadmium (1 mg $kg^{-1}$), but relatively high in strontium (20 g $kg^{-1}$) and arsenic (60 mg $kg^{-1}$). Fluorine pollution, originating from fertilizers and industry, is reported in the Baltic, Danube and Russian regions (Glazovskaya et al. 1991). Relatively high soil levels (450 mg $kg^{-1}$) were reported for Romania (Răută et al. 1987). Strontium and fluorine are mobilized under acid conditions, while arsenic is immobilized.

The number of livestock and the related manure production have increased enormously in the past decades with the highest rates in the Netherlands, Belgium, North Rhine-Westphalia and the Po area (RIVM et al. 1992). In the Netherlands, resulting ammonium emissions have led to increased acid rain. Localized hot-spots with high concentrations of nitrates in the groundwater occur in CEE due to manure dump sites and infiltration of sewage (Bollen et al., in press).

Europe is the largest agrochemical user in the world with major loads found in western Europe (Fig. 7.6). Highest pesticide loads are seen in the Netherlands, southeastern France and Azerbijan (>10 kg $ha^{-1}a^{-1}$). In Hungary, the Czech republic, Bulgaria and Azerbijan, pesticide use per ha is equal to western Europe (Bollen et al., in press).

Loads are generally relatively low (< 2 kg $ha^{-1}a^{-1}$) in the former Soviet Union, Poland and in the Scandinavian countries (RIVM et al. 1992). In general, herbicides are more important in CEE countries; fungicides in the EC countries. In

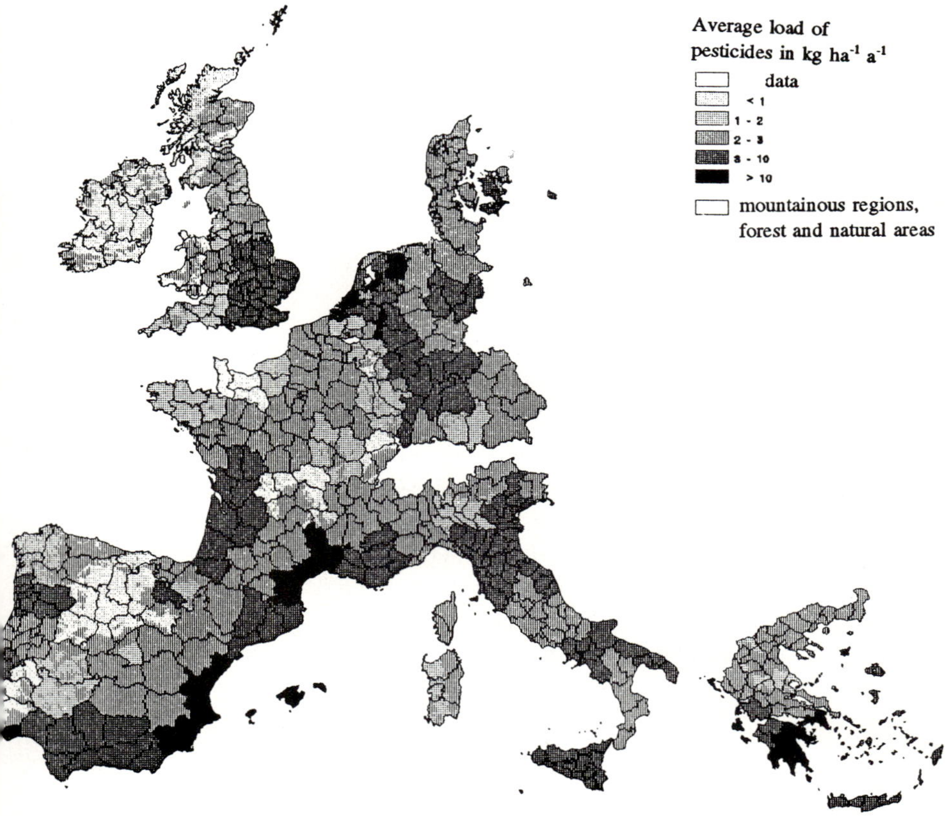


**Fig. 7.6.** Pesticide load on European soils: present situation (RIVM et al. 1992)

France, Italy and Spain fungicides are very important in viniculture; sulphur and copper compounds are still used in high quantities. During the past decades in western Europe the pesticides have become more potent and degradable, and the amount used has started to decrease. Inorganic compounds are being substituted with organic compounds.

### 7.2.6 General Trends in Europe

At the moment there is a shift to decreasing pollution input in most parts of Europe, particularly in the West. This "clean-up" action is most effective for water and air pollution. In southern Europe, decreases are fewer, as pollution has been less, and this area is under development. Pollution loading in CEE is expected to decrease during the next decades, but increased lead and fertilizer loading is expected in these regions.

Also observed is the shift in relative importance of diffuse and point sources. In the Rhine Basin in the period 1975–1985 the total load of cadmium was reduced from more than 140 t $a^{-1}$ to less than 30 t $a^{-1}$, but the relative contribution of urban and agricultural diffuse sources increased from less than 24 to more than 50% (Stigliani and Anderberg 1991). However, loads of persistent chemicals stored in soils and sediments will not decrease in the short term.

When comparing chemical loads in soils in Europe, the following trends can be observed:

- The highest population density is generally found in combination with large industrial centres, hence consumer and industrial pollution sources are coupled and concentrated.
- Agriculture is the second major pollution source, discharging similar substances to those found in industrial and consumer pollution (heavy metals, nitrate), but also specific substances like pesticides and ammonium.
- The highest pollution is observed in western Europe; the Black Triangle is the only part of CEE in which the pollution loads are comparable to western Europe. Severe pollution in CEE is mostly restricted to hot spots, like industrial and mining areas. One problem in this respect has been the difficulty of obtaining data from the former planned economy countries. A major problem in CEE is lignite burning, causing severe pollution and acidification. Another risk-bearing component is the waste dumps in CEE and southern Europe.
- Loads in Scandinavian and Mediterranean countries are generally lower, except for the industrial centres and specifically the Po Valley. Exceptions are special types of pesticides in southern Europe and mercury in Scandinavia.

## 7.3 Regional Differences in Storage Capacity in Soils

For a regional approach to soil sensitivity for delayed responses, storage capacities of soils should be assessed. This is to gain an impression of potentials for accumulation and leaching, especially for non-degradable compounds.

Storage capacities are mostly determined by the combination of soil type, climate and management. The combined weathering action of the climate on the parent rock, erosion and transport by glacial, fluvial and wind action produces the mineral particles which form the primary skeleton of the soil. Organic matter derived by decomposition processes from plant detritus interacts with these minerals to provide the solid phase of the soil. This mineral-organic mix determines greatly the storage capacity of the soil and the sensitivity of the soil to changes in capacity controlling properties (Swift et al. 1979). Soils with high storage capacities are generally undisturbed soils with high organic matter content, fine structure (clay), high Fe, Mn and Al (hydr)oxide content, low temperatures, low leaching, and low salt content and erodibility. In the following sections factors important for and possibly influential on storage will be assessed for different regions in Europe.

### 7.3.1 Climate

In Europe in general average temperatures show a typical increase going southwards, whereas temperature differences between winter and summer increase going eastward. Precipitation patterns are less straightforward: highest yearly precipitation (>1000 mm $a^{-1}$) is found in the Alp region and the northern Adriatic coast, and at the continental western coasts. Major dry areas (< 500mm $a^{-1}$) are found in Spain, the former southeast USSR, and the northern part of Scandinavia and former USSR. Leaching of contaminants increases with higher precipitation/evapotranspiration rates. Long lasting snow cover (>90 days snow per year) is found in most areas of the former USSR north of 50° latitude. The length of the snow-cover period is important for wet deposition; snowmelt may result in peak discharges in spring leading to the release of compounds in a relatively short time (Konsten et al. 1993). Permafrost, determining the active and drainable soil cover, is found in the tundra areas.

Similar soil types develop under similar climatic conditions. However, very similar soil types developed within widely differing climatic zones. This is the case for waterlogged soils such as peats, and soils derived from limestone such as rendzinas (Swift et al. 1979).

### 7.3.2 Organic Matter Content

Organic matter percentages do not only contribute to storage capacities for contaminants in soils, but also to robustness of soils against erosion and

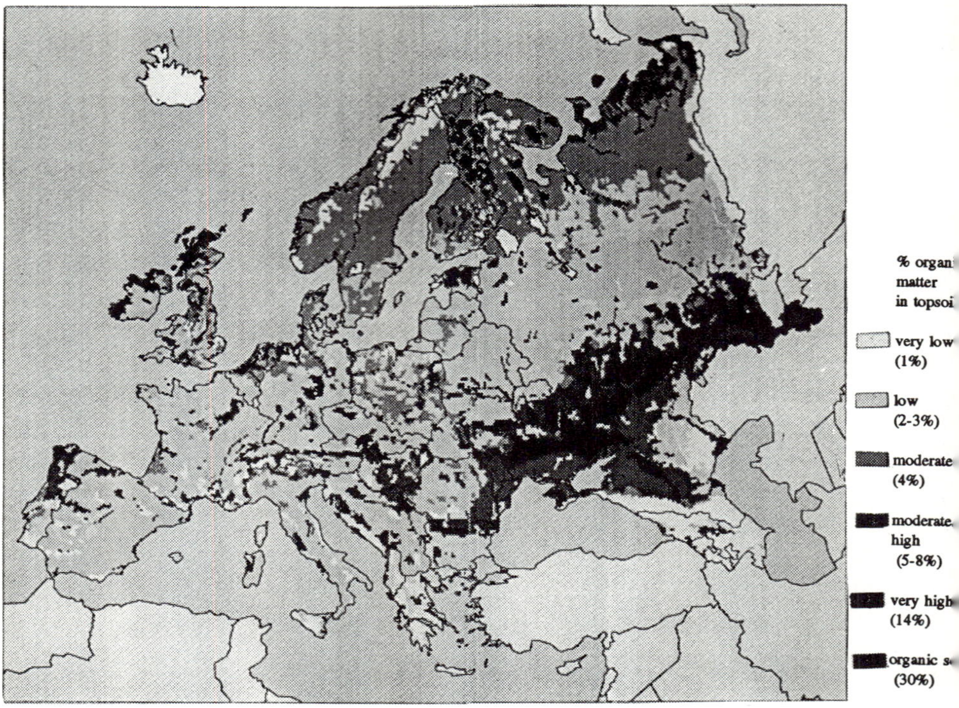


**Fig. 7.7.** Organic matter (%) in topsoils of arable land in Europe (Fraters et al. 1993)

acidification. Organic matter contents of more than 30% are found in peat soils. Peat bogs, important wetlands in Europe, are arranged along an oceanic-continental gradient and a temperate-subarctic gradient.

Figure 7.7 shows the organic matter content in arable land; the content in natural soils (forests) will be slightly higher. Major organic soils (organic matter > 8%) are found in the south-eastern part of the former Soviet Union, in the so-called chernozems and to a lesser extent in the kastanozems situated around the chernozems. The Nordic region has a moderate organic matter content (4%) in most of its soils. Most soils in Europe contain low to moderate organic matter content (2–3%); only in mountainous areas, some sites in the Mediterranean region (especially in Greece and the Adriatic coast) and a few places in Poland are there soils very low in organic matter (1%; Fraters et al. 1993).

### 7.3.3 Texture and Erosion

Texture classes can be determined from the FAO soil map of Europe, but as different texture classes occur within one grid, only very general comparisons of texture classes can be made on a regional scale. Generally, light sandy soils can be found in the north-west of the Netherlands, northern Germany and in Denmark. Poland contains a large region with light sandy soils. Scandinavian soils are characterized by a relatively thin soil cover, often overlying acidic bedrock. Heavy, clayey soils are found in Italy and Greece (FAO-UNESCO 1981).

Texture is a relatively stable soil characteristic but it can be changed by erosion, especially wind erosion, which only moves the smaller soil particles, leaving the larger ones.

Erosion is a widespread phenomenon in Europe, with emphasis on the south-eastern part of Europe and major parts of Russia. In the Russian Plain about 1200 ton $ha^{-1}$ of topsoil have been lost over the past 100 years due to erosion of sod-podsolic soils. Eroded land is observed in approximately 25% of the chernozem soils in the Ukraine (Karavayeva et al. 1991). Water erosion is by far the most common type of soil erosion, as 12% of the soils ($114 \times 10^6$ ha) show light to extreme erosion from water (Fig. 7.8a). For 10% of the soils, the erosion is moderate to strong. Wind erosion (Fig. 7.8b) is light to moderate for 5% of the European soils (RIVM et al. 1992). As pollutants are mostly tied to the fine fraction of the soil particles, comprising clay and organic matter particles, wind erosion is an important transport mechanism for contaminants (e.g. pesticides). The extensive and serious wind erosion in the south-eastern part of Europe (southern Russian Plain) can be partly explained by the dry continental climate and sensitive soils in combination with inappropriate farming practices (fertilization, compaction and salinization). Excessive drainage and overgrazing are prominent causes for conditions that favour wind erosion. In western Europe the fairly moist conditions and storm events often associated with heavy rain diminish the chances of wind erosion. However, some is observed in parts of NW Europe (Van Lynden 1993).

### 7.3.4 Vegetation

Ecosystem types (biomes) are correlated with the major soil types and environmental factors as shown in Fig. 7.9. Tundra soils range from stone deserts to waterlogged moss meadows. Taiga vegetation predominantly consists of evergreen conifers; the soils are strongly leachable podzols. Temperate grasslands with well-drained soils occur on the chernozems. The urban industrial society of western Europe is mainly located in areas previously dominated by temperate deciduous forest. The most intensive agricultural practice has also been developed within this biome. Under natural conditions soils vary from podzols to brown earth, podzols mostly under conifers and heather, and brown earth mostly under deciduous trees and grassland. Podzols are generally more acid than brown earth. A change from one vegetation type into the other can result in the respective change in soil type with correspondent soil characteristics (Swift et al. 1979). Climate change may shift vegetation belts northward. Binding of heavy metals, for instance, seems to be influenced by the vegetational background, probably due to the differences of physico-chemical properties (Krosshavn et al. 1993). This

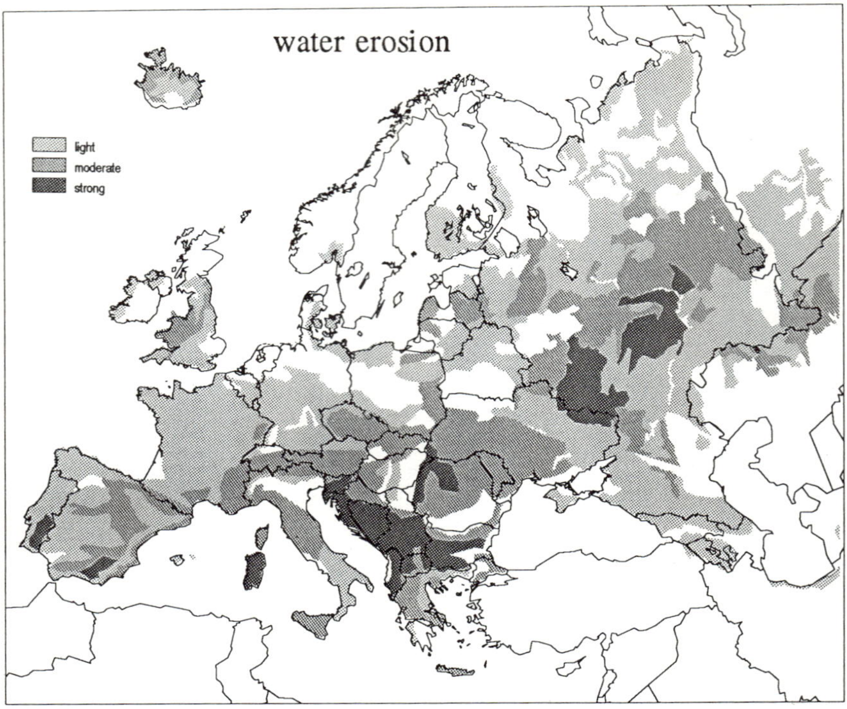


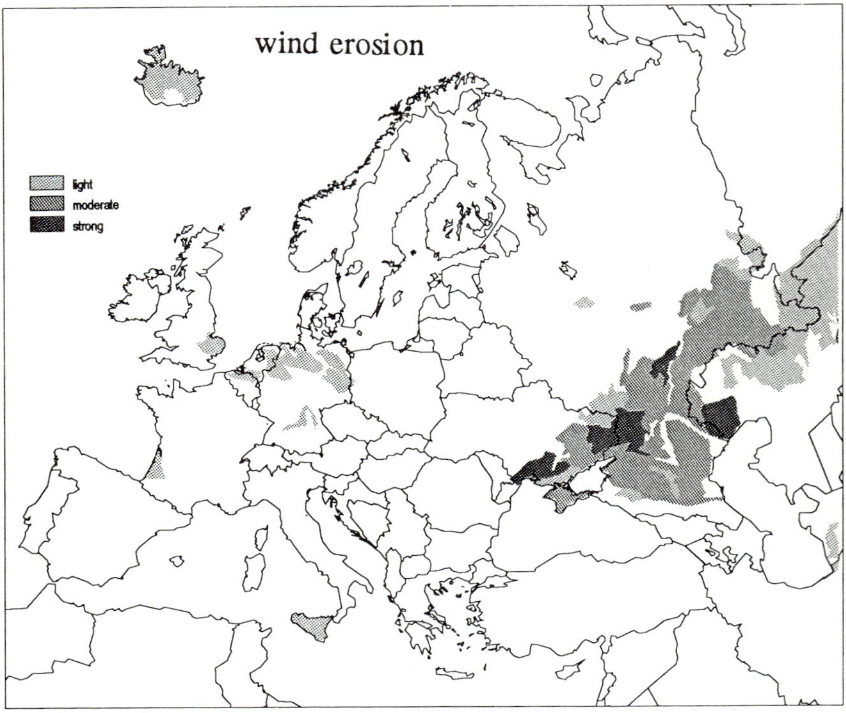


**Fig. 7.8 a–b.** Soil erosion in Europe. **a**: Water erosion; **b**: wind erosion (RIVM et al.1992)

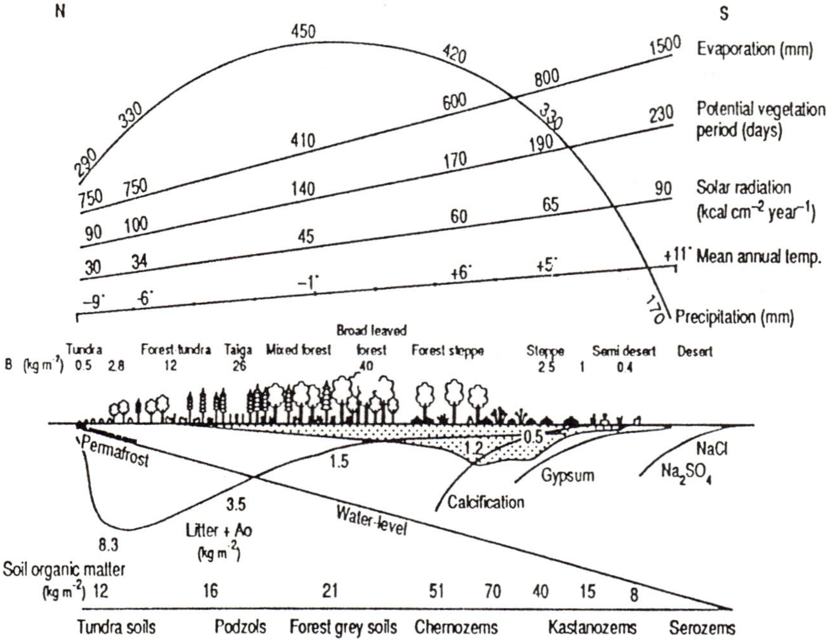


**Fig. 7.9.** The relation between biomes and environmental factors for the north–south succession of biomes in eastern Europe. Indicated are climate parameters, vegetation types, with B = biomass and NPP = net primary productivity; and soil parameters with litter + Ao = litter accumulation in upper soil horizon; gray = humus layer. (After Hekstra 1991)

means, that a change of vegetation types or land use may result in changes in the soils' storage capacities.

### 7.3.5 Sensitivity to Acidification

To determine soil acidification due to acid atmospheric deposition, calculations have been made for critical loads for acidity in European forest soils and the amounts by which the critical loads are exceeded by potential acidic deposition (the sum of nitrogen and sulphur deposition; De Vries et al. 1993; Hettelingh et al. 1991). Critical loads give an indication of the buffer capacity of soils. In many regions low critical loads coincide with low pH values, but this is not always the case.

Topsoils of forests in Europe show significant regional differences in critical loads and pH: The lowest current pH values (<4) are found in western Europe and in the podzols in the north part of CEE. Critical potential acid loads are low in the Nordic countries, Denmark, north-western Germany, the Netherlands and large parts of the former USSR; even remarkably low in Poland (200–

400 mol$_c$ ha$^{-1}$ a$^{-1}$) with its light sandy soils. In northern Scandinavia pH values are generally above 4.5, despite a thin soil cover overlying acid bedrock, and low critical loads. Critical loads and pH are much higher in the parts of Europe with calcareous soils: south of the Pyrenees-Alps-Carpathians-Caucasus line calcification has been more prolonged and more intensive than north of this line. This has resulted in regions with high soil pH values (>5) and high buffer capacities against acidification in the arid southern part of Europe, and in the chernozems and kastanozems in the former USSR. The same accounts for the Estonian rendzinas.

Forest soils often are relatively acid, most of the better soils being used for agriculture (Van Lynden 1993). Furthermore, agricultural soils are generally limed.

Critical acid loads are largely exceeded in western and central Europe, i.e. the Netherlands, Belgium, Germany, Poland and the Czech Republic. Major exceedences also occur in Byelorussia and in parts of the United Kingdom, Romania and Slowakia. In the Mediterranean countries, critical acid loads are exceeded in less than 20% of the area in Portugal and Italy, and the area exceeded is negligible (<5%) in Spain, former Yugoslavia, Bulgaria, Albania and Greece.

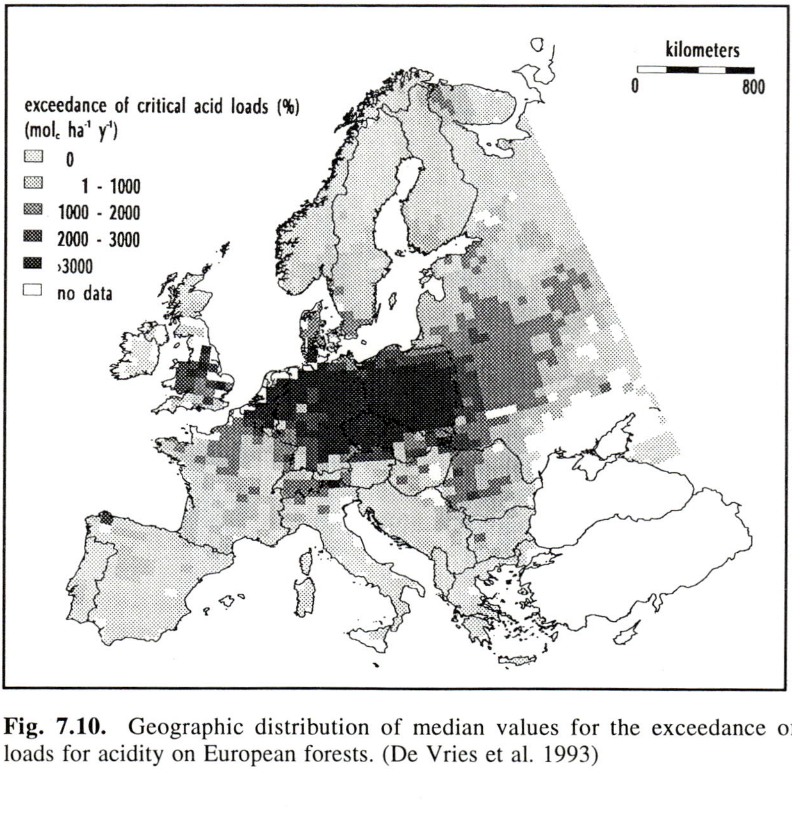


**Fig. 7.10.** Geographic distribution of median values for the exceedance of critical loads for acidity on European forests. (De Vries et al. 1993)

The same is true for most of Ireland and the Nordic countries (Fig. 7.10). Many publications report exceedance of critical loads in the Nordic region (Hettelingh et al. 1991) but in these observations not only forest soils, but also surface waters, have been included. These are the most vulnerable ecosystems for acidification in the Nordic region.

The areas with exceeded critical acid loads are subject to acidification; present soil pH values will decrease. Acidification is one of the strongest mobilizing factors in soil for many heavy metals. The areas showing acid deposition and with exceeded critical loads largely coincide with heavy metal deposition areas.

### 7.3.6 Sensitivity to Redox Changes

Waterlogging can result in low redox potentials and may be induced by flooding, thawing of permafrost, increased run-off from higher areas, or raised water-tables, for example as a result of irrigation. In drier areas, it also increases the risk of salinization though capillary rising of groundwater. Human-induced waterlogging occurs mainly in northern Russia and at a few scattered sites in the rest of Europe (foremost along the Black Sea coast and the lower Danube Valley). Inundation of land, as is planned for the Netherlands' river-forelands will also cause waterlogging. Waterlogging will stimulate sulphide binding, but will release contaminants adsorbed at (hydr)oxides (see Chap. 4 by Bourg and Loch).

Wetlands are widely distributed in Europe, but many have already suffered decline through drainage, water extraction, pollution and fisheries. Waterlogged soils in natural wetlands are often accumulation spots for sulphate and heavy metals. Drainage may lead to release of formerly anoxically stored contaminants. If the waterlogged soils also contain pyrite, drainage will lead to strong acidification (oxidation of sulphur to sulphuric acid) as happened in Lake Blamissusjon, now the most acidic lake in Sweden (Stigliani and Salomons 1991). This type of soils is not very widely distributed, but is found in some places in Europe.

Lowering of the groundwater table for agricultural purposes, resulting in excessive drainage and/or oxidation of organic soils, leads to shrinkage of organic soils in a large region in Byelorussia/Ukraine and south of St. Petersburg (Van Lynden 1993). Drainage of peat, leading to the breakdown of organic matter, occurs quite extensively in Finland (Konsten 1991).

### 7.3.7 Sensitivity to Salinization

Evapotranspiration of saline soil moisture, improper irrigation and drainage, and intrusion of salt water from the sea or from saline fossil sources, may all cause salinization. Salinization is mostly strongly tied to site-specific soil properties and climatic conditions; its distribution is restricted to the southern and eastern part of Europe where (semi-)arid conditions prevail (Szabolz 1991).

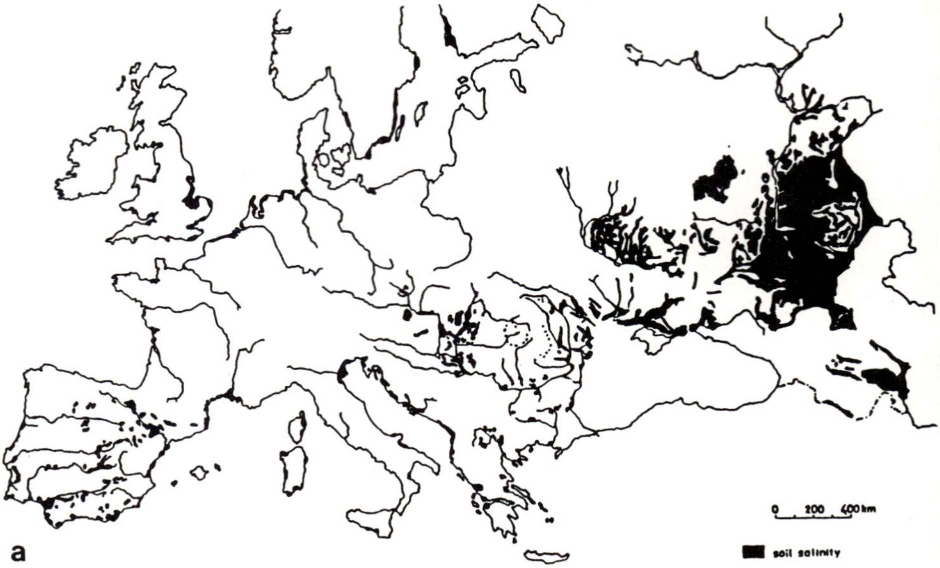


a

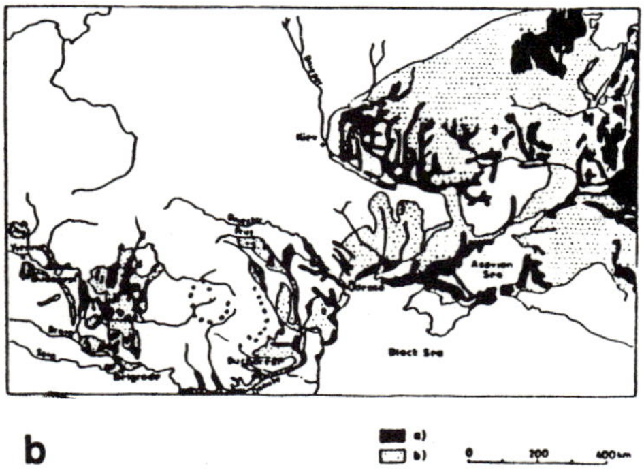


b

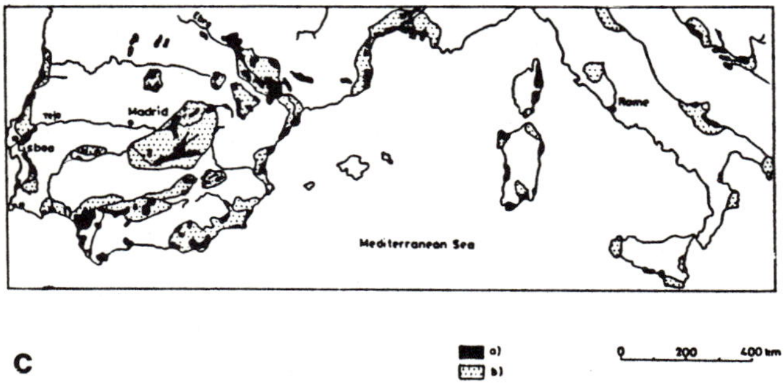


c

**Fig. 7.11a–c.** The present situation of **a** existing soil salinity in Europe ; **b** the territory of salt affected soils and potential soil salinity as a consequence of extended irrigation; **c** the territory of salt affected soils and potential salinity as a consequence of climatic changes. (Szabolz 1991)

Irrigation of agricultural lands is the main anthropogenic factor causing salinization in (semi-)arid areas. The effects of sea water intrusion can occur in all climatic zones (Van Lynden 1993). The total extent of salt affected soils in Europe surpasses 50 x $10^6$ ha (Fig. 7.11a). Three quarters of all salt affected soils of this continent occur in Spain, Hungary's Danube lowlands, and in the semi-arid steppe and the forest steppe regions of the former USSR. In several countries salt affected soils are found to a limited extent on the coastline.

The extent of potential salt affected soils has been estimated to exceed at least 2 to 3 times the acreage of existing saline soils. The rate of human-induced salinization is accelerating due to intensified farming combined with increased irrigation, domestic water use and climatic change. Szabolz (1991) showed that for a scenario of maximally doubling of irrigation until 2050 the territory of salt affected soils and potential soil salinity in mid-eastern Europe will be tripled (Fig. 7.11b). A scenario with an average temperature rise of 1°C in the next 50–70 years for the Mediterranean region showed an increase in the territory of salt affected soils of more than twofold, mostly taking place in Spain (Fig. 7.11c).

Increased irrigation will mostly take place on agricultural soils loaded with fertilizers and pesticides. Salinization due to sea level rise will take place in the lower coastal areas and deltas (e.g. the Netherlands, Po Delta) – areas often industrialized or used for agriculture. Consequently, salinization will often occur in places of high contaminant accumulation.

## 7.4 Delayed Effects due to Changes in Storage Capacity of Soils: Examples

### 7.4.1 Acidification and Release of Chemicals

Exceedances of the critical acid load imply that the critical Al concentration of 0.2 $mol_c\ m^{-3}$ is exceeded in a situation of equilibrium. Al leaching is predicted in the area stretching from the Netherlands to Russia, with major leaching (> 4.0 $mol_c\ ha^{-1}\ a^{-1}$) in the Black Triangle and parts of Germany, the Netherlands and Romania (De Vries et al. 1993). The most severe leaching with aluminium concentrations in groundwater exceeding 5 mg $l^{-1}$ is expected in southern Poland, north-western Germany and the eastern Netherlands. This is in accordance with the reports on declining forest vitality in these areas (RIVM et al. 1992).

Acid deposition in the Nordic region is relatively low, compared with most other regions in Europe. However, especially in the south of Norway, Sweden and Finland and in the neighbourhood of the Kola Peninsula, diffuse atmospheric deposition of acidic compounds and heavy metals is significant (Konsten et al. 1993). Soil cover is relatively thin and often overly-acidic bedrock. This results in small adsorption capacities for chemical pollutants, and low buffering capacities.

In southern Sweden, strong acidification of forest soils has already caused increased solubility of Al, Cd and Zn, resulting in high concentrations of these elements in surface waters (Konsten 1991). Consequently, however, this also means that in these areas the accumulation of chemicals will be relatively low, which in a way is favourable with respect to delayed responses.

In contrast with most towns in western Europe, some towns in the former Soviet Union experience alkalinization, caused by precipitation of carbonate-containing dust, as more typical than acidification. Some branches of industry (heat and energy production, cement production) cause strong alkalinization of the snow cover, soils and waters. For instance, the cement plant in Togliatti is responsible for a change in pH in the snow cover from 6–6.5 to 11. This constant alkalinization increases the pH of the surface layer by 2–3 values. An accompanying problem will be the release of alkaline melt water in spring. Being a centre of nitrogen fertilizer production, the town of Togliatti is a source of nitrogen, chromium, nickel, molybdenum, vanadium and lead (Kasimov and Moiseyenkov 1994). Especially for elements like molybdenum and vanadium this can cause increased mobilization, as these elements become more mobile in alkaline conditions.

The Danube countries generally do not suffer greatly from acidification. The soils in this region generally have high humic content (chernozems, cambizols) loamy/clayey (fluvisols and luvisols) and often contain high calcium contents (renzinas). They have large buffer capacities for acidic compounds and storage capacities for contaminants. Therefore, except for the very contaminated hot spots, effects of contaminants in the Danube basin are not as obvious as, for instance, in Poland. Acid and dust emissions are expected to decrease in line with international agreements (Haskoning 1994). The decrease of alkaline dust emissions might, just as in Poland, increase acidity.

Due to its calcareous soils and consequent high buffer capacities, the Mediterranean region does not experience much acidification either, except for small (industrial) areas (Fig. 7.10)

In Poland, sandy and light loamy soils cover about 60% of the arable land (Smal 1993). Acidification is the main soil degradation factor. About 58% of the cultivated soils are acid (pH below 5.5) in the surface layer and 25% in the lower layer (100–150 cm). The natural tendency to acidification is intensified by the atmospheric pollution, application of concentrated forms of mineral fertilizers with simultaneous decrease in organic fertilization and liming. (Turski and Glinsky 1991). Acid deposition in Poland is very high and it is not yet certain whether this is bound to decrease in the near future. Moreover, it has been observed that due to control measures resulting in a decline in alkaline dust emission, the acidity from deposition has increased (Smal and Salomons 1993). The highest proportion of acid soils is in western and south-western Poland, in spite of the fact that the average yearly dose of lime per ha in those regions has always been high (350 kg $ha^{-1}$ CaO in 1988/89). Decrease in liming enhances the acidification of agricultural soils, leading to increased uptake of metals in crops. In the most polluted areas such as Katowice and Krakow pH in more than 40% of the soils is below 5.5. The combined effects of heavy metal pollution and acidification have been exemplified in a study in Katowice Province on heavy metal pollution in arable soils. The study showed that the heavy metal levels in plants were so high that about 17% of the studied allotments and cultivated soils were not suitable for any crop production and should be absolutely excluded from it. In about 50% of the studied area certain crops with a high affinity for metal accumulation should be excluded from cultivation as well (Smal and Salomons 1993). From 1959 to 1990 the amount of arable land in Poland decreased by about 5%, partly due to an increase of forestation. This trend may become intensified by the recent economic and political changes in the country (Smal 1993). Afforestation will most likely specifically occur on the most polluted areas and increased acidification with coupled mobilization of contaminants can be expected.

### 7.4.2 Land-use Changes

The distribution of land use in Europe is 30% agricultural land, 16% permanent pasture, 37% forest and woodland and 26% other uses (Oldeman 1993). Forest percentages in the various countries vary from less than 1% in Ireland to more than 60% in Finland (De Vries et al. 1993).

Agricultural soils are very different from forest soils. They are generally highly manipulated: most soils are regularly mechanically mixed, which results in the "dilution" of the chemicals by uncontaminated deeper layers of the soil. The organic matter balance is disturbed (e.g. due to cropping and fertilizing) therefore abandonment often leads to a build-up of organic matter. As acidity decreases the productivity of soils, farmers lime soils to obtain a soil pH of about 6.

**Abandonment of Agricultural Soils.** During the past decades, farmland area in the whole of Europe has been decreased due to expansion of cities and roads. In western Europe the average rate was 0.5% per year during the last 20 years; in CEE the rate was much slower: 1% over 20 years (RIVM et al. 1992).

The area of agricultural land in the EC and EFTA is expected to decrease considerably due to the MacSharry plans, which propose a cutting back of about 15% of productive farmland. Probably this will mostly relate to marginal lands, which have been relatively heavily loaded with pesticides and chemical fertilizers. The areas will be abandoned, set aside, or turned into forest. The conversion of agricultural lands into forest has also been reported for several countries in CEE. In the Russian Plain, deforestation has taken place for the past 300 years, but the southern and western parts of the region have recently begun to reafforest abandoned agricultural land, stimulated by a state forest fund (Karavayeva et al. 1991). A Polish state forest fund stimulates reforestation of former agricultural lands in Poland. In Hungary arable land has recently decreased by almost 8% and grassland by 3%, in favour of forests and gardens.

Changing polluted agricultural land into, for example, uncultivated land, forests or gardens and the accompanying abandonment of liming practices may lead to acidification of the soils both in CEE and western Europe. This will have important ramifications with respect to the mobilization of chemicals, metals and pesticides in particular. Effects of abandonment of formerly intensively farmed land on cadmium mobilization have been modelled for an area in the south-eastern part of the Netherlands (the Kempen). In this region cadmium loads are elevated due to fertilizer application and deposition from zinc smelters. The soils are sandy and poor, and acid loads are high due to intensive lifestock farming. Cadmium has already been observed in the upper groundwater (RIVM 1992). According to EC policies, development of natural areas and afforestation is being considered for agricultural lands in this area. A simple model calculation for an abandonment scenario, showed a decrease in total cadmium, but a considerable increase in mobile cadmium for the next 50 years. This may lead to increased uptake of cadmium in the food web (Knoop and Traas 1993).

Abandonment of former intensively farmed soils may also lead to non-linear nitrate leaching. In Germany nearly 10% of former intensively farmed soils (6.2 million ha) have been abruptly abandoned or extensively managed because of the current structural change in agriculture. A field experiment showed for the first year of abandonment an additional wash out for nitrogen up to a maximum of 280% in relation to traditional farming. A considerable decrease of nitrogen wash out was observed in the second year of durable fallow ground (Meissner et al. 1993). On the other hand, during the last 50 years much previously uncultivated land has been brought into arable production: in some parts of eastern England permanent pasture has declined to about one-third of the area in 1940. When old grassland is ploughed, the organic matter content of the soil declines substantially due to mineralization into nitrogen. An experiment in Lincolnshire showed an soil organic matter loss of almost 4 ton $ha^{-1}$ N over 25 years. Half of the decline took place in the first 5.5 years (Whitmore et al. 1992). In both cases the non-linear nitrogen loads could turn out to be a considerable threat for the underground and surface waters.

Steep mountain areas of the Mediterranean have been terraced to reduce the probability of erosion. Ongoing depopulation of the mountain districts in most of the Mediterranean countries has led to abandonment, neglect and destruction of the terraces, with subsequent accelerated soil erosion. This relates mostly to marginal agricultural land. Between 1965 and 1988 arable land decreased by 25% in Portugal, 21% in Italy, 10% in Spain and 7% in Greece. Although in the past few decades forest destruction through fires has increased at an alarming rate for all southern European countries, in the period 1965–1988 forest and woodland in the Mediterranean region increased (Sanchez Diaz 1993).

**Drainage, Inundation and Deforestation.** The Mediterranean region is one of the few regions in Europe where rice is grown, for instance, in the Albufera Natural Park wetlands (Valencia, Spain). The park is contaminated by and sewage dumped into the irrigation ditches and the lagoon and by excessive use of fertilizers and pesticides. Average fertilizer consumption in the Valencian community is 192.23 kg $ha^{-1}$ $a^{-1}$, twice the average national consumption; and pesticide consumption (29.89 kg $ha^{-1}$ $a^{-1}$) is fivefold. A study on heavy metal levels in an anaerobic rice plot showed that, apart from the residual fraction for certain metals, the major metal fraction is precipitated in sulphide forms due to high sulphate contents of the flooding water. Therefore, the sulphate reduction rates in the soil are expected to be very high. The mobilization of accumulated heavy metals like cadmium, copper, lead and zinc can be triggered by an increase in redox potential due to changing land use. In the Albufera Natural Park there has been a continuous change in land use by the drying and draining of this area in order to adapt it to new types of crops (mainly orchards and citrus crops; Boluda et al., unpubl.).

In the Russian Plain chernozems are drastically transformed by rice cultivation. Inundation changes the redox regime and the chemical properties of the soils. In addition, the inundation may exacerbate compaction problems and may induce humus loss (Karavayeva et al. 1991).

In the Nordic region, peat areas are increasingly being drained for the planting of harvestable forests, resulting in oxidation of the peat, and loss of buffering and retention capacities. It has been observed that tillage of forest soils, a common measure in, for instance, Finland in the regeneration of clear-cut forests, caused a changed uptake in plants: elevated levels of Al, Mn and Hg were noted in bilberry (*Vaccinium myrtillus*) leaves growing on tilled areas; however, this was not the case for Fe, Cu and Cd (Nuorteva 1990). A similar phenomenon might occur in the change of forest into agricultural soil.

One of the results of the planned economic activity in the former "Post-Eastern Block" was the formation of large farming units comprising marginal soils (Várallyay 1994). An example is the catchment area of Ladoga Lake, close to St Petersburg. The Ladoga Lake watershed is covered with forests on podzol soils with a low pH (3.8–4.5), and organic content and marshes with pH levels from 2.6 to 4.2. The intensive deforestation and drying of marshes with resulting oxidation of organic matter, is considered the primary source of river-flow input of some elements into Lake Ladoga, comprising up to 70–80% for Cu, Fe and Zn. Efforts undertaken since the 1960s to improve agricultural capacities of the soils by intensive liming resulted in a decrease in the proportion of acid soils, but due to the severe decrease in lime use since 1990 rapid acidification of cultivated soils has taken place (Ikonnikov 1993).

Deforestation, reported from Romania, Bulgaria and the Ukraine causes increased erosion, especially on the Bulgarian banks of the Danube. Soils, washed away by erosion, have a serious impact on reservoirs in front of hydraulic structures like the recently built 69 dams in the Danube. Sediment accumulates in the reservoirs behind dams and other hydraulic structures, and reveals elevated concentration levels in heavy metals and agrochemicals. Breakage of the dams (for instance due to earth quakes) can cause a sudden release of the polluted sediment.

The construction of the dams also has implications for the wetlands on the Danube. Drainage of the wetlands (for agricultural reasons or due to the dam constructions) will lead to decreased filter functions and leaching of the accumulated contaminants (Haskoning 1994).

### 7.4.3 Intensive Agriculture

Agriculture and deforestation are the most important causative factors of soil degradation in Europe accounting for 38 and 29% respectively of the total degraded area (Oldeman et al. 1991). These factors have been affecting soils for a long time: as early as 590 BC in Greece the first environmental law was proclaimed to protect the soil on the steep slopes against soil loss due to cultivation (Amezaga and Hearns 1991). Agriculture and deforestation are the prime cause of soil erosion, have a major effect on compaction, waterlogging, loss of nutrients, and on contamination by pesticides, chemical fertilizers and manure. Agriculture-induced degradation occurs chiefly in western and northern Europe.

**Phosphate Saturation.** The Netherlands has the highest fertilizer application in Europe. This has resulted in phosphate-saturated soils in the sandy soils in the central, eastern and southern regions where animal waste surpluses from intensive lifestock farming are deposited. The region comprises about half of the Dutch sandy area (293 000 ha). Assuming current phosphate application rates, phosphate leaching is a distinct possibility over a time horizon of several decades to a century. The consequence is a time-delayed increase of eutrophication of surface waters. Vulnerable areas are especially those soils with high inputs, low phosphorus sorption capacity, high water-tables and low retention time of water. In the EC these may also be expected in Belgium (Flanders), Germany (Vechta region) and Italy (the Po region; Breeuwsma and Reijerink 1993). Increased phosphate leaching might be induced by raising water-tables or acidification as phosphate storage capacities appear to correlate well with iron and aluminium (hydr)oxide contents of soils. High groundwater levels explain why a considerable grassland area (43%) in the Netherlands is phosphate saturated, despite relatively low phosphate loads (Salomons and Stol 1992).

**Cadmium Saturation.** To assess the storage for cadmium as well as the potential of leaching, cadmium loads in agricultural soils in the European Community have been compared with the soil sensitivity to cadmium accumulation (Fraters and Van Beurden 1993). The cadmium-binding capacity was estimated using the soil acidity map, the soil organic matter map and the soil texture map. Soil pH was considered to be the most important parameter. The total loads for cadmium have been defined as the sum of cadmium applied with artificial fertilizer, animal manure and sewage sludge, and the load due to atmospheric deposition, minus the fraction removed due to harvesting and grazing. A distinction was made for arable land and grassland.

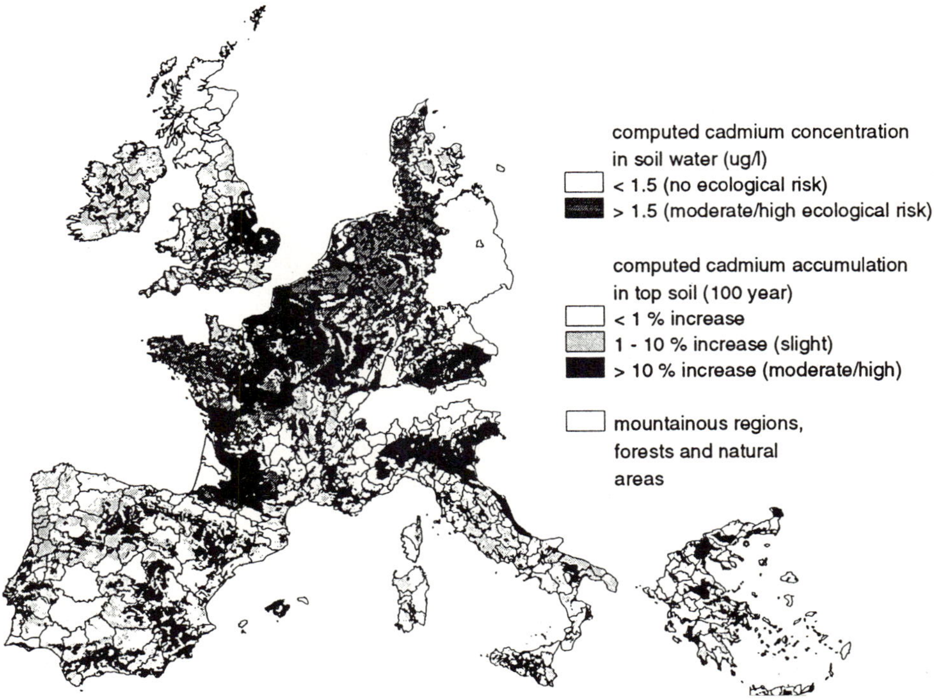


**Fig. 7.12.** Accumulation of cadmium in arable topsoils in the European Communities (Fraters and Van Beurden 1993)

The most sensitive soils for accumulation are found mainly in the eastern part of Spain, northern and south-eastern France, north-eastern and western Italy, considerable parts of Greece, and small areas in the United Kingdom and the Netherlands.

The cadmium accumulation rate (Fig. 7.12) shows continuing storage of chemicals in almost the whole of the EC. The affected areas are very relevant in respect of non-linear effects: in the regions where accumulation is combined with exceedance of critical acid loads in the natural soils, changing from limed agriculture into forest, for instance, may lead to mobilization. Similarly, in southern Spain, salinization in cadmium accumulation regions may lead to problems.

**Organic Matter Breakdown and Nitrate Leaching.** Accumulation of nitrogen occurs mostly by incorporation in organic material, such as humus. Nitrate is hardly adsorbed in soils, therefore it can leach out quickly, especially when groundwater levels and/or precipitation are high. Nitrate leaching will be more prominent in light soils, due to smaller water-holding capacities and lower denitrification rates. Anaerobic conditions are a prerequisite for denitrification and can be caused by water and the presence of easily degradable organic material. Denitrification and humus breakdown are higher at higher temperatures.

The central chernozem and Volga regions have been extensively farmed for a long period of time. This has resulted in decreases in humus content, mostly in the eastern part. In the Volga and Ural regions humus content has decreased from 13–16% to 7–10%, and even down to 4–7%. Under irrigation, the chernozem soils become increasingly prone to salinization, alkalinization and compaction (Karavayeva et al. 1991).

As many contaminants can be adsorbed by organic material, breakdown of organic matter may not only lead to increased leaching of nitrogen, but also to leaching of other contaminants. Copper sulphate is widely applied in viniculture as a pesticide to combat mildew. On non-irrigated land it is retained in the soil as a result of the high cation exchange capacity, favoured by a high organic matter content. This is particularly important in highly calcareous and silty Mediterranean soils that rely on organic matter for nutrient retention and soil structural stability. Desertification and reducing soil organic levels lower the storage capacity of the soil, and may release, mobilize and erode copper into colluvium and alluvium where it is concentrated (Imeson and ter Meulen 1991).

### 7.4.4 Climate Change

Northern Europe is expected to experience the greatest warming (>5°C at a doubling of $CO_2$) with increased soil wetness, which will lead to massive mineralization of litter and subsequent nitrogen leaching (Stigliani and Salomons 1991). This phenomenon will also take place in the highly organic chernozem and chestnut soils. Organic matter loss reduces CEC, and may cause mobilization of pesticides, organic micropollutants and heavy metals stored as a result of agricultural practices and deposition. It will also lead to increased vulnerability to erosion, which may be in turn be enhanced by climate-change-driven increases in extreme thunderstorm events. Both phenomena will occur mostly in erosion-sensitive areas with low organic matter content, such as the Mediterranean region. The Mediterranean region will also suffer increases in incidents of fire. Forest fire causes deterioration of physical structure, decrease in organic matter, loss of nutrients in smoke and erosion of ash by wind and water. Post-fire erosion and transport has the effect of mobilizing and concentrating the colloidal sediments that retain chemicals.

A Polish study suggests that the net results of predicted climate changes would speed up the hydrological cycle in the freshwater wetlands in the investigated area. In other areas, conditions might be reversed, with increased drought and disappearance of wetlands (Markham et al. 1993). Peat bogs exist in a delicate balance with precipitation and evapotranspiration. Climate changes will lead to desiccation and changes in composition of many peat bogs resulting from decreased precipitation and increasing salinity due to sea level rise (especially in the Netherlands; Markham et al. 1993). Resulting degradation of peat may lead to enhanced mobility of contaminants.

Forest soils in Sweden contain high amounts of mercury, especially from pesticides and industrial emissions. Although mercury loading was drastically reduced since the 1970s, it has been estimated that at the end of the 1980s fish in over 10 000 Swedish lakes had muscle wet weight mercury concentrations over 1 mg/kg (SEPA 1991). Moreover, in areas studied the average methyl mercury transport to the lakes only forms less than 0.2% of the total methyl mercury store in soils. Mercury is bound mainly to organic matter. Higher temperatures due to climate change may decrease the amount of organic matter in the soils, and hence may cause a sudden release of high amounts of mercury, which can be transformed into mobile methyl mercury (Lee et al. 1993). Toxic implications of methyl mercury have been shown in the Minamata disease in Japan, where the consumption of methyl mercury polluted fish had lethal consequences for a fishing community.

Decrease in frost may replace the spring acid pulse by melting acid snow through a series of more frequent acid pulses throughout the winter, or by chronic acid water all winter (Markham et al. 1993). In the tundra permafrost areas will be reduced and permafrost levels in the soils will retreat to deeper layers. Permafrost determines the active and drainable soil cover and thawing will increase soil reduction and leaching (Loch et al. 1993). Furthermore, deepening of the permafrost level may lead to delayed mobilization of chemicals stored in the more or less closed depression on the permafrost front under towns, resulting from the towns' heat production. These so-called kryopegs contain wastes from sewage, street dust and industrial pollution (Nekrasov 1991).

### 7.4.5 Delayed Effects of Past and Present Mining Activities

Mining activities the wastes generated can lead to pollution processes, not only directly from the contaminants, but also as a result of changes in storage properties like pH or redox potential. One of the major side effects of mining are the changes in hydrogeological systems (Rybicka 1994). Mining exploitation causes drainage of the surrounding areas by dewatering of mines and excessive salinity of surface flows by waters hoisted from mining works. On the other hand, it may cause progressive lowering of the terrain and flooding of large areas. In Katowice Province 690 ha of the land are permanently flooded and 2733 ha are in the range of local "depression cones" where groundwater tables are considerably lowered. For the sake of three lignite mines in North-Rhine Westphalia $1.2 \times 10^9$ $m^3$ of groundwater is pumped into the Rhine yearly. This kind of actions may greatly affect the redox status in the area, and may cause a change in storage capacities of the subsoil, for instance, due to effects on pyritic layers.

One of the most dangerous hazards of mining is the pollution of surface and groundwater by soluble solids leached from the wastes due to the formation of highly acid waters ($pH<4$) resulting from sulphide oxidation. This appears to be a common process. The leaching time of sulphides from fine-grained, oxygenated spoils is estimated to be about 11 years on average. In dumps, this process is much slower due to the coarse fraction of spoils and limited volumes of meteoric waters available. Some spoil dumps can be the persistent sources of pollution containing products of sulphide oxidation, which may affect the environment for dozens of years (Rybicka 1994). Examples are the waste rocks of sulphide-containing coal fields in the Preurals and around Moscow. Sulphuric acid and Fe, Mn and Al sulphates penetrate into the surroundings with the seepage waters of the mine heaps ($pH < 2$) or with surface run-off. Due to disturbance by mining activities, large areas adjacent to the mine heaps are composed of loose layers of varying depth and/or composition overlaying natural soils. The soils are inhomogeneous and consist of areas which may differ in redox potential. In this kind of area, the migrating sulphates are transformed into secondary sulphides, stable in conditions with low redox potential. Hence numerous "secondary storage areas" with a high intensity of toxic pollutants are formed as a kind of islands sometimes far from the waste heaps. Although these islands are rather stable, natural or artificial drainage with subsequent lowering of the groundwater table can cause remobilization, further dispersion by rivers, and storage at river estuaries or artificial reservoirs. An example is the Kama reservoir, which is polluted by the Kizelovskyi coal field in one of the main tributaries of the Volga (Solntseva and Gerasimova 1993; Solntseva 1994).

Alkalinization due to mining activities is another potential for non-linear releases. north-eastern Estonia contains the largest commercially exploited oil shale deposit of the world. Estonia's fuel consumption is mostly covered by this locally mined, low-grade fuel, 90% fired in power plants and 10% processed into oil and chemical products (Nurste 1991). The combustion of pulverized oil shale causes high levels of pollutant emissions, such as tin, arsenic, and to a lesser extent, uranium and cadmium. In contrast to wet depositions in Europe, due to contact with alkaline fly ash, rain and snow turn alkaline (pH changes from 8.5 near power stations to 6–7 further away). High Ca and Mg levels result in extremely alkaline ash plateaus. The water leaching from the stored ash has a pH of over 12 and is polluted by phenols and heavy metals (Vallner and Sepp 1993). In the vicinity of two big thermal power plants in the Narva region the pollution load for sulphates extends up to 125 kg $ha^{-1}$ $a^{-1}$ and for nitrogen up to 40–50 kg $ha^{-1}$ $a^{-1}$. But in this area large basic depositions of limestone and concrete dust (from concrete production complexes) and oil shale ash from the power plants also occur. These million tons of alkaline material per year can neutralize acid depositions and cause changes in some natural ecosystems like high moors and coniferous forests on podzolic soils. For most of agricultural areas such kinds of alkaline depositions avoid uptake of heavy metals (the lead content in some soils of this area reaches 500 mg $kg^{-1}$) in plants or organisms. Arsenic, however, can be mobilized.

## 7.5 Summary

Stored pollutants and finite storage capacities of soils are present in Europe everywhere. Storage capacities can be exceeded by continuous input, or by manipulation of the land. When comparing pollution concentrations with storage capacities of soils, it appears that the main pollution sources are often situated in or in the neighbourhood of vulnerable soils. This is most obvious for the poor soils in the Netherlands (the Rhine Delta) and in the Black Triangle.

When comparing regions in Europe, two trends are the most observable:

- An east–west trend, which is mainly anthropogenic: point sources of pollution can be much higher in CEE, but the diffuse loading is generally lower than in western Europe. This might be connected to lower population densities. In CEE especially wastes are a large problem, due to their nature, magnitude and the processing potentials. The use of lignite in the CEE is another difference between East and West. A soil-connected difference is found in the occurrence of the soils with high organic content, the chernozems and kastanozoms in CCE.
- A north–south trend is mainly related to climate and corresponding soil types. Problems connected to high evapotranspiration/precipitation rates, erosion and salinization/alkalinization, are mostly found in the south and south-eastern part of Europe. Acidification is most prominent in the industrialized and densely populated central-western part of Europe; loads in the North and South are lower, moreover, the soils in the south have much more buffer capacity due to their calcareous character.

Land-use changes occur all over Europe. Wetlands are endangered everywhere, be it by drainage (Finland), dams (Hungary), or climate change. Most land-use changes in Europe, however, are connected to changes into or out of agricultural use. In all of Europe the trend of taking agricultural soils out of production is prevalent. The fact that these manipulations can cause exceedances of storage capacities of soils, indicates the urgency of a long-term view on how to manage these moderately to lightly contaminated areas in order to achieve their sustainable use.

It appears that long-term risk assessment for chemicals in soils is a complicated process; it should not only include loads, loadings and storage capacity of soils, but also should take into account linkage with biogeochemical cycles. Changes in these cycles, be it due to natural changes, changes in loading or changes in management, can influence the CCP determining the behaviour of chemicals in soils. Politics can not be excluded from the risk assessment, as they have major influence on loadings and management. Therefore the idea of "sustainable development", as defined by the Brundtland commission, is not just a idealistic concept, but very realistic within the framework of future soil potentials.

## References

Amezaga JM, Hearns G (1991) Chemical time bombs in the Mediterranean region. Foundation for ecodevelopment, Hoofddorp, the Netherlands

Boels D, Fleming G (1993) Chemical time bombs from landfills: appraisal and modelling. Land Degradation and Rehabilitation 4:399-407

Bollen J, Hettelingh JP, Maas R (1993) Scenarios for economy and environment in Central and Eastern Europe. Document prepared for the Worldbank, RIVM, the Netherlands

Boluda R, Gimeno-Garcia E, Sanchez J (unpublished) An example of a "hot spot" for a chemical time bomb: Albufera Natural Park (Valencia, Spain). Preliminary studies

Breeuwsma A, Reijerink JGA (1993) Phosphate-saturated soils, a "new" environmental issue. In: Meulen GRB ter, Stigliani WM, Salomons W, Bridges EM, Imeson AC (eds) Chemical time bombs. Proceedings of the European state-of-the-art conference on delayed effects of chemicals in soils and sediments, Veldhoven, the Netherlands, 2-5 Sept 1992. Foundation for Ecodevelopment, Hoofddorp, the Netherlands, pp 79-86

Csikós I (1991) Long term environmental risks for soils, groundwater and sediments in the Danube catchment area. Report of the workshop, Budapest, December 1990. Foundation for Ecodevelopment, Hoofddorp, the Netherlands

Diercke Weltatlas (1988) Georg Westermann Verlag

FAO-UNESCO (1981) Soil map of the world 1:5 000 000 vol V Europe, UNESCO, Paris

Fleming G (1992) Landfills and contaminated land: time delayed responses. Foundation for Ecodevelopment, Hoofddorp, the Netherlands

Fraters D, Beurden AUCJ van (1993) Cadmium mobility and accumulation on soils of the European communities, RIVM-report 481505005. RIVM, Bilthoven, the Netherlands

Fraters D, Bouwman AF, Thewessen TJM (1993) Estimates of soil organic matter content of the topsoil of FAO-Unesco soil units, RIVM report 481505004. RIVM, Bilthoven, the Netherlands

Glazovskaya MA, Gerasimova MI, Kremlenkova NP, Bogdanova MD (1991) Mapping of soil vulnerability to acidification and fluorine accumulation. In: Batjes NH, Bridges EM (eds) Mapping of soil and terrain vulnerability to specified chemical compounds in Europe at a scale of 1:5 M. Proceedings of an International Workshop held at Wageningen, the Netherlands, 20-23 March 1991. ISRIC, Wageningen

Grenon M, Batisse M (eds) (1989) Futures for the Mediterranean basin. The blue plan. Oxford University Press, Oxford

Haskoning (1994) Environmental programme for the Danube river basin; Danube integrated environmental study. Report phase 1

Hekstra GP (1991) Climate change and land-use impact in Europe. In: Brouwer FM, Thomas AJ, Chadwick MJ (eds) Land use changes in Europe, processes of change, environmental transformations and future patterns. Kluwer Academic Publishers, Dordrecht, pp 351-377

Hettelingh JP, Downing RJ, Smet PAM de (1991) Mapping critical loads for Europe.RIVM, Coordination Centre for Effects, Bilthoven, the Netherlands. Technical report no 1

Ikonnikov V (1993) Ladoga lake catchment area and gulf of Finland sediment pollution: influencing factors and processes. Land Degradation and Rehabilitation. Vol 4 no 4:297-307

Imeson AC, Meulen-Smidt GRB ter (1991) Developing strategies to manage the effects of environmental change and desertification on sudden and delayed changes in the buffering and storage capacities of Mediterranean soils. Paper prepared for the European Symposium on Desertification in the Mediterranean Area, EC-EPOCH, Florence 24-26 June 1991

Jaarsveld JA van Atmospheric deposition of cadmium, copper, lead, benzo(a)pyrene and lindane over Europe and its surrounding marine areas. RIVM, Bilthoven, the Netherlands. In press

Karavayeva NA, Nefedova TG, Targulian VO (1991) Historical land use changes and soil degradation on the Russian Plain. In: Brouwer FM, Thomas AJ, Chadwick MJ (eds) Land use changes in Europe, processes of change, environmental transformations and future patterns. Kluwer Academic Publishers, Dordrecht, pp 351-377

Kasimov NS, Moiseyenkov OV (1994) Agents of pollutants behaviour in urban environment. In: Gerasimova MI (ed) Report of the international workshop on long term environmental risks for soils, sediments and groundwater in the Volga catchment area, Moscow, 23-27 May 1992. Foundation for Ecodevelopment, Hoofddorp, the Netherlands

Knoop J, Traas ThP (1993) Acidification, changing land use, and cadmium mobilization: a modelling approach. In: Meulen GRB ter, Stigliani WM, Salomons W, Bridges EM, Imeson AC (eds) Chemical time bombs. Proceedings of the European state-of-the-art conference on delayed effects of chemicals in soils and sediments, Veldhoven, the Netherlands, 2-5 September 1992. Foundation for Ecodevelopment, Hoofddorp, the Netherlands, pp 107-118

Konsten CJM (1991) Report of the international workshop on delayed environmental effects of chemical pollution with focus on the Nordic countries, Uppsala, Sweden, September 14,15, 1991. Foundation for Ecodevelopment, Hoofddorp, the Netherlands

Konsten CJM, Meulen-Smidt GRB ter, Stigliani WM, Salomons W, Eijsackers H (1993) Summary of the workshop on delayed effects of chemicals in soils and sediments (chemical time bombs), with emphasis on the Scandinavian region. Applied Geochemistry, Suppl Issue no 2, pp 295-299

Krosshavn M, Steinnes E, Varskog P (1993) Binding of Cd, Cu, Pb and Zn in soils organic matter with different vegetational background. Water, Air and Soil Pollution 71:185-193

Lee, Y-H, Borg, G, Iverfeldt, Å, Hultberg, H (1993) Fluxes and turnover of methylmercury: mercury pools in forest soils. In: Watras and Huckabee (eds) Mercury as a global pollutant – toward integration and synthesis. Lewis Publ, Chelsea, Michigan, in print

Loch JPG, Berg GA van den, Weijden CH van der (1993) The stability of Fe-, Mn, and Al-(hydr)oxides and their binding capacity for contaminants in changing soil and sediment environments. In: Meulen GRB ter, Stigliani WM, Salomons W, Bridges EM, Imeson AC (eds) Chemical time bombs. Proceedings of the European state-of-the-art conference on delayed effects of chemicals in soils and sediments, Veldhoven, the Netherlands, 2-5 September 1992. Foundation for Ecodevelopment, Hoofddorp, the Netherlands, pp 45-52

Lynden GWJ van (1993) Handbook of soil conservation in Europe. Council of Europe, ISRIC, Wageningen, the Netherlands

Markham A, Dudley N, Stolton S (1993) Some like it hot. WWF International, CH-1196 Gland, Switzerland

Mayer R (1993) Chemical time bombs related to forestry practice: distribution and behaviour of pollutants on forest soils. Land Degradation and Rehabilitation 4:275-281

Meissner R, Seeger J, Rupp H, Schonert P (1993) Der Einfluss von Flächenstillegung und Extensivierung auf den Stickstoffaustrag mit dem Sickerwasser. Vom Wasser 81, 197-215

Meulen GRB ter, Stigliani WM, Salomons W, Bridges EM, Imeson AC (eds) (1993) Chemical time bombs. Proceedings of the European state-of-the-art conference on delayed effects of chemicals in soils and sediments, Veldhoven, the Netherlands, 2-5 September 1992. Foundation for Ecodevelopment, Hoofddorp, the Netherlands

Nekrasov IA (1991) Vechna li vechnaya merzlota? Nedra, Moscow, 128 pp

Nuorteva P (1990) Metal distribution patterns and forest decline: seeking Achilles' heel for metals in Finnish forest biocoenoses. Publ Dept Environ Conserv, Univ of Helsinki, no 11, Helsinki

Nurste H (1991) Estonian energy situation and pollutant emission. In: Abstracts of the 3rd International conference on system analysis approach to environment, energy, and natural resource management in the Baltic region,session 1, Copenhagen, Denmark, May 7-10 1991.

Oldeman LR (1993) An international methodology for an assessment of soil degradation and georeferenced soils and terrain database. Paper presented for the "Third Expert Consultation of the Asian Network of Problem Soils", Bangkok, Thailand, 25-29 October 1993

Oldeman LR, Hakkeling RTA, Sombroek WG (1991) World map of the status of human-induced soil degradation; an explanatory note, 2nd revised edn. GLASOD, ISRIC, Wageningen and UNEP, Nairobi

Răută C, Cârstea S, Mihăilescu A (1987) Influence of some pollutants on agricultural soil in Romania. Arch Ochrony Srod 1-2:33-37

Renberg I, Persson MW, Emteryd O (1994) Pre-industrial atmospheric lead contamination detected in Swedish lake sediments. Nature 368:323-326

RIVM (1992) Milieudiagnose 1991 I. Integrale rapportage lucht-, bodem- en grondwaterkwaliteit. "pilot studie", RIVM Rep 724801004, Bilthoven, the Netherlands

RIVM, RIZA (1991) Sustainable use of groundwater. RIVM/RIZA Rep 600025001, RIVM Bilthoven, the Netherlands

RIVM, ISRIC, RIZA, DGW, WHO (1992) The environment in Europe: a global perspective. RIVM Rep 481505001, Bilthoven, the Netherlands

RIVM (1994) Nationale milieuverkenning 3, 1993-2015. Samson HD, Tjeenk Willink bv Alphen aan den Rijn, the Netherlands

Rybicka EH (1994) Environmental impact of mining industry in Poland. In: Salomons W, Mader P, Förstner U (eds) Heavy metals, problems and solutions. Springer, Berlin Heidelberg New York

Salomons W, Stol BEM (1992) Diffuse bronnen in de landbouw. IB-DLO & DLO, Haren, the Netherlands

Sanchez Diaz H (1993) Soil use in the Mediterranean basin and its relation to chemical time bombs. In: Meulen GBR ter, Stigliani WM, Salomons W, Bridges EM, Imeson AC (eds) Chemical time bombs.Proc Eur State-of-the-art Conf on Delayed effects of chemicals in soils and sediments, Veldhoven, the Netherlands, 2-5 Sept 1992. Foundation for Ecodevelopment, Hoofddorp, the Netherlands, pp 147-156

SEPA-informs (1991) 'Mercury in the environment, problems and remedial measures in Sweden.' Tryckeri Balder AB, Stockholm

Smal H (1993) Some environmental conditions and chemical time bombs hazards in Poland. In: Soil degradation and land rehabilitation, 4:269-275

Smal H, Salomons W (1993) Heavy metals and acidification in Poland. In: Meulen GRB ter, Stigliani WM, Salomons W, Bridges EM, Imeson AC (eds) Chemical time bombs. Proceedings of the European state-of-the-art conference on delayed effects of

Smal H, Salomons W (1993) Heavy metals and acidification in Poland. In: Meulen GRB ter, Stigliani WM, Salomons W, Bridges EM, Imeson AC (eds) Chemical time bombs. Proceedings of the European state-of-the-art conference on delayed effects of chemicals in soils and sediments, Veldhoven, the Netherlands, 2-5 September 1992. Foundation for Ecodevelopment, Hoofddorp, the Netherlands, pp 97-106

Solntseva NP (1994) CTBs in the coalfields of the Volga catchment area. In: Gerasimova MA (ed) Report of the international workshop on long term environmental risks for soils, sediments and groundwater in the Volga catchment area (Chemical time bombs in the Volga basin). Moscow, 23-27 May 1992

Solntseva NP, Gerasimova M (1993) Sediment-derived CTBs in coal mining areas: "fusion" processes and interaction with the environment. In: Meulen GRB ter, Stigliani WM, Salomons W, Bridges EM, Imeson AC (eds) Chemical time bombs. Proceedings of the European state-of-the-art conference on delayed effects of chemicals in soils and sediments, Veldhoven, the Netherlands, 2-5 September 1992. Foundation for Ecodevelopment, Hoofddorp, the Netherlands, pp 163-166

Stigliani B, Salomons W (1991) Pollutants and some not impossible environmental problems caused by climate change. Foundation for Ecodevelopment, Hoofddorp, the Netherlands

Stigliani WM (1994) A proposed research agenda for sustainable use of materials and long-term maintenance of soil quality. Foundation for Ecodevelopment, Hoofddorp, the Netherlands

Stigliani WM, Anderberg S (1991) Industrial metabolism and the Rhine basin. Options, Sept 1991, IIASA

Swift MJ, Heal OW, Anderson JM (1979) Decomposition in terrestrial ecosystems. Studies in ecology, vol 5. Blackwell Scientific Publications, Oxford

Szabolz I (1991) Salinization potential of European soils. In: Brouwer FM, Thomas AJ, Chadwick MJ (eds) Land use changes in Europe, processes of change, environmental transformations and future patterns. Kluwer Academic Publishers, Dordrecht, pp 293-315

Turski R, Glinski J (1991) Long term environmental risks for soils, sediments and groundwater in the eastern part of the Baltic sea catchment area. In: Krutikow M (ed) International workshop on long-term environmental risks for soils, sediments and groundwater in the Baltic catchment area: contributions. Foundation for Ecodevelopment, Hoofddorp, the Netherlands

Vallner L, Sepp K (1993) Oil shale mining pollution effects in Estonia. Soil Degradation and Land Rehabilitation 4:381-387

Várallyay Gy (1994) Soil management and environmental relationships in Central and Eastern Europe. Trans 15th World Congr of Soil Science, Acapulco, Mexico July 10-16, 1994. 7a:34-58

Vries W de, Posch M, Reinds GJ, Kämäri J (1993) Critical loads and their exceedance on forest soils in Europe. Report 58, revised version SC-DLO, WRI, Wageningen, the Netherlands

Vries W de, Reinds GJ, Posch M (1994) Assessment of critical loads and their exceedance on European forests using a one-layer steady-state model. Water, Air Soil Pollution 72: 357-394

Whitmore AP, Bradbury NJ, Johnson PA (1992) Potential contribution of ploughed grassland to nitrate leaching. Agriculture, Ecosystems and Environment 39:221-233

# 8 Capacity Controlling Parameters and Their Impact on Metal Toxicity in Soil Invertebrates

*C. A. M. van Gestel, M. C. J. Rademaker, and N. M. van Straalen*

## 8.1 Introduction

The behaviour of a chemical in a soil or sediment depends on its physical-chemical properties as well as on the environmental conditions prevalent in its immediate surroundings. In soils and sediments, contaminant behaviour may be determined by the content of organic matter, clay, and hydroxides of aluminium and iron and by pH (Alleyway 1990; Bolt and Bruggenwert 1976; Van Riemsdijk and Hiemstra 1993; Yong et al. 1992). These properties determine the storage capacity of soils for heavy metals and other pollutants, and are therefore called Capacity Controlling Parameters (CCP) (ter Meulen et al. 1993). The same properties are believed to determine bioavailability of contaminants for organisms living in soil or sediment.

Changed land use may lead to considerable changes in the properties of soil and sediment. Due to acidification, the pH of terrestrial and aquatic ecosystems has shown a decrease over the last decades. The reclamation of sea or lake sediment to land leads to a change from anaerobic to aerobic conditions. And the abandonment of agricultural land may lead to losses of organic matter and a decrease in soil pH. All such changes may significantly affect the behaviour and bioavailability of contaminants in soil and sediments. For a proper assessment of the long-term risks of contaminants in soils and sediments, a quantification of the contribution of the different CCP is needed.

Ecotoxicological data for terrestrial and aquatic organisms form the basis for the risk assessment of chemicals in terrestrial and aquatic ecosystems. To derive "safe" levels of chemicals in these environmental compartments, statistical or other more straightforward methods are applied (Aldenberg and Slob 1993;

US-EPA 1984; OECD 1992; Suter 1993; Van Straalen and Denneman 1989; Wagner and Løkke 1991). The bioavailability of chemicals in soil and sediment will, however, also depend on the physico-chemical properties (CCP) of these compartments. Attempts are therefore made to include the influence of these CCP in the risk assessment procedure (Denneman and Van Gestel 1990; Van de Meent et al. 1990; MILBOWA 1991).

The aim of Chapter 8 is to give an overview of the present knowledge of the influence of CCP on the bioavailability of chemicals for soil organisms. The description will be restricted to soil invertebrates. For most non-charged or non-dissociated organic chemicals, soil organic matter content is the main CCP affecting behaviour, and sufficient evidence is available to predict changes in toxicity due to changes in organic matter content (Van Gestel, 1992). For heavy metals, many more factors influence behaviour and their role in determining bioavailability is difficult to predict (Van Riemsdijk and Hiemstra 1993; Yong et al. 1992). In this chapter, the emphasis will therefore be put on heavy metals as contaminants.

## 8.2 Toxicity Tests and Routes of Exposure in Soil

### 8.2.1 Toxicity Tests

Toxicity tests in soil are usually carried out in the laboratory. An overview of test methods with different soil organisms, most of which are still in development, can be found in Van Straalen and Van Gestel (1993) and Van Gestel and Van Straalen (1994).

In test methods described in the literature, a great variety of test substrates are used. Besides natural and artificial soils, a number of artificial substrates, such as quartz sand, wet filter paper, silica gel, agar, water or nutrient solutions may be used. The relevancy of the latter substrates to natural soils is limited, however, as it is difficult to translate results obtained in such substrates to effect levels in soil. For the purpose of this chapter, and also from the point of view of ecological realism, only tests in real or artificial soils are relevant. For the sake of standardization, and to allow for a comparison of tests with different organisms, the use of artificial or standardized natural soils is recommended. An example of such a standardized soil is the artificial soil used for earthworm and Collembola toxicity testing. This soil is composed in the laboratory by mixing quartz-sand (70%), kaolin clay (20%) and *Sphagnum* peat (1 mm sieved, 10%), and adjusted to a pH of 6.0 by adding some calcium carbonate powder (OECD 1984).

In toxicity tests, usually a range of test concentrations is prepared by mixing different amounts of the test chemical through portions of soil. Also a non-treated control is included. Organisms, preferably cultured in the laboratory and having the same age and size, are exposed to the different test concentrations under standardized laboratory conditions. Test conditions are chosen which are about optimal for the test organism used. During and/or at the end of the exposure period, effects on survival, growth and/or reproduction are determined.

For effects on survival, $LC_{50}$ is used, indicating the concentration causing 50% mortality among the test organisms. For expressing sublethal effects, $EC_{50}$ or NOEC (no-observed effect concentration) may be used, indicating the concentration at which the measured parameter is 50% inhibited compared to the control and the NOEC, respectively. $LC_{50}$ and $EC_{50}$ values can be calculated from the concentration-response relationship obtained, whereas the NOEC can be derived by comparing the performance (e.g. growth or reproduction) at the different test concentrations with that in the control group. For calculating $LC_{50}$ and $EC_{50}$, several probit and logistic models are available; for estimating NOEC, normal statistical tests (e.g. Students t test, Williams test, Dunnetts test) are applied.

### 8.2.2 Routes of Exposure

Organisms can be exposed to pollutants present in the soil by two major routes:

1. Ingestion and oral uptake of soil particles
2. Dermal uptake of pollutants from the soil pore water

Also exposure by inhalation of pollutants present in the soils' air phase may occur; no data are available on the relevance of this route of exposure compared to the others. For organisms with a firm exoskeleton, such as many arthropods, direct dermal uptake of pollutants from soil may be less important than oral uptake. Everts et al. (1991) however, found that dermal contact of spiders with deltamethrin contaminated soil was an important uptake route. This finding is supported by studies with the collembolan *Folsomia candida* in our laboratory. Much higher concentrations of cadmium had to be added to the food than to the soil to obtain the same internal concentrations in the organisms (Van Gestel, in prep.).

Soft-bodied organisms, living in close contact with the soil, such as earthworms and enchytraeids, will mainly be exposed via soil pore water. For the acute toxicity for earthworms of a number of selected organic chemicals, Van Gestel and Ma (1988, 1990) demonstrated that the equilibrium partitioning concept developed in aquatic ecotoxicology (US EPA 1989) may be valid. This concept is shown in Fig. 8.1. This means that, for earthworms, the toxicity of chemicals in soil is mainly determined bythe pore water concentration, which can be derived from the total concentration by means of sorption data. A similar finding was reported for nematodes (Houx and Aben 1993).

For organic chemicals, at low concentrations, a linear relationship may be found between the concentration in the soil and that in the pore water. For metals, this relationship is not linear and dependent on the properties of soil and pore water: pH, organic matter and clay content, cation exchange capacity and ionic strength of the pore water may all affect sorption behaviour of metals in soil (Van Riemsdijk and Hiemstra 1993). Passive uptake of metals via the skin may occur in earthworms, as is suggested by Streit (1984). It is however, assumed that the uptake of metals from the pore water will also be affected by the properties of the pore water. This was shown by Kiewiet and Ma (1991), who found that the uptake of lead and cadmium from water by the earthworm *Lumbricus rubellus* was dependent on the pH and calcium concentration of the solution. Nederlof et al. (1993) found a stronger increase of adsorption of copper on yeast cells with increasing pH than for the adsorption on soil particles, resulting in higher copper concentrations in the yeast cells than in the soil. A similar finding was reported for copper in maize roots.

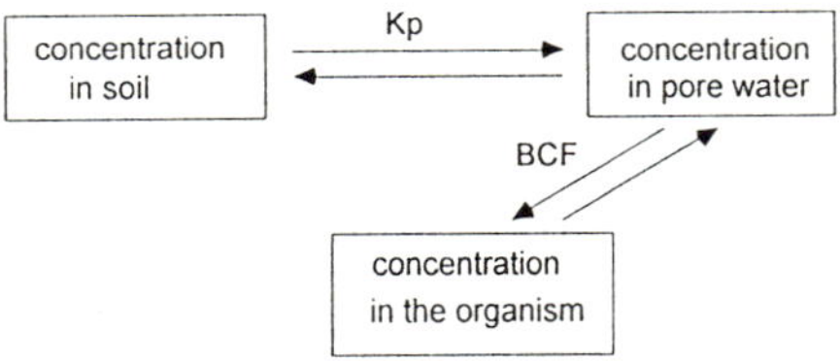


**Fig. 8.1.** Equilibrium partitioning concept: the concentration of chemicals in soil organisms is considered to be related to the concentration in the pore water, which in turn is dependent on the sorption behaviour of the chemicals. *Kp* = adsorption coefficient; *BCF* = bioconcentration factor (= ratio of concentration in organism and concentration in pore water)

Van Gestel et al. (1992a) incubated earthworms (*Eisenia fetida*) for 4 weeks in soils from 19 different nature reserve areas and related metal concentrations in the earthworms to measured pore water concentrations. For cadmium, lead and zinc, earthworm concentrations appeared to increase with increasing pH; for copper no such relationship was found. This finding is in agreement with those of Nederlof et al. (1993) and Kiewiet and Ma (1991).

## 8.3 The Influence of Soil Properties on the Toxicity of Heavy Metals for Soil Invertebrates

The data on the influence of soil properties on the toxicity of heavy metals for soil invertebrates found in the literature, only concern cadmium, lead and copper and are restricted to earthworms and Collembola as test organisms.

### 8.3.1 Cadmium

The toxicity of cadmium for the earthworm *Eisenia andrei* in three soils was influenced by soil pH and organic matter content, with pH having the greatest influence (Van Gestel and Van Dis 1988). Bengtsson et al. (1986) found a decrease of the cocoon production of the earthworm *Dendrobaena rubida* by a factor of 2 when pH of the sandy soil decreased from 6.5 to 4.5. Hatchability of cocoons, however, also decreased with decreasing pH and seemed to be higher upon exposure to 100 mg Cd/kg in the soil. These studies do not however, allow for a quantification of the effect of pH or organic matter content on toxicity.

Crommentuijn et al. (in prep.) studied the influence of pH and organic matter content on the toxicity of cadmium for the Collembola *Folsomia candida* using artificial soils composed of quartz sand, kaolin clay and *Sphagnum* peat. By using different amounts of *Sphagnum* peat, the organic matter content of these soils was varied; by adding different amounts of calcium carbonate, soils with different pH were obtained. Uptake of cadmium by the Collembola decreased with increasing pH and organic matter content, and appeared to be related to the water-soluble cadmium fraction in soil. Water-soluble cadmium concentrations could not, however, explain toxicity, probably because the response of the organisms was also affected by other soil properties.

In another study with *Folsomia candida* using both artificial and natural soils, Wohlgemuth et al. (1990) found no correlation of cadmium toxicity with soil pH nor with organic matter content. Toxicity could, however, be related to the cation exchange capacity (CEC) of the soils, which was calculated from clay and organic matter contents. This is shown in Fig. 8.2. From this figure, it can be concluded that a reduction of soil CEC by a factor of 3 may increase cadmium toxicity for *Folsomia candida* by a factor of 3–4.

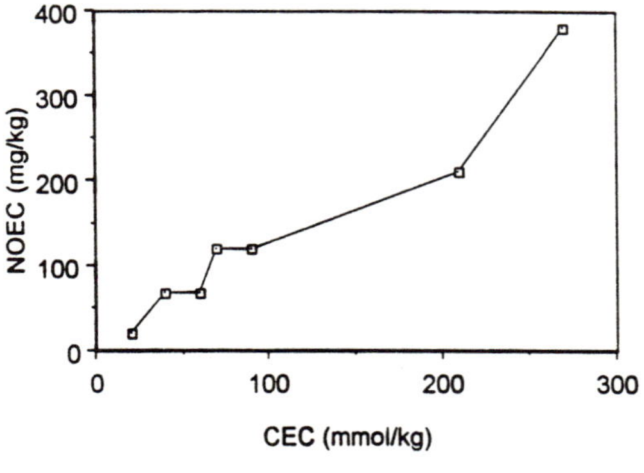


**Fig. 8.2.** Effect of cation exchange capacity (CEC) on the toxicity of Cd for the Collembola *Folsomia candida* in a number of natural and artificial soils. Toxicity is expressed as the NOEC (no-observed effect concentration) for the effect on the number of juveniles produced after a 4-week exposure period (adapted from Wohlgemuth et al. 1990)

### 8.3.2 Copper

Jäggy and Streit (1982) found the following correlation between copper toxicity for the earthworm *Octolasium cyaneum* and soil organic matter content:

$$\log (LC_{50}) = 1.476 + 1.053 * \log(\%OM) \quad (n=3; r=0.9997; p<0.01)$$

This relationship suggests that copper toxicity decreases by a factor of 2 when the organic matter content is doubled. It should be noted, however, that the toxicity tests lasted for only 4 days, which is rather short, and that only three different soils were used in this study.

Ma (1984) found that cocoon production of the earthworm *Lumbricus rubellus* was significantly reduced by 131 mg Cu/kg in a loamy sand (pH 4.8, 5.7% OM, 2% clay) and by 63 mg Cu/kg in a sandy loam (pH 7.3, 3.4% OM, 17% clay). This difference could be attributed to the lower organic matter content in the latter soil, but indirect effects due to the reduced activity of this earthworm in the low organic soil cannot be ruled out.

From another study by Ma (1988), it can be concluded that CCP not only affect bioavailability of copper, but may also affect the sensitivity of earthworms. In the latter study, the effect of copper on cocoon production by *Aporrectodea caliginosa* appeared not to be affected by soil pH, whereas the sensitivity of *Lumbricus rubellus* to copper was increased by a factor of 2 when soil pH was 2 units lower.

Bengtsson et al. (1986) found a reduced survival of the earthworm *Dendrobaena rubida* with decreasing soil pH and cocoon production was halved when pH dropped from 6.5 to 4.5. Extractability of copper with 1 M ammonium acetate was not affected by soil pH, but copper concentrations in the earthworm cocoons were twice as high at pH 6.5 than at pH 5.5. As for cadmium, cocoon hatchability was decreased at low soil pH, but increased upon exposure to 100 mg Cu/kg dry soil. From this study, no conclusions can be drawn on the influence of soil pH on the toxicity of copper.

### 8.3.3 Lead

The only study on the influence of soil properties on lead toxicity for earthworms was carried out by Bengtsson et al. (1986). Extractability of lead with 1 M ammonium acetate increased with decreasing soil pH, but, as in the studies with copper and cadmium described above, no conclusions can be drawn on the influence of soil pH on the toxicity of lead for *Dendrobaena rubida*.

## 8.4 The Influence of Soil Properties on the Bioaccumulation of Heavy Metals in Soil Invertebrates

From the Section 8.3, it is obvious that only few data are available concerning the influence of CCP on the toxicity of heavy metals for soil invertebrates. More data were found on concentrations of heavy metals in these organisms in relation to a CCP. As toxicity will be determined by internal concentrations in the organism, we therefore consider data on this relationship also useful for gaining insight into the possible influence of CCP on the toxicity of heavy metals for soil invertebrates. Furthermore, bioaccumulation may be relevant for organisms feeding on soil invertebrates, and data on the influence of CCP may also be used to predict effects on organisms higher in the food-chain.

For cadmium, chromium, copper, lead, nickel and zinc, literature data were found on the influence of CCP on the concentrations in earthworms. For other organisms only few data were found.

Concentrations of metals in earthworms seem to be determined by a number of factors: food source, structure of digestive system and of the storage and detoxification mechanisms present (Bengtsson and Tranvik 1989). Some metals, such as cadmium and zinc, are known to accumulate in earthworms, whereas others do not (Beyer and Cromartie 1987; Wright and Stringer 1980; Mueller *et al.*, 1993; Bengtsson and Tranvik 1989). Ma (1982) found the following order for uptake of metals in earthworms: Cr<Ni<Pb, Cu<Zn<Cd.

In Section 8.4, a number of studies on the influence of CCP on metal concentrations in earthworms will be discussed. In an attempt to quantify the influence of CCP, relationships determined by different authors and additional data from the literature will be integrated in alternative relationships.

### 8.4.1 Relationships Found in the Literature

In Table 8.1, a number of relationships describing metal concentrations in earthworms as a function of soil concentrations and CCP are shown. First, the studies which have led to these relationships will be described.

Beyer and Cromartie (1987) studied the influence of soil properties on the concentrations of metals in ten earthworm species present in four soils having a different degrees of metal contamination: natural (non-polluted), mine, industrial and near a galvanic factory. A negative correlation was found between cadmium concentrations in the earthworms and soil pH. Earthworm concentrations of lead were also negatively correlated with soil pH, but positively with soil organic matter content. For copper and zinc no such relationships could be found. It should be noted that for the regression analysis, data for the different earthworm species were taken together.

Ma (1982) studied the influence of CCP on the uptake of metals (cadmium, chromium, copper, iron, manganese, nickel, lead en zinc) in the earthworm species *Lumbricus rubellus* in six soils. The pH of these sandy and loamy soils was between 4.7 and 7.1, organic matter content between 2.8 and 12.4% and clay content between 5 and 40%. Soils were treated annually with compost at rates of 0.20 or 40 tonnes/ha. From his results, it appears that BSAF values (BSAF = biota to soil accumulation factor = ratio of concentration in earthworms and soil concentration) are higher on the sandy soils. For cadmium, zinc and lead, BSAF values appeared to be negatively correlated with soil pH, and it appeared that a decrease of soil pH from 6.5 to 4.5 leads to an increase of BSAF with a factor of about 5. For all metals, except for chromium, BSAF was negatively correlated with CEC. And for copper, a negative relationship was found between BSAF and soil organic matter content.

In a field study, Ma (1983) studied the influence of soil pH and organic matter content on the uptake of cadmium, lead, zinc and copper in *Lumbricus rubellus* in a gradient near a zinc smelter. Soil pH varied between 3.5 and 6.1 and organic matter content between 2.2 and 8.6%. pH appeared to have a significant influence on the earthworm concentrations of cadmium, lead and zinc, with higher values being found at low soil pH. For copper, no such pH effect was noticed. Organic matter content influenced the uptake of lead in earthworms, but did not affect the uptake of the other metals.

Morgan and Morgan (1988b) studied the influence of soil properties on the metal concentrations in *Lumbricus rubellus* and *Dendrodrilus rubidus* in 16 soils from old mining areas. A positive linear relationship was found between concentrations in earthworms and those in soil. Only for cadmium BSAF was >1. For copper and zinc, which are essential elements, earthworms seemed to be capable of regulating uptake. Organic matter content and CEC did not have much influence on earthworm concentrations of cadmium, zinc and copper. Only at high organic matter content did copper concentrations in earthworms seem to be lower. Earthworm concentrations of lead appeared to be related to soil pH (and CEC) and also to calcium concentration in soil (Morgan and Morgan 1988a)

From a study of Corp and Morgan (1991) it appears that natural populations of *Lumbricus rubellus* accumulate more cadmium and zinc from seven polluted soils than newly-introduced earthworms of the same species. For lead and copper no such differences were found, and concentrations of these metals in earthworms increased with increasing soil concentrations.

Van Gestel et al. (1992a) studied the influence of CCP on the bioavailability of metals for *Eisenia fetida* in 19 non-polluted soils from nature reserve areas in the Netherlands. After 4 weeks of exposure to these soils, the earthworms were analyzed. Metal concentrations in the earthworms were higher on soils having low organic matter and clay contents and a low pH. In the case of copper and zinc, no relationship with soil properties was found: levels in earthworms of these metals appeared not to be affected by soil pH. For cadmium, concentrations in the earthworms seemed to increase rather than to decrease with increasing soil pH, but this might be an artefact.

In the studies described here, the authors derived relationships between concentrations in the earthworms and CCP. In Table 8.1, these relationships are summarized for cadmium, copper, lead, and zinc. For other metals no such relationships were found. In this table, the goodness of fit of the relationship is indicated with $R^2$, and the significance of the relationship with $p$ (if no value for $p$ is given, the authors did not mention it). When more relationships were given in a study, only the one with the highest $R^2$ is given in Table 8.1.

Some of the relationships depicted in Table 8.1 use log or ln transformed values to increase homogeneity of the data. Application of such a transformation to pH values, which are already log-transformed, does not seem to make much sense. From Table 8.1, it can be concluded that *cadmium* concentrations in earthworms are negatively correlated with soil pH, except for one study in which earthworm concentrations seem to increase with increasing soil pH. In all cases, cadmium concentrations in earthworms seem to increase with decreasing soil organic matter content. Clay content was included in only one study; from this it appears that cadmium concentrations in earthworms increase with decreasing clay content of the soil. From all relationships shown, the impression can be obtained that soil pH and organic matter content are the most important factors determining cadmium uptake in earthworms.

For *lead*, all relationships shown in Table 8.1 indicate an increase of earthworm concentrations with decreasing soil pH. There is no consistent relationship with soil organic matter: in two studies a negative correlation between lead concentrations in earthworms and organic matter content was found, whereas in one study the correlation was positive. In three studies, earthworm concentrations were negatively correlated with CEC. Finally, in one study clay content was included, showing a negative correlation with earthworm concentrations of lead. From the relationships shown in Table 8.1, it can be concluded that soil pH seems to be the most important factor determining the bioavailability of lead for earthworms. At low pH, organic matter content and CEC may, however, also play a role. Bioavailability may increase with decreasing CEC and organic matter content.

For *zinc* and *copper* each, six relations could be found in the literature, but most of these were having fairly low $R^2$ values indicating that for these metals it is more difficult to establish consistent relationships between earthworm concentrations and CCP. This is not so strange, when it is realized that these metals are essential elements, the concentrations of which can, at least to a certain extent, be regulated by many living organisms. This was most clearly demonstrated by Van Gestel et al. (1992a), who found zinc levels in earthworms *Eisenia fetida* between 88 and 111 mg/kg dry weight when exposed for 4 weeks to 19 natural soils containing between 5.2 and 127 mg Zn/kg dry weight. In the same study copper concentrations in the earthworm ranged between 5.1 and 17 mg/kg dry weight, while soils contained 0.39–31 mg Cu/kg dry soil.

**Table 8.1.** Relationships between concentrations in earthworms and capacity controlling parameters of soils as found in the literature. See text for a description of the different studies

| Relationships[a] | $R^2$[b] (%) | p | Species[c] | Ref.[d] |
|---|---|---|---|---|
| Cd | | | | |
| $\log C_w = 1.14 - 0.079*pH$ | 12.0 | - | E | 1 |
| $\log C_w = 2.37 + 0.519*\log C_s - 0.57*\log pH - 0.585*\log OC$ | 90.1 | <0.05 | L.r. | 2 |
| $\ln C_w = 6.018 + 0.787*\ln C_s - 0.402*pH - 0.106*OM$ | 51.5 | - | L.r. | 4 |
| $\log C_w = 1.417 + 0.492*\log C_s - 0.181*\log OM$ | 87.2 | <0.01 | L.r. | 5 |
| $\log C_w = 1.207 + 0.618*\log C_s - 0.194*\log OM$ | 82.8 | <0.01 | D.r. | 5 |
| $C_w = -5.0 + 0.8*C_s - 0.08*clay - 0.03*OC + 2.1*pH$ | 79.4 | <0.01 | E.f | 6 |
| Cu | | | | |
| $\log C_w = 1.48 + 0.194*\log C_s - 0.310*\log OC$ | 82.0 | <0.05 | L.r. | 2 |
| $C_w = 31.7 - 0.44*C_s - 0.99*CEC$ | - | - | L.r. | 3 |
| $C_w = 20.57 + 0.350*C_s + 0.238*pH - 1.301*OM$ | 58.4 | - | L.r. | 4 |
| $\log C_w = 0.726 + 0.229*\log C_s$ | 10.6 | <0.01 | L.r. | 5 |
| $\log C_w = 0.327 + 0.487*\log C_s$ | 31.9 | <0.01 | D.r. | 5 |
| $C_w = 9.1 + 0.09*clay$ | 18.0 | <0.05 | E.f | 6 |
| Pb | | | | |
| $\log C_w = 1.2 + 0.035*C_s - 0.17*pH + 0.10*OM - 0.19*[Mg]$ | 64.0 | - | E | 1 |
| $\log C_w = 2.65 + 0.897*\log C_s - 3.56*\log pH$ | 93.3 | <0.05 | L.r. | 2 |
| $C_w = 165 + 0.54*C_s - 2.64*CEC - 17.9*pH$ | - | - | L.r. | 3 |
| $\ln C_w = 4.157 + 1.131*\ln C_s - 0.746*pH - 0.176*OM$ | 66.5 | - | L.r. | 4 |
| $\log C_w = 3.690 + 1.660*\log C_s - 6.878*\log pH - 1.010*\log CEC$ | 81.7 | <0.01 | L.r. | 5 |
| $\log C_w = 4.154 + 1.530*\log C_s - 6.657*\log pH - 0.898*\log CEC$ | 75.5 | <0.01 | D.r. | 5 |
| $C_w = 8.8 + 0.3*C_s - 0.2*clay - 0.4*OC - 2.2*pH$ | 76.4 | <0.01 | E.f | 6 |
| Zn | | | | |
| $\log C_w = 1.86 + 0.250*\log C_s - 0.643*\log pH$ | 74.2 | <0.05 | L.r. | 2 |
| $C_w = 4918 + 2.74*C_s - 40.2*CEC - 536*pH$ | - | - | L.r. | 3 |
| $\ln C_w = 6.878 + 0.439*\ln C_s - 0.298*pH - 0.088*OM$ | 49.0 | - | L.r. | 4 |
| $\log C_w = 2.024 + 0.313*\log C_s$ | 58.0 | <0.01 | L.r. | 5 |
| $\log C_w = 2.134 + 0.213*\log C_s$ | 52.7 | <0.01 | D.r. | 5 |
| $C_w = 95 + 0.3*OC$ | 35.1 | <0.01 | E.f | 6 |

[a] $C_w$ = concentration in earthworms (mg $kg^{-1}$ dr. wt.); $C_s$ = concentration in soil (mg $kg^{-1}$ dry soil); OM = organic matter content (%); OC = organic carbon content (%); clay = clay content (%); CEC = cation exchange capacity (meq/100 g); pH = pH-$H_20$, except for references 3 and 4 in which pH = pH-KCl

[b] $R^2$ = goodness of fit of the relationships between earthworms and CCP

[c] D.r. = *Dendrodrilus rubidus*; E = *Eisenoides* spp.; E.f. = *Eisenia fetida*; L.r. = *Lumbricus rubellus*

[d] 1 = Beyer and Cromartie (1987); 2 = Corp and Morgan (1991); 3 = Ma (1982); 4 = Ma (1983); 5 = Morgan and Morgan (1988b); 6 = Van Gestel et al. (1992a)

In three studies, earthworm concentrations of *zinc* appeared to be negatively correlated with soil pH, in two studies a negative correlation was found with organic matter content and in one study a negative correlation with CEC was reported. Also for zinc, soil pH seems to have the greatest effect on bioavailability. From the study of Ma (1983) it appears that at low pH bioavailability of zinc is also affected by soil organic matter content.

From the six studies with *copper* given in Table 8.1, it is difficult to derive a consistent relationship with CCP. Only in one study a relationship with soil pH was found: in this study, earthworm concentrations increased with increasing pH. In two studies, copper concentrations in earthworms were negatively correlated with soil organic matter content, and in one case they were negatively correlated with CEC. From the relationships described in Table 8.1, it can be concluded that bioavailability of copper for earthworms is more influenced by soil organic matter and clay content than by soil pH.

### 8.4.2 Other Data

Besides these studies described here, reporting on relationships between metal concentrations in earthworms and CCP, a number of data was found in the literature on the same issue but not leading to such relationships. These studies may, be useful however, also for the purpose of this chapter, and therefore a brief description will be given below. The full data of these and the previously described studies are provided in the original papers.

Bengtsson and Rundgren (1992) studied the influence of soil pH on the seasonality of lead accumulation in *Lumbricus terrestris*. Soil was artificially polluted with $PbNO_3$. Lead concentrations in the earthworms were linearly correlated with those in soil and appeared not to be affected by soil pH, suggesting active uptake of lead through feeding.

From a study by Streit (1984), on the uptake of copper by the earthworm *Octolasium cyaneum* using a brown earth (pH 4.78; 3.3% organic carbon) and a peaty soil (pH 4.5; 42.6% organic carbon) treated with $CuSO_4$, it can be concluded that uptake is strongly correlated with organic matter content of the soils. In the peaty soil, no effects were observed at 400 and 1200 mg Cu/kg soil, whereas earthworms did not survive these concentrations in brown earth. From this study it can also be concluded that earthworms die when the internal body concentration exceeds a value of 100–120 mg Cu/kg dry weight.

Van Gestel et al. (1992b) and Adema et al. (1987) determined the uptake of heavy metals by *Eisenia fetida* in polluted and decontaminated soils, using the same bioassay procedure as applied by Van Gestel et al. (1992a). Metal concentrations in the earthworms were determined after 4 weeks exposure to the soils. During the extractive decontamination process, most of the fine soil particles (organic matter, clay) were washed away, making the decontaminated soils very sandy and almost unsuitable for earthworms. The soils were therefore amended with some peat or cow dung to improve structure and to provide the earthworms with food. This appeared not to affect bioavailability. From the BSAF values determined in both studies, it can be concluded that bioavailability of metals may increase due to decontamination, in some cases even leading to increased metal concentrations in earthworms kept on decontaminated soils.

Beyer et al. (1982) determined metal concentrations in earthworms sampled from soils amended with sewage sludge. Dominant species on the loamy and sandy loam soils included in this study were *Allolobophora longa, Allolobophora turgida, Lumbricus terrestris* and *Aporrectodea tuberculata.*

Stafford and McGrath (1986) exposed *Lumbricus terrestris* for 2 weeks to soils from different polluted sites, and analyzed them afterwards for zinc, lead, copper and cadmium. It should be noted that an exposure period of 2 weeks is rather short, and probably not long enough to reach full equilibrium between earthworm and soil concentrations.

Benninger-Truax and Taylor (1993) determined metal concentrations in plants, earthworms (*Lumbricus* spp.), crickets and spiders sampled from soils treated with sewage sludge. Liming significantly affected metal concentrations in these organisms.

### 8.4.3 Alternative Relationships

Applying multi-regression analysis, using the software package SYSTAT on a Macintosh computer, attempts were made to assess the influence of CCP on the bioavailability of metals for earthworms by integrating all data available. For this purpose, all original data given by Adema et al. (1987), Benninger-Truax and Taylor (1993), Beyer et al. (1982), Ma (1982), Stafford and McGrath (1986) and Van Gestel et al. (1992a,b) were used, including data on concentrations in earthworms and soils and physico-chemical properties of the soils. For the analysis, data from different soil types, different earthworm species, various soil concentrations, different exposure times and from laboratory and field studies are combined. Only those data were used, for which sufficient information on soil properties were available. For the analysis, BSAF values were calculated as the ratio of the concentration in earthworms divided by the concentration in soil. This ratio is a relative measure of the bioavailability of metals for earthworms. In the analysis, BSAF and log BSAF values were related to all possible combinations of soil pH, clay and organic matter content and CEC (as CEC is dependent on clay and organic matter content, no relationships were tested in which CEC and clay or organic matter content were combined). Obvious outliers, such as extremely high or low BSAF values, were omitted from the regression analysis. Table 8.2 shows all resulting relationships with $p<0.05$. In case both BSAF and log BSAF resulted in significant relationships with the same parameter(s), only the best fitting one is given in Table 8.2.

For *cadmium*, BSAF values >50 and one very low value of 0.54 obtained on a high organic soil had to be excluded from the regression analysis; these data were obvious outliers. From Table 8.2, it can be concluded that BSAF values for cadmium are negatively correlated with soil pH, organic matter and clay content and CEC. Of these CCP, soil pH seems to have the greatest influence on bioavailability of cadmium for earthworms. A significant relationship between BSAF and pH is found. This relationship significantly improves when log BSAF is taken instead of BSAF and organic matter content is included, while inclusion of clay content does not add much to the correlation. Inclusion of CEC instead of organic matter content further improves the relationship. In Fig. 8.3, the relation between log BSAF and soil pH is shown for soils with a different CEC, using the relationship with these two parameters.

From the relationships in Table 8.2 and from Fig. 8.3, the following conclusions can be drawn: at low CEC a change in pH has a greater effect on BSAF than at high pH, and the effect of a pH reduction seems to be higher in low pH soils than in high pH soils. As the relationship of BSAF with organic matter content was similar to that with CEC, the same conclusions can be drawn for changes in organic matter content.

In the analysis of *copper*, BSAF values >10 were excluded. From Table 8.2, it can be concluded that BSAF values for copper are negatively correlated with all CCP included in the analysis. BSAF was significantly correlated with soil pH. This relationship improved when organic matter content or CEC was included. A significant negative correlation of log BSAF was found with clay content, which slightly improved by adding organic matter content and significantly improved when pH was included. The most significant correlation was found when pH, organic matter and clay content were combined. Of these factors, pH has the greatest impact on bioavailability, followed by clay content. The relationship between log BSAF and pH for different clay and organic matter content is shown in Fig. 8.4.

From the relationships in Table 8.2 and from Fig. 8.4, it can be concluded that clay has the greatest impact on the bioavailability of copper for earthworms. At low clay content, a decrease of soil pH leads to a greater increase of BSAF than at high soil pH. A change in organic matter content has only a slight influence on BSAF.

**Table 8.2.** Relationships between BSAF values for the uptake of heavy metals in earthworms and Capacity Controlling Parameters obtained by combining data from various studies. See text for further information [a]

| Relationship [b] | | n | p | $r^2$ |
|---|---|---|---|---|
| Cd | | | | |
| BSAF | = 31.2 - 3.11*pH | 31 | 0.032 | 0.127 |
| log BSAF | = 2.0 - 0.14*pH - 0.013*CEC | 21 | 0.005 | 0.443 |
| log BSAF | = 1.95 - 0.14*pH - 0.018*OM | 31 | 0.001 | 0.415 |
| log BSAF | = 1.95 - 0.14*pH - 0.018*OM + 0.001*clay | 31 | 0.002 | 0.416 |
| Cu | | | | |
| BSAF | = 3.08 - 0.32*pH | 47 | 0.005 | 0.162 |
| BSAF | = 4.14 - 0.49*pH - 0.021*CEC | 37 | 0.003 | 0.296 |
| log BSAF | = 0.12 - 0.015*clay | 38 | 0.002 | 0.246 |
| log BSAF | = 0.65 - 0.10*pH - 0.014* clay | 38 | 0.001 | 0.346 |
| BSAF | = 3.79 - 0.42*pH - 0.022*OM | 38 | 0.015 | 0.213 |
| log BSAF | = 0.11 - 0.016*clay + 0.002*OM | 38 | 0.007 | 0.249 |
| log BSAF | = 0.74 - 0.11*pH - 0.003*OM - 0.014*clay | 38 | 0.002 | 0.354 |
| Ni | | | | |
| BSAF | = -0.22 + 0.093*pH | 28 | 0.024 | 0.181 |
| log BSAF | = -0.64 - 0.011*CEC | 28 | 0.026 | 0.177 |
| BSAF | = -0.079 + 0.085*pH - 0.005*CEC | 28 | 0.012 | 0.297 |
| log BSAF | = -0.84 - 0.011*clay | 24 | 0.026 | 0.206 |
| Pb | | | | |
| log BSAF | = 0.048 - 0.18*pH | 48 | 0.001 | 0.228 |
| log BSAF | = 0.15 - 0.18*pH - 0.007*CEC | 38 | 0.020 | 0.201 |
| BSAF | = 0.14 - 0.0026*clay | 37 | 0.049 | 0.106 |
| BSAF | = 0.41 - 0.052*pH - 0.002*OM | 37 | <0.001 | 0.485 |
| log BSAF | = 0.035 - 0.19*pH - 0.013*clay | 37 | <0.001 | 0.566 |
| log BSAF | = 0.07 - 0.19*pH - 0.001*OM - 0.012*clay | 37 | <0.001 | 0.567 |
| Zn | | | | |
| log BSAF | = 2.09 - 0.34*pH | 49 | <0.001 | 0.361 |
| BSAF | = 22.3 - 2.29*pH - 0.13*CEC | 39 | <0.001 | 0.408 |
| log BSAF | = 2.08 - 0.36*pH - 0.001*clay | 39 | <0.001 | 0.430 |
| log BSAF | = 2.18 - 0.37*pH - 0.004*OM | 39 | <0.001 | 0.433 |
| log BSAF | = 2.18 - 0.37*pH - 0.004*OM + 0.000*clay | 39 | <0.001 | 0.433 |

[a] n = the number of data; p = significance of the relationship; $r^2$= correlation coefficient
[b] BSAF = Biota Soil Bioaccumulation Factor; pH = pH-$H_2O$ (sometimes pH-KCl); OM = organic matter content (%); clay = clay content (%); CEC = cation exchange capacity (meq/100 g)

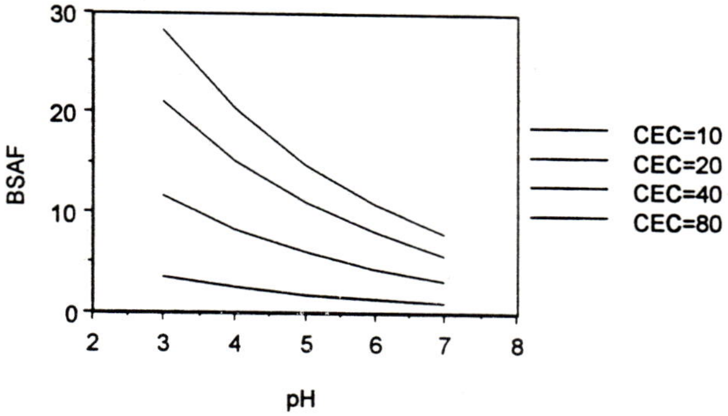


**Fig. 8.3.** log BSAF values for the accumulation of cadmium in earthworms in relation to pH for soils with different CEC (in mEq./100 g), according to the relationship log BSAF = 2.0 - 0.14*pH - 0.013*CEC (see Table 8.2)

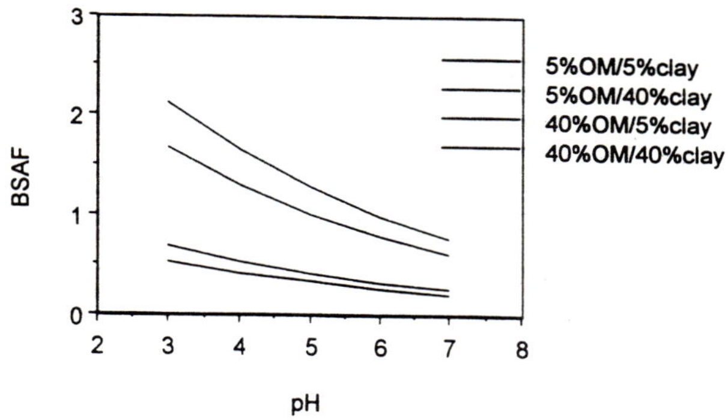


**Fig. 8.4.** log BSAF values for the accumulation of copper in earthworms in relation to pH for soils with different clay and organic matter contents, according to the relationship log BSAF = 0.74 - 0.11*pH - 0.014*clay - 0.003*OM (see Table 8.2)

In the analysis for *lead*, BSAF values >1 and some values obtained by Ma (1982) appeared to be outliers and were therefore omitted. Uptake of lead by earthworms appears to be negatively related with soil pH, CEC, organic matter and clay content. For lead, pH appears to be the most important factor determining bioavailability, and a significant relationship was found between log BSAF and soil pH. This relationship did not improve by the inclusion of CEC, but it did when clay content was included. This relationship was only slightly improved when organic matter content was added. A significant correlation was found between BSAF and clay content, and also between BSAF and the combination of pH and organic matter content. The effect of soil pH on log BSAF values resulting from the relationship with pH, clay and organic matter content is shown in Fig. 8.5.

From Table 8.2 and Fig. 8.5, it can be concluded that also for lead , after soil pH, clay content is an important factor determining bioavailability for earthworms. A change in clay content has great influence on the bioavailability of lead, whereas it is hardly affected by a change in organic matter content. At low clay content, a reduction in soil pH may lead to a strong increase in BSAF. At high clay content, the effect of pH reduction on BSAF seems to be somewhat smaller.

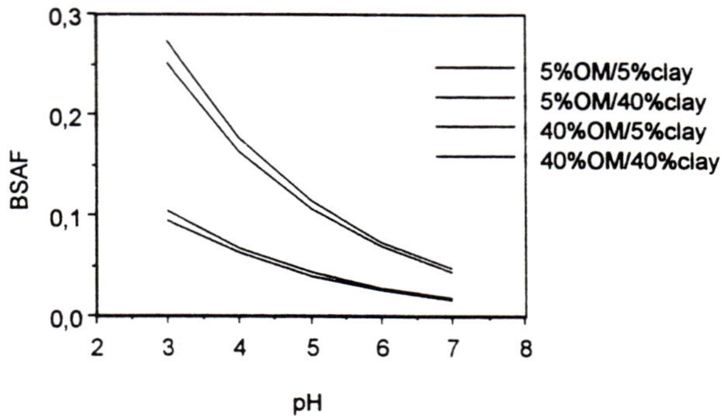


**Fig. 8.5.** log BSAF values for the accumulation of lead in earthworms in relation to pH for soils with different organic matter and contents, according to the relationship log BSAF = 0.07 - 0.19∗pH - 0.001∗OM - 0.012∗clay (see Table 8.2)

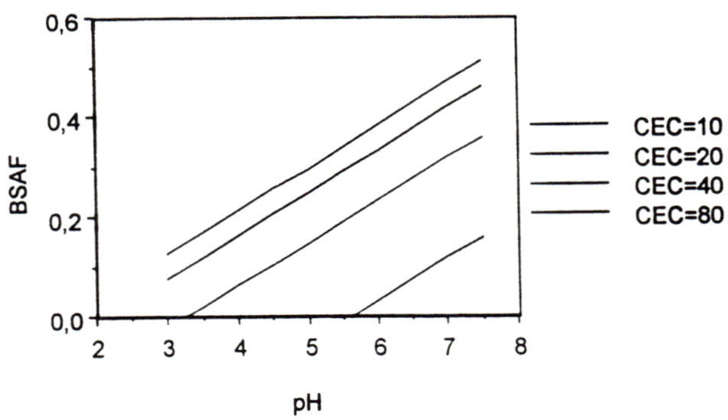


**Fig. 8.6.** BSAF values for the accumulation of nickel in earthworms in relation to pH for soils with different CEC (in mEq./100 g), according to the relationship BSAF = -0.079 + 0.085∗pH - 0.005∗CEC (see Table 8.2)

For *nickel* BSAF values >1 were excluded from the regression analysis. In this case, BSAF values were positively correlated with soil pH, suggesting that bioavailability of nickel increases with increasing soil pH. Apparently, the uptake of nickel by earthworms becomes greater than adsorption to soil with increasing

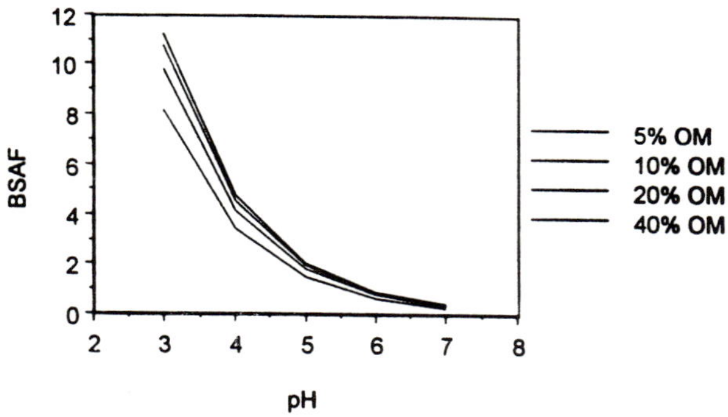


**Fig. 8.7.** log BSAF values for the accumulation of zinc in earthworms in relation to pH for soils with different CEC (in mEq./100 g), according to the relationship log BSAF = 2.18 - 0.37∗pH - 0.004∗OM (see Table 8.2)

soil pH. This may be similar to the finding of increased copper accumulation in yeast cells with increasing pH reported by Nederlof et al. (1993). BSAF values for nickel were negatively correlated with CEC and clay content of the soils, and not with organic matter content. Soil pH seems to have the greatest effect on bioavailability, and the most significant relationship was obtained between BSAF and the combination of pH and CEC. This relationship is elaborated in Fig. 8.6.

From the relationships shown in Table 8.2 and from Fig. 8.6, it can be concluded that there is a linear relationship of the BSAF for nickel in earthworms and soil pH. The slope of the lines is not affected by changes in CEC. It therefore has to be concluded that a change of soil pH will have about the same effect on BSAF in all soils. It should be realized, however, that for nickel only a limited number of data is available, not allowing for further conclusions.

BSAF values >25 were not included in the regression analysis for *zinc*. For this metal, BSAF values are negatively correlated with soil pH, CEC and organic matter and clay content. Bioavailability of zinc is most affected by soil pH, followed by organic matter content and CEC. A significant relationship between log BSAF and pH is obtained, which improves when BSAF values are used and CEC is included. Further improvement is obtained by the inclusion of organic matter or clay content. Clay content does not add much to these relationships, suggesting that bioavailability of zinc is only slightly affected by clay content. In Fig. 8.7, the relationship of log BSAF with soil pH is elaborated for soils with different organic matter content.

For zinc, from the relationships in Table 8.2 and from Fig. 8.7, the following conclusions can be draw: a decrease of soil pH from 7.0 to 3.0 leads to an increase of BSAF by a factor of >10, and this effect of pH seems to be almost independent of soil organic matter content. As already indicated, clay content does not have any effect on the bioavailability of zinc. So, bioavailability of zinc seems to be mainly dependent on soil pH.

## 8.5 Discussion and Conclusions

### 8.5.1 Toxicity

Only few data on the toxicity of metals for soil invertebrates were found that could be related to CCP. Most studies lacked a proper description of all relevant soil characteristics. Useful data found were restricted to two groups of organisms: Collembola and earthworms.

Toxicity of cadmium for soil invertebrates seems to be influenced by both soil pH and organic matter content with pH being the most important. CEC was shown to be a useful descriptor of cadmium toxicity for Collembola. Toxicity of copper in earthworms seems to be most affected by soil pH followed by CEC. The study of Ma (1988) also showed that the choice of the test species may have great influence on such conclusions. Lead toxicity for earthworms seems to be affected by soil pH. None of the studies reviewed in this chapter however, allows for a quantification of the influence of CCP on the toxicity of heavy metals for soil invertebrates.

### 8.5.2 Bioaccumulation

As toxicity will be determined by the amount of metal taken up by an organism, in this chapter also data on the bioaccumulation of metals in soil invertebrates were reviewed. Again, many of the studies considered lack a proper description of the relevant soil characteristics and most studies were restricted to earthworms.

In the literature, a number of studies came to the derivation of relationships between earthworm concentrations of metals and CCP. These relationships have been described in Table 8.1. It can be concluded that there are great differences between the results of the different studies. These differences may be caused by differences in test conditions (field versus laboratory), source and degree of contamination and the consequent differences in metal speciation, different earthworm species used and different exposure times. It is therefore difficult to compare the results of different studies. Nevertheless, an attempt has been made to unify all data found into one relationship. The results have been described in Table 8.2 and Figs. 8.3 through 8.7.

From this, it appears that for all metals soil pH is the main factor dominating uptake in earthworms. For lead and copper, also clay content seems to be an important factor, while for cadmium and nickel bioavailability is also related to CEC. For lead and copper, the effect of soil pH on BSAF seems to be greater at a low clay content, whereas the effect of soil pH on the bioavailability of cadmium is higher at low CEC. It can be concluded that in poor soils, having low CEC and clay content, acidification may lead to a greater increase of bioavailability of heavy metals than in heavy soils.

### 8.5.3 Conclusions

Only few data are available in the literature concerning the influence of CCP on the toxicity and bioavailability of heavy metals for soil invertebrates. Most of these data are restricted to earthworms, and many studies lack a proper description of relevant soil characteristics. From these data, it is therefore not possible to quantify the impact of the different CCP on toxicity.

Using the data from different studies with earthworms, attempts have been made to derive quantitative relationships between BSAF values and CCP. It should be stressed that for this purpose data from different origin have been pooled, which strictly speaking is not valued and results in a large variation in the data. Nevertheless, from this analysis it can be concluded that for each metal a different combination of CCP governs bioavailability for earthworms. Uptake in earthworms seems to be most affected by soil pH, followed by CEC, clay or organic matter content.

For a proper risk assessment of chemicals in soils and for a prediction of the potential risks of changed land use for soil ecosystems, a quantification of the effect of CCP on the toxicity and bioavailability of metals for soil invertebrates is needed. This study has made an attempt to arrive at such a quantification and may be useful as a starting point for further studies, which are no doubt needed considering the great lack of appropriate data. Considering the complexity of this subject, such research should be performed in a joint effort between ecotoxicologists and soil chemists.

### References

Adema DMM, Boer JLM de, Van Gestel CAM, De Jong P (1987) De invloed van bodemreiniging op de biologische beschikbaarheid van metalen, TNO-Rapport nr R87/120. Delft, The Netherlands

Aldenberg T, Slob W(1993) Confidence limits for hazardous concentrations based on logistically distributed NOEC toxicity data. Ecotox Environ Safety 25:48-63

Alloway BJ (ed)(1990) Heavy metals in soils. Blackie and Son Ltd, Glasgow

Bengtsson, G, Rundgren S (1992) Seasonal variation of lead uptake in the earthworm *Lumbricus terrestris* and the influence of soil liming and acidification. Arch Environ Contam Toxicol 23:198-205

Bengtsson G, Tranvik L (1989) Critical metal concentrations for forest soil invertebrates. Water Air Soil Pollut 47:381-417

Bengtsson G, Gunnarsson T, Rundgren S (1986) Effect of metal pollution on the earthworm *Dendrobaena rubida* (sav.) in acidified soil. Water Air Soil Pollut 28:361-383

Benninger-Truax M, Taylor DH (1993) Municipal sludge metal contamination of old-field ecosystems; do liming and tilling affect remediation? Environ Toxicol Chem 12 1931-1943

Beyer WN, Cromartie EJ (1987) A survey of Pb, Cu, Zn, Cd, As and Se in earthworms and soil from diverse sites. Environ Monit Assess 8:27-36

Beyer WN, Chaney RL, Mulhern BM (1982) Heavy metal concentrations in earthworms from soil amended with sewage sludge. J Environ Qual 11:381-385

Bolt GH, Bruggenwert MGM (1976) Soil chemistry, A. Basic elements. developments in soil science, vol 5A. Elseviers, Amsterdam

Corp N, Morgan AJ (1991) Accumulation of heavy metals from polluted soils by the earthworm *Lumbricus rubellus*: a laboratory exposure of control worms reduce biomonitoring problems. Environ Pollut 74:39-52

Denneman CAJ, Gestel CAM van (1990) Bodemverontreiniging en bodemecosystemen: voorstel voor C-(toetsings)waarden op basis van ecotoxicologische risico's, Reportnr 725201001. National Institute for Public Health and Environmental Protection, Bilthoven, The Netherlands

Everts JW, Aukema B, Mullié WC, Van Gemerden A, Rottier A, Van Katz R, Gestel CAM van (1991) Exposure of the ground dwelling spider *Oedothorax apicatus* (Blackwall) (Erigonidae) to spray and residues of deltamethrin. Arch Environ Contam Toxicol 20:13-19

Gestel CAM van(1992) The influence of soil characteristics on the toxicity of chemicals for earthworms: a review. In: Greig-Smith PW, Becker H, Edwards PJ, Heimbach F (eds) Ecotoxicology of earthworms. Intercept Ltd, Andover, Hants, pp 44-54

Gestel CAM van, Ma W (1988) Toxicity and bioaccumulation of chlorophenols in earthworms, in relation to bioavailability in soil. Ecotox Environ Safety 15:289-297

Gestel CAM van, Ma W (1990) An approach to quantitative structure-activity relationships in terrestrial ecotoxicology: earthworm toxicity studies. Chemosphere 21:1023-1033

Gestel CAM van, Van Dis WA (1988) The influence of soil characteristics on the toxicity of four chemicals for the earthworm *Eisenia fetida andrei* (Oligochaeta). Biol Fertil Soils 6:262-265

Gestel CAM van, Van Straalen NM (1994) Ecotoxicological test systems for terrestrial invertebrates. In: Donker MH, Eijsackers H, Heimbach F (eds) Ecotoxicology of soil organisms. Lewis Publ, Chelsea, Michigan, pp 205-228

Gestel CAM van, Dirven-Van Breemen EM, Kamerman JW (1992a) Beoordeling van gereinigde grond, IV. Toepassing van bioassays met planten en regenwormen op referentiegronden, Reportnr 216402004. National Institute for Public Health and Environmental Protection, Bilthoven, The Netherlands.

Gestel CAM van, Dirven-Van Breemen EM, Kamerman JW (1992b) Beoordeling van gereinigde grond, V. Toepassing van bioasays met planten en regenwormen op verontreinigde en gereinigde gronden, Reportnr 216402005. National Institute for Public Health and Environmental Protection, Bilthoven, The Netherlands

Houx NWH, Aben WJM (1993) Bioavailability of pollutants to soil organisms via the soil solution. Sci Total Environ Suppl pp 387-395

Jäggy A, Streit B (1982) Toxic effect of soluble copper on *Octolasium cyaneum* Sav. (Lumbricidae). Revue Suisse Zool 89:881-889

Kiewiet AT, Ma W (1991) Effect of pH and calcium on lead and cadmium uptake by earthworms in water. Ecotox Environ Safety 21:32-27

Ma W (1982) The influence of soil properties and worm-related factors on the concentration of heavy metals in earthworms. Pedobiologia 24:109-119

Ma W (1983) Regenwormen als bio-indicators van bodemverontreiniging. Bodembescherming 15. Staatsuitgeverij, The Hague, The Netherlands

Ma W (1984) Sublethal toxic effects of copper on growth, reproduction and litter break down activity in the earthworm *Lumbricus rubellus* with observations on the influence of temperature and soil pH. Environ Pollut (Series A) 33:207-219

Ma W (1988) Toxicity of copper to Lumbricid earthworms in sandy agricultural soils amended with Cu-enriched organic-waste material. Ecol Bull 39:53-56

Meulen GRB ter, Stigliani WM, Salomons W, Bridges EM, Imeson AC (eds)(1993) Chemical time bombs. Proceedings of the European state-of-the-art conference on delayed effects of chemicals in soils and sediments. Foundation for Ecodevelopment "Stichting Mondiaal Alternatief", Hoofddorp, The Netherlands

MILBOWA (1991) Milieukwaliteitsdoelstellingen bodem en water. Ministry of Housing, Physical planning and the Environment, Tweede Kamer, The Hague 21990(1):1990-1991

Morgan JE, Morgan AJ (1988a) Calcium-lead interactions involving earthworms. Part 1: the effect of exogenous calcium on lead accumulation by earthworms under field and laboratory conditions. Environ Pollut 54:41-53

Morgan JE, Morgan AJ (1988b) Earthworm as biological monitors of cadmium, copper, lead and zinc in metalliferous soils. Environ Pollut 54:123-138

Mueller BR, Roth M, Rittner P (1993) Influence of compost and lime on population structure and element concentrations of forest soil invertebrates. Biol Fertil Soil 15:165-173

Nederlof M, Van Riemsdijk WH, De Haan FAM (1993) Effect on pH on the bioavailability of metals in soil. In: Eijsacker HJP, Hamers T (eds) Integrated soil and sediment research: a basis for proper protection. Kluwer, Dordrecht, pp 215-219

OECD (1984) Guideline for testing of chemicals no 207: earthworm, acute toxicity tests. Adopted 4 April 1984, Organization for Economic Co-operation and Development, Paris

OECD (1992) Report of the OECD workshop on the extrapolation of laboratory aquatic toxicity data to the real environment. Organization for Economic Co-operation and Development, Paris

Stafford EA, McGrath SP (1986) The use of acid insoluble residue to correct for the presence of soil-derived metals in the gut of earthworms used as bio-indicator organisms. Environ Pollut (Series A) 42:233-246

Streit B (1984) Effects of high copper concentrations on soil invertebrates (earthworms and oribated mites): experimental results and a model. Oecologia 64:381-388

Suter GW II (1993) New concepts in the ecological aspects of stress: the problem of extrapolation. Sci Total Environ Suppl 63-76

US-EPA (1984) Estimating concern levels for concentrations of chemical substances in the environment. Environmental Effects Branch, US Environmental Protection Agency, Washington, DC

US-EPA (1989) EPA Office of Water Regulations and Standards, Report number 440/5-89-002. US Environmental Protection Agency, Washington, DC

Van de Meent D, Aldenberg T, Canton JH, Gestel CAM van, Slooff W (1990) Streven naar waarden. Achtergrondstudie ten behoeve van de nota "Milieukwaliteitsnormering water en bodem", Reportnr 670101001. National Institute for Public Health and Environmental Protection, Bilthoven, The Netherlands

Van Riemsdijk WH, Hiemstra T (1993) Adsorption to heterogeneous surfaces. In: Allen HE, Perdue EM, Brown DS (eds) Metals in groundwater. Lewis Publ, Chelsea, Michigan, pp 1-36

Van Straalen NM, Denneman CAJ (1989) Ecotoxicological evaluation of soil quality criteria. Ecotox Environ Safety 18:241-251

Van Straalen NM, Gestel CAM van (1993) Soil invertebrates and micro-organisms. In: Calow P (ed) Handbook of ecotoxicology. Blackwell Sci Publ, Oxford, 1:251-277

Wagner C, Løkke H (1991) Estimation of ecotoxicogical protection levels from NOEC toxicity data. Water Res 25:1237-1242

Wohlgemuth D, Kratz W, Weigmann G (1990) The influence of soil characteristics on the toxicity of an environmental chemical (cadmium) on the newly developed mono-species test with the springtail *Folsomia candida* (Willem). In: Barcelo J (ed) Environmental Contamination 4th International Conference, Barcelona. CEP-press, Edinburgh, pp 260-262

Wright MA, Stringer A (1980) Lead, zinc and cadmium content of earthworms from pasture in the vicinity of an industrial smelting complex. Environ Pollut (Series A) 23:313-321

Yong RN, Mohamed AMO, Warkentin BP (1992) Principles of contaminant transport in soils. Developments in Geotechnical Engineering, 73. Elsevier, Amsterdam

# 9 Acidification and its Long-term Impact on Metal Mobility

*H. Smal and W. Salomons*

## 9.1 Introduction

In the past, there was a general belief among scientists that the buffering ability of the soil would suffice to counteract acid rain. That is now known to be untrue. Throughout much of southern Scandinavia the pH values of soils have fallen by 0.3 to 1 unit, and in some instances by as much as 1.5 in the last few decades (Hanneberg 1993). Soils of southern and central Europe developed from more weathered rocks with higher lime content, that were better able to neutralize acids than were Scandinavian soils. Even so, in recent decades similar trends in acidification have appeared in this region as well.

Emissions of sulfur dioxide and nitrogen oxides (the main causes of anthropogenic soil acidification) increased rapidly after the Second World War as a result of rapid industrialization and urbanization. Development of industry has caused not only a high pollution of the environment with acids but with various contaminants including heavy metals as well. Consequently there are areas now in Europe and other parts of the world where polluted soils and acidification of the environment are the serious problems. For instance, in the United Kingdom, Germany, Poland, and Italy past mining and industrial activities have lead to a widespread level of contamination (sometimes very high) of Cd, Pb, Zn, Cu and Ni ( Angelone and Bini 1992, Davis 1987; Marchwińska and Kucharski 1988, Dudka and Sajdak 1992; Leita et al. 1988; Leita et al. 1989). There is no area in western Europe where the local balance for most heavy metals in the ecosystem is not strongly influenced by atmospheric pollution (Angelone and Bini 1992).

Chapter 9 is a literature review on environment acidification and on two areas with metals problems in Poland as an example. In the final sections of this chapter an attempt is made to combine both environmental impacts in a discussion of future effects.

## 9.2 Some Basic Geographical Data on Poland

Poland covers an area of 312 677 $km^2$ located between the 55°and 49° latitude North and 14° and 24° longitude East (of Greenwich). The territory is in the zone influenced by continental and marine climates. The mean temperatures are -2.5°C in January, ranging to +18°C in July and the annual mean is +7.5°C. Average annual precipitation ranges from 500 to 700 mm with substantial differentiation among the regions of the country. The heaviest rainfalls above 1000 mm occur in the Carpathian and Sudeten Mountains and the lowest, less than 500 mm, in the central and midwestern regions.

The water resources in Poland are one of the poorest in Europe. Its outflow indices on the basis of $km^2$ and amount per capita, place the country in the 22nd and 20th positions, respectively, among the 27 European nations ( GUS 1990 ). Poland's main mineral resources are: pit coal, brown coal, sulfur, iron and nonferrous ores (lead, zinc, copper). Their exploitation ( except iron ore ) and processing increased rapidly after World War II. In this period the country has changed from primarily an agricultural country into an increasingly industrialized one. In 1987 the population of Poland was 37.8 million. About 40% of the population lives in rural areas, and 28% are employed in agriculture.

## 9.3 Acidification of the Environment

In 1988, Poland was third in Europe in emissions of $SO_2$ and $NO_x$, contributing about 10% and 8% respectively to the global emissions. While emission of $SO_2$ in Europe in the 1980s decreased by 20%, in Poland emissions were maintained at the same high level of 4.1–4.2 million tons per year in 1980–1985 (GUS 1990). However, approximately 45% of total yearly deposit of $SO_2$ in Poland comes from its own sources and 55% from import. For example, the largest pollution contributors in 1988 (according to the Institute of Environmental Protection, after GUS 1990) were: former East Germany (486 000 t), Czechoslovakia (246 000 t), West Germany (78 000 t) and Hungary (54 000 t). In turn, Poland exported mainly to the former USSR (518 000 t), Czechoslovakia (126 000 t), Romania (68 000 t) and Sweden (56 000 t). Power production, necessary for all economic activities, is one of the main causes of atmosphere pollution. In 1989, 52% of total emission of $SO_2$ and 32% of $NO_x$ were contributed from that industry (GUS 1990). In the last decade increasing energy needs were covered by increased output of brown coal (from 37 million t in 1980 to 72 million t in 1989). High amounts of $SO_2$ and $NO_x$ emitted to the atmosphere cause "acid rain", considered as a main source of acidity.

Calculations performed by the Institute of Meteorology and Water Management in Warsaw showed that one-third of Poland's area is situated within the area of most acidic rainfall in Europe (pH 4.1) and that the remaining area has an average pH of rainwater of about 4.3. The average yearly pH in Suwałki (east–north) was equal to 4.4, and in the Snieżka mountains (south-west) it was lower than 4.0 (Chudzyńska 1985 after Strzyszcz 1990). A high variability of the acidity of rain (pH 2.8–8.0) was recorded in Warsaw and its vicinity. It has also been found that occurrence of rain with pH below 4.0 has been increasing systematically in that region.

The most polluted areas of Poland are the southern and southwestern parts, the regions called Upper and Lower Silesia respectively. In the Karkonosze mountain chain (Lower Silesia) in the years 1983–1986 the pH of rainfall ranged from 3.0 to 6.3 (Wasilewski and Twarowski), and was only slightly changed in 1991–1993 from 3.2 to 6.6 (Twarowski and Gendolla). Generally, in the whole region of Lower Silesia rainfall was acidic and in 1991 its pH was below 6.0 (Twarowski and Gendolla 1993). In that year, the lowest pH was in the Karkonosze mountains (3.7–6.1). It was also very low in Legnica-Głogów (area of copper mining and processing) where it ranged from 4.2 to 6.0.

Interesting observations of the response of rainfall were made by Czyż et al. (1968, after Strzyszcz 1990) in the Upper Silesia region. Although, the acidity of the rain there ranged from pH 3.7 to pH 7.7 no relationship between the pH and $SO_2$ concentrations in the air or dust was found nor between the concentrations of sulfates in rainwater and pH. Authors assume that this lack of correlation was caused by another contributing factor which is the emission of alkaline compounds into the air. Acidic reaction of rain was monitored on the borders of the Silesia region where overall pollution was lower. The higher pH in regions with the greater pollution was also shown by the study at the Zabrze area (Strzyszcz 1990). Recent data on pH of rain in Sosnowiec station, located in Upper Silesia, from July 1992 to March 1993 showed that 32% of the rainfall was of pH below 4.1; 48% from 4.1 to 4.6 and only 2% showed pH from 6.1 to 6.5 (Leśniok and Opałka-Gądek 1993). No rain with pH above 6.5 was stated.

## 9.4 Soil Conditions: Vulnerability

### 9.4.1 In General

Light (with regard to granulometric composition), not very fertile soils cover about 60% of the agriculturally utilized land in Poland (Gliński 1987). There are mainly podzolic (Leptic and Ortic podzols) and brown soils (Eutric Cambisols) developed from various sands and loams. These soils have a low cation exchange capacity (10–100 mmol(+) $kg^{-1}$). Organic matter content ranges from 1% to 2% in mineral soils and only locally from 4% to 5% (Dobrzański and Zawadzki 1981).

About 28% of agricultural and forest land are subject to wind, water and surface erosion (GUS 1991). About 36% of soils are seasonally or permanently too dry and about 20% too wet (Gliński 1987).

### 9.4.2 Acidity

Acidification of soils is considered to be the main problem of soil degradation in Poland (Gliński 1987; Kern 1987; Siuta 1990) An extensive soil survey carried out in the years 1954–1965 showed that about 58% of all cultivated soils were acidic (pH <5.5) in the surface layer, 25% weakly acidic (pH 5.6–6.5) and 17% were neutral and alkaline soils (pH >6.5; Włodarczyk 1983). The second survey (1966–1977) did not show any substantial changes in soil pH (Kern 1987). It should be added that the average dose of lime application has been increased from 12.4 kg $ha^{-1}$ CaO in the late 1950s to 202 kg $ha^{-1}$ in the late 1980s (GUS 1991). The majority of very acidic soils (with pH<4.5) are sandy or light loamy soils with very weak sorptive and buffer capacities (Fotyma 1987).

A high proportion of acid soils can be seen in the western and southwestern parts of Poland, although the average yearly dose of lime fertilizers per ha in those regions has always been much higher (350 kg CaO in 1988–1989) compared to the average for the country. The most polluted areas such as Katowice, Kraków, Legnica, Jelenia Góra, Łódź voivodeships have soils with pH<5.5 in more than 40% of their areas (Włodarczyk 1983). Many soils in Poland are acidified throughout the whole soil profile. This is an important property as it affects metals mobility and leaching into the deeper layers. According to Kern (1987), 61%, 39% and 26% of arable soils were acidic (pH<5.5) in the layers from 0–50 cm; 50–100 cm and 100–150 cm respectively.

### 9.4.3 Vulnerability to Degradation

In Poland, a map of soil resistance to degradation has been constructed. (Fig. 9.1). The authors took into account the buffering capacity of soils in respect to acidification determined by the concentration of base cations, e.g., Ca, Mg, K, Na (Siuta 1976). This index can be used for nearly all mineral soils, the exception being those containing $CaCO_3$ as it greatly increases resistance to acidification. Six classes of soil resistance were defined., ranging from first class soils – with very weak resistance to degradation (containing from 0 to 30 mEq. $kg^{-1}$ ) – to the sixth class with very high resistance (above 160 mEq. $kg^{-1}$; Siuta 1976).

It has been estimated that in ca. 51.3% of the total area, the soils range from very weak to weak resistance to degradation, 30.4% range from medium resistance to resistant, and only 18.4% comprise highly resistant soils. The regions with severe industrial pollution have soils weakly resistant to chemical contamination. Vulnerable soils cover almost all of the western part of the country, and they are also common in the southern and other parts as well.

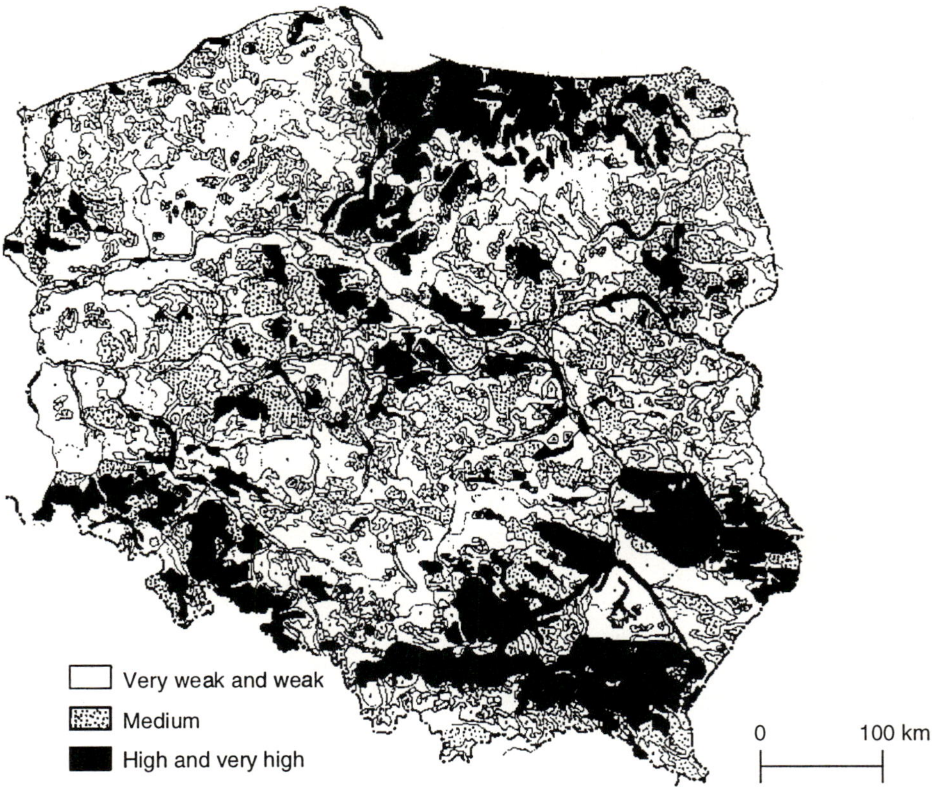


**Fig. 9.1.** A generalized map of resistance of soils to degradation. Weak and very weak resistant soils combine first and second, medium-third, high and very high-fourth and fifth classes respectively. (modified from Siuta 1990)

## 9.5 Heavy Metals

### 9.5.1 Geochemical Background Concentrations

In the 1950s and early 1960s little data on trace metal content in soils were available (Boratyński et al. 1972). In the last two decades more studies have been conducted to determine their geochemical background as well as concentrations in polluted areas.

The natural concentrations of heavy metals in soils are related to the amounts in parent materials and to the soil-forming processes (Kabata-Pendias and Pendias

1993). Table 9.1. shows means of total concentration of heavy metals in unpolluted soils representing typical soils from all over the country. The data may be considered as background values of those metals in Polish soils. They are generally lower than the means of central eastern Europe and of the world soils.

**Table 9.1.** Mean trace metal concentrations in mg $kg^{-1}$ in unpolluted cultivated soils in Poland. Samples of surface layer ( 0–20 cm)

| Reference | Mn | Zn | Cr | Pb | Cu | Ni | Co | Cd |
|---|---|---|---|---|---|---|---|---|
| Kabata-Pendias (1981) | 563 | 59 | - | 28.0 | 12.0 | - | - | 0.33 |
| Andruszczak and Czuba (1984) [a] | 362 | 55.4 | - | - | 8.7 | - | - | - |
| Gworek (1985a) [a] | 321 | 39 | 24.5 | 9.0 | 6.6 | 7.9 | 3.2 | 0.27 |
| Gworek (1985b) | 461 | 29 | 22.0 | 12.0 | 3.9 | 4.7 | 4.5 | 0.52 |
| Czarnowska and Gworek (1987) | 322 | 36 | 30.0 | 10.3 | 8.8 | 12.7 | 4.8 | 0.27 |
| Czarnowska ( 1989) [a] | 370 | 32 | 28.0 | 14.5 | 10.5 | - | - | - |
| Dudka (1992) [b] | 304 | 40.3 | 16.8 | 18.8 | 6.3 | 7.4 | 3.6 | 0.40 |
| Mean of means | 386 | 41.5 | 24.2 | 15.4 | 8.1 | 8.2 | 4.0 | 0.35 |
| n = | 721 | 721 | 232 | 283 | 721 | 222 | 222 | 273 |
| Central-eastern Europe (Kabata-Pendias et al. 1992) | 505 | 50 | 75 | 35 | 16 | 22 | - | 0.5 |
| World (Ure and Berrow 1982) | 761 | 59.8 | 84.0 | 29.2 | 25.8 | 33.7 | 12.0 | 0.62 |
| n = | 8354 | 7402 | 3290 | 4970 | 7819 | 4625 | 5504 | 1642 |
| World (Kabata-Pendias and Pendias 1993) | 440 | 64 | 54 | 30 | 20 | 22 | 7.9 | 0.5 |

[a] modified
[b] geometric mean

### 9.5.2 Inputs

There are many pathways for trace metals to enter soils. In the mass balance by Kabata-Pendias and Pendias (Table 9.2) for slightly contaminated Polish soils the main source of Cd, Pb and Zn was atmospheric precipitation. Its share in total input was 34.4%, 71.4%, 76.2% for Cd, Pb and Zn respectively. Substantial amounts were also added to the soils with manuring. Loss of metals occurs mainly due to removal via plant biomass and leaching. As shown in the balance, these two ways were equally important in removing of Cd and Pb from the soils. In the case of Zn, losses via leaching were much higher than via plant uptake.

The problem of soil contamination by heavy metals in Poland occurs locally and concerns only some of the so called ecologically threatened areas, of which there are 27 in Poland. These areas were defined in 1983. Until now they have not

**Table 9.2.** Balance of Cd, Zn, Pb in slightly contaminated Polish soils (mean values from measurements in 1980–1985; Kabata-Pendias, Pendias 1993) given in g $ha^{-1}$ $a^{-1}$ and in percentage of total input/output

| Input: | Cd | % | Pb | % | Zn | % |
|---|---|---|---|---|---|---|
| Mineral fertilization | 1.0 | 6.9 | 2.0 | 0.7 | 10.0 | 1.4 |
| Liming | 1.5 | 10.9 | 5.0 | 1.8 | 4.0 | 0.6 |
| Manure | 2.5 | 17.2 | 38.0 | 13.6 | 75.0 | 10.6 |
| Sewage sludge | 1.5 | 10.4 | 15.0 | 5.4 | 30.0 | 4.2 |
| Plant residues | 3.0 | 20.6 | 20.0 | 7.1 | 50.0 | 7.0 |
| Precipitation | 5.0 | 34.4 | 200.0 | 71.4 | 540.0 | 76.2 |
| Input in total | 14.5 | 100 | 280.0 | 100 | 709.0 | 100 |
| Output: | | | | | | |
| Yield | 3.0 | 50.0 | 40.0 | 50 | 100.0 | 35.7 |
| Leaching | 3.0 | 50.0 | 40.0 | 50 | 180.0 | 64.3 |
| Output in total | 6.0 | 100 | 80.0 | 100 | 280.0 | 100 |
| Yearly balance | | + 8.5 | | + 200.0 | | + 429.0 |

**Table 9.3.** Annual emission of heavy metals into the air: totally in Poland and in "ecologically threatened areas" in 1989 from " particularly hazardous industrial plants"[a] (source: GUS 1990)

| Element | Poland | | Areas of ecological threat | | |
|---|---|---|---|---|---|
| | ton | kg $km^{-2}$ | ton | % | kg $km^{-2}$ |
| As | 33 | 0.11 | 27 | 81.8 | 0.79 |
| Cr | 407 | 1.30 | 393 | 96.6 | 11.45 |
| Cd | 12 | 0.04 | 11 | 91.7 | 0.32 |
| Cu | 352 | 1.13 | 350 | 99.4 | 10.18 |
| Pb | 610 | 2.00 | 560 | 91.8 | 16.29 |
| Hg | 12 | 0.04 | 3 | 25.0 | 0.09 |

[a] There are 1612 of these kinds of plants. Their contribution to the total emission of air pollutants in Poland is estimated to be 60–70%. Out of these 1612 industrial plants, 712 (44%) are located in ecologically threatened areas.

been verified, although in the opinion of experts, some of them are not actually so endangered as originally thought, although there are perhaps other areas which should be included in the threatened category. They occupy 11.2% of the total area and are inhibited by 35.4% of the population (GUS 1991).

As shown in Table 9.3, more than 90% of total industrial emissions of Cu, Cr, Pb, Cd in the country arise from those areas.

### 9.5.3 Soil Pollution

**Generally.** Very recently a map of soil pollution by heavy metals in Poland has been constructed (on the scale 1:500 000) at the Institute of Soil and Plant Cultivation in Puławy. However, it is not yet available for publication. The map is based on a large amount of data, and shows that in Poland pollution of soils with heavy metals is a problem primarily in some heavily industrialized areas, and that generally the rest of the soils have concentrations within the natural concentration of trace metals.

The areas in the southern and southwestern parts of the country are the most polluted with metals. Among them Upper Silesia is classified as the worse polluted area. This region has without any doubt not only the most degraded soil environment but degraded air and water as well (GUS 1990). Legnica-Głogów (southwest) is another region with heavy metal problems.

Dudka (1992) has compared the concentration of trace elements in soils of southwestern Poland with the northeastern part, which is not polluted. The concentrations of *Cd, Pb, and Zn* in soils in the first case were significantly higher than in the soils of the other regions. Especially high concentrations of these elements were found in the soils of Katowice voivodeship (Upper Silesia). The author stated that this situation was due to anthropogenic factors, and not due to the natural properties of the soils. There were no substantial differences in the concentrations of other trace elements in the regional comparison.

**Upper Silesia: Sources of Metals.** There are natural and anthropogenic sources of heavy metals in Upper Silesia. This is the area of Poland with the largest deposits of lead and zinc ore. In some places ore-bearing Triassic limestones lie very close to the surface or occur as outcroppings that may naturally contaminate soils of the region. It is an old mining area dating back to the twelfth century. In the region of Tarnowskie Góry alone, some 15 000 shallow pits have been documented, which were used for mining ores containing silver, lead, zinc and cadmium. These locations must have caused contamination of neighboring soils with those metals (Marchwińska et al.,1989).

Although these soils must have been exposed to heavy metals for centuries, a substantial increase must have occurred in the last decades due to expanding industrialization. For example, the output of lead and zinc ores in 1988 was 2.5 times that in 1961 (GUS 1990).

Nonferrous metal production caused very high emissions of those metals into the air. In 1985 the lowest mean concentration of lead monitored in towns of Upper Silesia exceeded the allowed standard (Fig. 9.2.), and maximum concentrations were 17 times higher (Bytom). As a result, mean annual concentrations were several times higher as compared to the allowed standard (0.2 $\mu g\ m^{-3}$).

In recent years in Upper Silesia, a decrease in heavy metal concentrations in the air has been recorded due to more active environmental protection and prod uction decreases (Table 9.4). However, maximum yearly concentrations of Cd

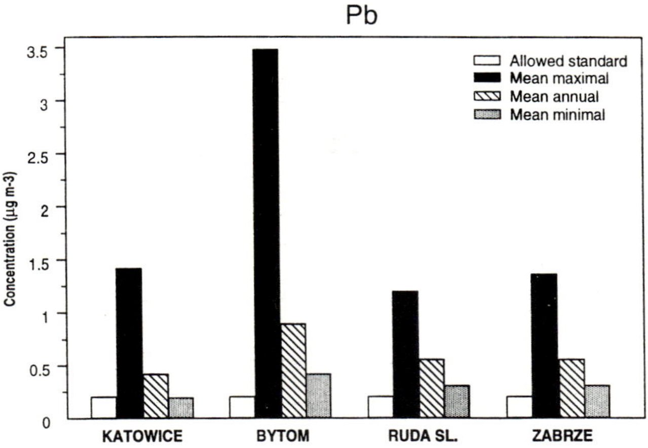


**Fig. 9.2.** Mean annual concentration of lead in the air of some towns in Upper Silesia in 1985 by Sanitary-Epidemiology Station data (Sądelski 1989)

and Pb in the Katowice voivodeship still often exceed allowed standards by several fold.

Polluted atmospheres are one of the main sources of heavy metals for soils. According to Marchwińska and Kucharski (1988) the amount of lead deposited with dust in the soils in Katowice voivodeship ranges from 27 to 2687 kg $km^{-2} a^{-1}$, and in case of cadmium from 0.6 to 74.1 kg $km^{-2} a^{-1}$. In 1990, according to the WSSE report (1991), the fallout of lead in the voivodeship area ranged from 2.0 to 2822 kg $km^{-2} a^{-1}$, and for Cd from 0.0 to 87.5 kg $km^{-2} a^{-1}$. The standards (since they were established in 1990) for airborne deposition of lead and cadmium in Poland are 100 kg $km^{-2} a^{-1}$ and 10 kg $km^{-2} a^{-1}$, respectively.

**Table 9.4.** Annual concentration of Pb and Cd in the air in Katowice voivodeship in years 1991–1992 by Sanitary-Epidemiology Station data (WUS – Katowice 1993)

| Element | Yearly concentration | | | | Standard |
|---|---|---|---|---|---|
| | Minimal | | Maximal | | |
| | 1991 | 1992 | 1991 | 1992 | |
| Pb [μg $m^{-3}$] | 0.11 | 0.09 | 1.50 | 1.07 | 0.20 |
| Cd [ng $m^{-3}$] | 2.10 | 1.10 | 72.00 | 49.50 | 10.00 |

**Upper Silesia: Soil and Plant Contaminations.** Such high inflows of heavy metals to the environment have caused their continuous accumulation in soils. However, until the 1970s very few studies had been carried out in these areas, and there are no data available to show changes over decades.

**Table 9.5.** The concentration of trace metals [mg $kg^{-1}$] and pH in arable soils (n=2270) of Katowice voivodeship (Dudka and Sajdak 1992)

| Element or pH | Geometric mean | Range | Geometric SD |
|---|---|---|---|
| Cd | 3.2 | 0.7–143 | 2.6 |
| Zn | 270.2 | 5.0–13 250 | 3.0 |
| Pb | 102.3 | 4.0–8200 | 2.6 |
| Cr | 5.1 | 0.7–54 | 2.6 |
| Cu | 8.8 | 0.7–250 | 2.1 |
| Ni | 6.7 | 0.7–75 | 2.4 |
| pH[a] | 6.1 | 3.2–7.8 | |

[a] arithmetic mean

A comprehensive survey on heavy metal concentrations in Katowice voivodeship was conducted in the 1980s. Until now 70% of arable soils have been tested. Also included are a large number of allotments (private gardens). The study showed that accumulation of heavy metals is a major factor for chemical degradation of soils in that area (Marchwińska and Kucharski 1988; Gzyl 1990; Dudka and Sajdak 1992). Very high soil contamination by cadmium, lead and zinc was found, and these three metals are considered as the main problem, although there is a slight disagreement between authors on their ranking. Namely, Marchwińska and Kucharski (1984) have found that soils are most contaminated by Cd and Pb, and to a less extend Zn, whereas Dudka and Sajdak (1992) claim that the worse pollutants are Cd and Zn and then Pb. In the opinion of Dudka and Sajdak, the difference probably results in part from the way of calculating mean values (e.g., using the geometric rather than the arithmetic mean), and in part from differing interpretations of the data. Nevertheless, there is no dispute that the contamination of soils with these three metals is the major cause of soil degradation.

As shown in Table 9.5 mean concentrations of Zn and Pb in the soils were about seven times higher than background values and in case of Cd about ten times higher as compared with the data in Table 9.1. Generally, soil contamination of the Katowice region by Cr, Cu and Ni has not been stated (Dudka and Sajdak 1992).

Some of the results for allotments are shown in Table 9.6. The lowest values of Pb and Cd concentration in the soils where the plants were grown were higher than the mean values in unpolluted soils. Mean Pb and Cd soil concentrations were about 15 and 24 times higher respectively.

In Poland maximum allowable (critical) concentrations are for cadmium: 3, lead: 50–100, zinc: 200–300 mg $kg^{-1}$, whereas the lower values refer to light textured soils and the higher values to the heavy textured soils (Monitor Polski 1986). However, most frequent values by various authors and recommended by European Community are for Cd: 1–8 (3); Pb: 100–500 (100); Zn: 150–400

**Table 9.6.** Lead and cadmium concentrations in soils and vegetables in 126 allotments of Upper Silesia (Gzyl 1990); concentrations [mg $kg^{-1}$ dr. wt][a]

| | | Soil | Vegetables | | | | | pH |
|---|---|---|---|---|---|---|---|---|
| | | | Parsley leaves | Parsley roots | Celery leaves | Celery roots | Carrot roots | |
| Pb | minimum | 16.8 | 1.3 | 1.0 | 3.5 | 1.0 | 1.0 | 5.0 |
| | maximum | 1640.8 | 49.5 | 9.6 | 47.3 | 10.6 | 9.4 | 7.5 |
| | mean | 221.4 | 7.9 | 2.7 | 9.5 | 3.2 | 2.7 | 7.0 |
| Cd | minimum | 1.2 | 0.3 | 0.2 | 0.21 | 0.70 | 0.70 | |
| | maximum | 51.7 | 13.5 | 2.35 | 19.33 | 10.88 | 5.25 | |
| | mean | 8.3 | 0.97 | 0.57 | 2.82 | 2.21 | 1.0 | |

[a] Allowable concentrations in plants for human consumption are, in mg $kg^{-1}$ dr. wt: Cd <0.15; Pb <1.0 (Kabata-Pendias et al. 1993)

(300) mg $kg^{-1}$ (Kabata-Pendias and Pendias 1993; values in brackets are mean values). Dudka and Sajdak (1992) in evaluation of Katowice voivodeship soils accepted the following allowable concentrations of trace metals (in mg $kg^{-1}$): Cd 3–5; Pb 100–400; Zn 300–500. They stated that, the exceedance of the allowable values for soil samples of Katowice voivodeship was 27% for Cd, 5% for Pb, and 40% for Zn.

Also a classification of polluted soils from 0 (unpolluted soil) to V (heavily polluted soil) has been proposed by Kabata Pendias et al. (1993) for agricultural soils in Poland. The classification is based on the total concentrations of Pb, Zn, Cu, Ni, Cd in soils. Moreover, three soil categories (by pH and soil texture) in each class are distinguished. The values of the heavy metal concentrations in the class II of this classification (slightly contaminated soil) are: for Cd in very light (independent of pH) and light-textured soils (pH<5.5) 2 mg $kg^{-1}$, and in heavier and heavy-textured soils (pH>5.5) 5 mg $kg^{-1}$; for Pb of 100 and 500 mg $kg^{-1}$, and for Zn of 300 to 1000 mg $kg^{-1}$ for given textural classes, respectively. As can be seen the values for class II could be related to the critical allowable concentrations. Soils classified in class I are below the critical levels in metal concentrations.

High concentration of heavy metals in the air and soils has a direct influence on their concentrations in cultivated plants as shown in Table 9.6. The results were compared also with tolerable concentrations on the basis of fresh weight, which are in Poland 0.3 mg $kg^{-1}$ for lead and 0.03 mg $kg^{-1}$ for cadmium. Out of 756 vegetable samples studied, only 170 met the standard for lead and 17 for cadmium. Other research showed high concentrations of lead and cadmium not only in vegetables, but also in potatoes (Marchwińska et al. 1984) and pasture plants such as grass, legumes, perco and corn ( Kucharski et al. 1989).

The survey showed clearly that soil pollution by heavy metals in Upper Silesia is a major problem not only in "sanitary zones" but also for much larger areas. In Katowice voivodeship arable soils occupy 47.1% of its area (313 400 ha). Results of the heavy metal survey showed that about 17% of the studied allotments and cultivated soils are unsuitable for any crop production (Marchwińska et al. 1989), and agricultural activities should be forbidden. About 50% of the studied area contains soils where risk of plant contamination by heavy metals above allowed standards is less than in the case of Katowice, but certain crops with a high affinity for metals accumulation should be excluded from cultivation as well. Remaining areas are considered to be of lower risk regarding cultivation of edible plants.

**Lubin-Głogów Area.** Situated in southwestern Poland, the Lubin-Głogów region is the main area for copper ore resources (95% of national deposits). The ores also contain substantial amounts of lead, silver, cobalt, and nickel. Mining and processing have a long history in this region. Additionally, new deposits of copper ores were discovered in late 1950s. The new mines were dug and first exploited 1968. Subsequently, smelters (Głogów I and II) were built, and reached full production in 1977. Within 1 year the deteriorating effect of sulfur pollution was noticed (Roszyk and Roszyk 1975), and in the second year the accumulation of heavy metals in soils was documented (Roszyk and Roszyk 1976). Emission of gaseous pollutants, as well as dust containing copper, arsenic, lead, nickel, zinc, and mercury increased accordingly with increasing outputs of ores and copper production. These activities have caused the continuous addition of metals into the soils (Drozd and Kowaliński 1977; Roszyk and Szerszeń 1988a; Roszyk and Szerszeń 1988b).

Heavy metals in soils of that region have not been studied on such a broad scale as in the case of Upper Silesia. Nevertheless, available data show that their

**Table 9.7.** Heavy metal concentrations (mg $kg^{-1}$) in surface soils in the vicinity of two copper works: Legnica and Głogów (source: Roszyk and Szerszeń 1988a,b) and in soils of Legnica voivodeship according to Greinert et al. (1987)

| Legnica-Głogów | | | Legnica voivodeship | |
|---|---|---|---|---|
| Element | Range[a] | Mean of means [a] n=807 | Range | Mean n=45 |
| Mn | 40.00 – 4800.0 | 557.0 | – | – |
| Zn | 25.00 – 4000.0 | 103.0 | 5–116 | 38 |
| Pb | 30.00 –18 400.0 | 175.0 | 6–220 | 43 |
| Cu | 25.00 – 9800.0 | 338.0 | 3–620 | 54 |
| Ni | 3.00 – 450.0 | 27.0 | 2– 22 | 9 |
| Cd | 0.02 – 11.0 | 0.7 | – | – |
| Co | 1.30 – 38.3 | 6.5 | – | – |
| As | 0.30 – 145.0 | 7.6 | – | – |
| Hg | 0.01 – 15.0 | 0.5 | – | – |

[a] modified

accumulation in the soil environment is very high. In Table 9.7 one can see an example of soil contamination in the vicinity (9000 ha) of two copper works. Copper, lead and zinc contents were extremely high there. Also Andruszczak et al.(1986) found levels of Cu ranging from 71 to 1244 mg $kg^{-1}$ and Pb ranging from 95–645 mg $kg^{-1}$ at a distance of 600 m to 2550 m from the copper works.

An attempt to assess the contamination by heavy metals of Legnica voivodeship soils showed that copper and lead concentration in soils were much higher than in unpolluted arable soils in Poland. However, this data may be over- or underestimated because of the low number of samples and the wide range of concentrations.

The above two examples for heavy metals in polluted soils do not show the full extent of the problem. There are other areas with higher-than-background concentrations of heavy metals in soils, but they are more local, rather than regional in scale. And the concentrations of metals in these localities often do not exceed threshold values. There are other affected areas in the close vicinity of industrial plants, as well as in some urban areas (Czarnowska and Gworek 1988; Czarnowska and Walczak 1988).

## 9.6 Effects on Humans

Sensitivity of organisms to deficiencies or excesses of trace elements varies according to age, sex, and health conditions. The analyses of blood, urine, hair or nails for content of trace elements can be used as an indicator of their possible effects regarding deterioration of function of organisms and organs.

Contamination of the environment (air, soils, waters) with trace elements has direct or indirect (via the food chain) effects on human health. The health of the inhabitants in heavily polluted regions is generally worse than those in relatively unpolluted areas (Jonderko 1990; Sroczyński 1989).

The concentrations of Cd and Pb in blood are integral indices for intake of these elements by respiratory and alimentary pathways. Table 9.8 shows the

**Table 9.8.** Average concentrations (x) of cadmium in blood (Cd-B), of lead in blood (Pb-B), and of cadmium in urine (Cd-U) in µg $dm^{-3}$ in inhabitants of Szopienice and Łódź (industrial) and Suwałki (rural) areas in Poland (Jakubowski et al. 1991)

| Parameter | Szopienice | | Łódź | | Suwałki | |
|---|---|---|---|---|---|---|
| | n | x | n | x | n | x |
| Cd-B | 251 | 3.87 | 294 | 3.01 | 173 | 2.22 |
| Cd-U | 248 | 2.28 | 279 | 1.12 | 129 | 0.78 |
| Pb-B | 94 | 101.70 | 289 | 66.30 | 168 | 56.10 |

comparison of blood concentrations of cadmium and lead and urinary cadmium in inhabitants of Szopienice area, Katowice voivodeship (nonferrous metals industry) of Łódź (textile and electric industry) and Suwałki (rural, unpolluted region in the northeastern part). The highest average concentrations of Pb and Cd in Szopienice were significantly higher than for the other two areas. In the case of Pb these average concentrations in blood were in the range of values allowable by the European Community, and were lower or comparable to the average concentrations in other countries (Jakubowski et al. 1991). In contrast to Pb, the Cd concentrations in blood were higher than in other countries, indicating an apparently higher intake of Cd in the Polish regions. Other studies showed no regional effect (industrial or rural) on the concentration of Pb in the ribs of inhabitants of four compared districts: Katowice, Kraków, Gdańsk and Białystok (Jaworowski et al. 1988). The observed values were three times lower than for the standard "reference man" as defined by the International Commission on Radiological Protection.

Generally, studies show that the population of southern Poland today is less contaminated with Pb than it was between the eleventh and nineteenth centuries. In the other countries as well, the lowering of Pb concentrations in bones relative to past centuries has been observed in the contemporary population. In contrast to lead, the concentration of Cd in bones of Europeans has increased since the last century by about one order of magnitude (Jaworowski et al. 1988). The level of Cd in tissues of inhabitants of Katowice was two times higher compared to the three other areas, but lower than levels in the standard "reference man".

## 9.7 What May Happen with Heavy Metals Accumulated in the Soils?

Soils in polluted areas of Poland are prone to conditions favoring metal mobilization. In Katowice voivodeship the dominating soils are Podzols, Cambisols and Luvisols comprising about 68% of the agricultural land (Langhamer 1990). They were formed mostly from various types of sand a smaller part was formed from semiloess, and only a very small percentage was formed from silts, loams or clay rocks. These soils are acidic, low in humus content (up to 2%) and sorptive capacity, and highly permeable – all factors that promote ion migration. Similar types of soils are found in southwestern Poland.

The main factor determining metal behavior in soils is pH, which directly affects sorption and solubility (mobility) of metals. Small shifts in pH can cause a sharp increase or decrease in dissolved metal levels (Salomons 1988). Again in the affected regions, soils are high in acidity. About 51% of the soils in Katowice voivodeship (Langhamer 1990), and 61% in Legnica (Siuta 1990) are acidic, e.g., of $pH<5.5$.

About 22% of the soils in Katowice voivodeship are neutral and alkaline. These are rendzinas and some soils developed from semiloess, which are naturally neutral and alkaline. However, some soils in these areas are also artificially neutral, especially in the central part of the voivodeship (most industrialized). There, soils have been effected by emissions of dust containing alkaline metal oxides and in these areas the pH of the plough layer is usually neutral and decreases to acidic values in deeper layers (Langhamer 1990).

Generally, about 42% of agriculturally used land of Katowice voivodeship requires very intensive liming, and liming is strongly recommended on about another 30% (Langhamer 1990). However, as shown in Table 9.9, lime application in this area has decreased substantially in recent years. The same trend has been observed in the region of Lower Silesia and generally in the entire country.

Decreases in lime application to the soils causes their further acidification. Additionally, it should be pointed out that the soils in areas with high heavy metal loads, are naturally more subject to acidification than soils in the other parts of the country. It is not only that these soils have developed from acidic rocks, but that they are subject to the stronger leaching of bases as well. This is so because

**Table 9.9.** Changes in lime applications in Poland, Katowice and Legnica voivodeships and in the southwestern region in the years 1960–1992. (in kg of CaO per ha of agriculturally used land per year; source: GUS-Statistical Annuals for 1961–1993)

| Area/year | 59/60 | 68/69 | 79/80 | 84/85 | 89/90 | 90/91 | 91/92 |
|---|---|---|---|---|---|---|---|
| Poland | 12.4 | 87.8 | 159.7 | 139.4 | 182.4 | 139.0 | 117.2 |
| Katowice | 39.2 | 105.4 | 166.1 | 170.4 | 216.0 | 196.8 | 172.1 |
| Legnica | 25.5 | 157.2 | 205.7 | 163.9 | 338.0 | 296.6 | 125.0 |
| SW region [a] | 28.3 | 166.1 | 201.1 | 175.1 | 265.2 | 210.5 | 146.8 |

[a] mean values for voivodeships: Opole, Wroclaw, Walbrzych, Jelenia Góra, Legnica

**Table 9.10.** Air pollution in Upper Silesia and Katowice Province in 1981 and 1987 by dust (t $km^{-2} a^{-1}$); $SO_2$, $NO_x$ and Pb concentrations [$\mu g\ m^{-3}$]. (Marchwińska et al. 1989)

| | Dust deposit | | $SO_2$ | | $NO_x$ | | Pb | |
|---|---|---|---|---|---|---|---|---|
| | 1981 | 1987 | 1981 | 1987 | 1981 | 1987 | 1981 | 1987 |
| Upper Silesia | 284 | 154 | 46 | 52 | 74 | 98 | 1.02 | 0.38 |
| Central part of Silesia | 365 | 253 | 55 | 73 | 87 | 118 | 2.30 | 0.53 |
| Katowice voivodeship | 150 | 112 | 41 | 40 | 63 | 88 | 0.50 | 0.31 |
| Allowed standard [a] | 250 | | 64 | | 22 | | 0.2 | |

[a] refers to standards at that time; currently they are: 200 t $km^{-2} a^{-1}$ for dust; 32 $\mu g\ m^{-3}$ for annual concentration of $SO_2$; and 50 $\mu g\ m^{-3}$ for annual concentration of $NO_x$ (as $NO_2$)

precipitation is relatively high there as compared to the other parts of Poland, and the positive water balance is one of the highest in the country (Rojek 1987).

As already mentioned, in some areas acid pollutants in the past were neutralized to some extent by deposition of alkaline dust. In the last decade, however, deposition of dust has decreased, but the concentration of $SO_2$ and $NO_x$ have increased significantly (Table 9.10) particularly in the case of nitrogen oxides. Data by the Sanitary-Epidemiology Station (WSSE 1991) show that the annual average concentration in years 1986–1990 measured at 27 stations in Katowice voivodeship in the late 1980s was still high and ranged from 31 to 81 and from 62 to 141 $\mu g\ m^{-3}$ for $SO_2$ and $NO_x$ respectively.

The present accumulation of heavy metals in soils of industrial regions is already very high. Therefore, decreasing inputs will not solve the problem because metals once accumulated will circulate in the soils for a very long time. According to Marchwińska and Kucharski (1988), assuming that deposition of Pb and Cd will not exceed the allowed standards, enrichment of soils with these metals in Katowice voivodeship will amount to 0.76 mg of Pb and 0.01 mg of Cd per kg of soil per year. The amount of metals removed with biomass and leaching may decrease the Pb content in the soil surface by about 0.08 and the Cd by about 0.0025 mg $kg^{-1}a^{-1}$. Hence, the authors state that even with very intensive cultivation of these soils combined with a hypothetical lowering of dust deposition to zero, decreasing the metal concentrations in the soils to allowable standards would require many generations.

Another factor which may change soil properties and metals mobility is surface deformation caused by intensive industrial activity, especially mining of pit coal and nonferrous ores. It is estimated that in Katowice voivodeship, 46 403 ha (7.4%) of land are affected by mining activities (Marchwińska et al. 1989). About 690 ha of the land are permanently flooded and 2733 ha are in the range of local "depression cones" (landforms caused by subsidence due to underground mining activities). The average subsidence amounts to 0.2 m per year (Konstantynowicz 1989). By the year 2010, some 12% of the district area could be affected by the mining industry. Land deformations cause changes in the soil hydrology and soil water properties. Any flooding will cause decreases in redox potential, which in turn will increase the mobility of some metals.

It must be remembered that, in most cases, these soils will remain contaminated for hundreds of years and that bioavailability as well as mobility will vary during this time as a result of changes in soil properties. Although "clean-up" procedures may be possible in some circumstances they are unlikely to be practical for large areas of contaminated land in Poland, mainly because of the very high costs of soil cleaning. This means that other control measures are necessary, such as increased use of liming and controlling the hydrology of a region (methods of geochemical engineering, as suggested by Salomons and Förstner 1988)

## 9.8 Concluding Remarks

In Poland we encounter a number of simultaneous problems which reinforce each other with regard to their environmental impact. The area is still subject to $SO_2$ and $NO_x$ emissions to the air. Acid deposition will continue as well as the breakdown of the buffering capacity of soils against pH changes. The resulting lowering of the pH in areas of high metals accumulation in soils, when no control measures are undertaken, will cause an increase in their leaching and an increase in their bioavailability.

Very polluted areas such as Upper Silesia have also experienced what can be called a surprise factor. The control measures which caused a decline in the dust emissions are on the other hand responsible for an increase in acidic deposition, since acids are less neutralized by the reduced amounts of alkaline dust particles.

The mining activity has caused changes in hydrology in some areas, and, as a consequence, changes in the redox conditions in the soil, thus affecting metal solubility changes.

To summarize one can state that all conditions in the areas of heavy metal pollution will exert a tendency toward delayed responses to the environment. Even in those cases where the input of metals will decrease, increases in acid precipitation and/or changes in hydrology will effect the mobility of the metals already accumulated in the soils.

The discussion presented here on prospective future problems has been rather qualitative, since more knowledge is urgently required on the kinetics of long-term effects of acidification on metal behavior in soils of contaminated regions such as Upper and Lower Silesia. The study should comprise the complex laboratory and field studies combined with elaboration of predicting models to assess/simulate time dependent changes in heavy metals status due to environment/soil acidification. It is needed not only from a scientific point of view but from a practical standpoint as well, since it has considerable practical implications for agriculture, the food chain, and human health.

### References

Andruszczak E, Czuba R (1984) Preliminary characteristics of the total content of macro and microelements in Polish soils.(in Polish) Roczniki Gleboznawcze 2:61-78

Andruszczak E, Straczyński S, Czerniawska W, Radwan (1986) Content of some elements in soils and plants as influenced by emissions of the copper works.(in Polish) Roczniki Gleboznawcze 4:47-66

Angelone M, Bini C (1992) Trace elements concentrations in soils and plants of western Europe. In: Adriano DC (ed) Biogeochemistry of trace metals. Lewis Publ, Chelsea, Michigan, pp 19-60

Boratyński K, Roszyk E, Ziętecka M (1972) Survey of investigations on trace elements carried out in Poland, part II. Zinc, Molybdenium, Cobalt, Titanium, Nickel, Chromium and other elements.(in Polish) Roczniki Gleboznawcze 1:285-333

Czarnowska K (1989) Content of some heavy metals in soils developed from silty formations (in Polish). Roczniki Gleboznawcze 2:107-117
Czarnowska K, Gworek B (1987) Heavy metals in some soils of the central and northern parts of Poland (in Polish). Roczniki Gleboznawcze 3:41-57
Czarnowska K, Gworek B (1988) Contamination of Warsaw soils with cadmium (in Polish). Roczniki Gleboznawcze 4:129-133
Czarnowska K,Walczak J (1988) Distribution of zinc, lead and manganese in soils of Łódź city. Roczniki Gleboznawcze 1:19-27
Davis BE (1987) Consequences of environmental contamination by lead mining in Wales. Hydrobiologia 149:213-220
Dobrzański B, Zawadzki S (1981) Soil Science (in Polish). Państwowe Wydawnictwo Rolnicze i Lesńe, Warsaw
Drozd J, Kowaliński S (1977) Changes of certain soil properties as influenced by pollutants emitted by copper smelter works "Legnica" (in Polish). Roczniki Gleboznawcze 2:49-75
Dudka S (1992) Establishing baseline concentrations of major and trace elements in surface soils of Poland (in Polish). Institute of Soil Science and Plant Cultivation, Puławy, pp 1-48
Dudka S, Sajdak G (1992) Evaluation of concentrations of some trace metals in soils of Katowice voivodeship. Archiwum Ochrony Środowiska 2:125-134
Fotyma M (1987) Reaction and acidity of soils. In: Foytma M., Mercik S, Faber A (eds) Chemical principles of soil fertility and fertilization. Panstwowe Wydawnictwo Rolnicze i Lesne (State Agricultural and Forest Publisher), Warsaw, pp 34-56 (in Polish)
Gliński J (1987) Problems of soil degradation in Poland. Zeszyty Problemowe Postępów Nauk Rolniczych 344:7-15
Greinert H, Gajewski K, Drab M (1987) Assessement of the contamination with heavy metals of Legnica province soils (in Polish). Archiwum Ochrony Środowiska 1-2:135-142
GUS – The Main Statistical Office, Warsaw (1989) Statistical materials 68. Environment protection (in Polish). Zakład Wydawnictw Statystycznych pp 64
GUS – The Main Statistical Office, Warsaw (1990) Report on state endangerment and environment protection (in Polish). Zakład Wydawnictw Statystycznych pp 1-357
GUS – The Main Statistical Office, Warsaw (1991) Environment protection 1991 (in Polish). Zakład Wydawnictw Statystycznych
GUS – The Main Statistical Office, Warsaw. Yearbooks: 1961; 1970; 1981; 1986; 1991; 1992; 1993
Gworek B (1985a) Trace elements (Mn, Zn, Cr, Cu, Ni, Co, Pb and Cd) in cultivated soils developed from boulder loams and silty formations of the northeastern region of Poland (in Polish). Roczniki Gleboznawcze 2:43-59
Gworek B (1985b) Microelements in soils developed from silty formations in the northeastern part of Poland (in Polish). Roczniki Gleboznawcze 3:41-50
Gzyl J (1990) Lead and cadmium contamination of soil and vegetables in the Upper Silesia region of Poland. Sci Total Environ 96:199-209
Hanneberg P (1993) Acidification and air pollution: a brief guide. Swedish Environmental Protection Agency, 35 pp
Jakubowski M, Razniewska G, Trzcinka-Ochocka M, Trojanowska B (1991) Pb and Cd levels among inhabitants of three regions in Poland (in Polish). Archiwum Ochrony Środowiska 3/4:165-176
Jaworowski Z, Bilkiewicz J, Pietrzak-Flis Z, Sulpinska M (1988) Influence of industry on contamination of Polish population with natural radionuclides and heavy metals (in Polish). Archiwum Ochrony Środowiska 1/2:101-112

Jonderko G (1990) Directions and outcome of investigations on the influence of environmental pollution upon the human health of the population in the Upper Silesian industry region. Archiwum Ochrony Scrodowiska 3/4:7-15

Kabata-Pendias A (1981) Heavy metal concentrations in arable soils of Poland (in Polish). Pamiętnik Puławski 74:101-111

Kabata-Pendias A, Pendias H (1993) Biogeochemistry of trace elements (in Polish). Wydawnictwo Naukowe PWN, Warsaw, 364 pp

Kabata-Pendias A, Piotrowska M, Witek T (1993) Evaluation of the possibilities for agricultural use of soils contaminated with heavy metals. In: Kabata-Pendias A, Matowicka-Terelak T, Piotrowska M, Terelak H, Witek T (1993) Evaluation of soils and plants contamination with heavy metals and sulfur. Instructions to agriculture (in Polish). IUNG, Pulawy, pp 5-14

Kabata-Pendias A, Dudka S, Chłopecka A, Gawinowska T (1992) Background levels and environmental influences on trace metals in soils of the temperate humid zone of Europe. In: Adriano DC (ed) Biogeochemistry of trace metals. Lewis Publishers, Chelsea, Michigan, pp 61-84

Kern H (1987) Acidity and CaCO3 content in soils of the agricultural areas of Poland. Zeszyty Problemowe Postepów Nauk Rolniczych 344:43-58

Konstantynowicz E (1989) The influence of development of resources base in Katowice District on the state of its environment. Conference materials: functioning barriers of Upper Silesia. Regional Group for Protection and Management of Environment in Upper Silesia, Katowice (in Polish), pp 65-90

Kucharski R, Marchwińska E, Piesak Z, Nikodemska E, Witala BW (1989) Pollution of pasture plants with lead and cadmium in chosen regions of Katowice province (in Polish). Veterinary Medicine 3:162-166

Langhamer L (1990) Natural conditions for Agricultural Production. Katowice voivodeship (in Polish). IUNG, Pulawy, pp 1- 40

Leita L, De Nobili M, Sequi P (1988) Content of heavy metals in soils and plants near Cave del Predil, Udine, Italy. Agrochimica 32:94-97

Leita L, De Nobili M, Pardini G, Ferrari F , Sequi P (1989) Anomalous contents of heavy metals in soils and vegetation of a mine area in SW Sardinia, Italy. Water Air Soil Pollut 48:423-433

Leśniok M, Opałka-Gądek J (1993) Rainfall contamination in transboundary area of pollutants flow in southern Poland. Materials of 6th Conference on chemistry of rainfalls, surface and ground waters. Łódź, 16-18 Sept 19, pp 37

Marchwińska E, Kucharski R (1988) The state of arable soils pollution with heavy metals in Katowice voivodeship and possibilities to conteract it (in Polish). Materials for the conference: agriculture in industrial agglomeration of Upper Silesia, Katowice, September 1988. pp 43-53

Marchwińska E, Kucharski R, Gzyl J(1984) Cadmium and lead concentrations in samples of potatoes from various regions of Poland (in Polish). Roczniki PZH 2:113-118

Marchwińska E, Bukowy E, Fudala J, Grzbiela Z, Hanus K, Hławiczka S, Kosarewicz O, Kucharski R, Pastuszka J, Sieja L, Wysokińska E, Suschka J (1989) Information on the state of environment in Katowice district (in Polish). Institute of Environment Protection, Katowice, November1989, pp 1-68

Rojek M (1987) The time and spatial distribution of climatic and agricultural-climatic water balances in territory of Poland (in Polish). Agricultural University in Wroclaw, pp 1-61

Roszyk E, Roszyk S (1975) The influence of copper metallurgy on certain soil properties and chemical composition of cultivated plants, part I. First year of emission (in Polish). Roczniki Gleboznawcze 3:277-291

Roszyk E, Roszyk S (1976) The influence of copper metallurgy on certain soil properties and chemical composition of cultivated plants, part II. Second year emission (in Polish). Roczniki Gleboznawcze 4:57-58

Roszyk E, Szerszeń L (1988a) Accumulation of heavy metals in the arable layer of soils of the sanitary protection zone in the vicinity of copper metallurgic plants, part I. The "Legnica" metallurgic plant (in Polish). Roczn Glebozn 4:135-146

Roszyk E, Szerszeń L (1988b) Accumulation of heavy metals in the arable layer of soils of the sanitary protection zone in the vicinity of copper metallurgic plants, part II. "Głogów" (in Polish). Roczn Glebozn 4:147-158

Ryszkowski L, Balazy S (1988) Environmental hazards in Poland and a strategy for conservation of living natural resources. INTECOL Bull 16:17-28

Sądelski M (1989) Actual state and possibilities of improvement of air quality in Katowice province. Conference materials: functioning barriers of Upper Silesia. Regional group for protection and management of environment in Upper Silesia, Katowice (in Polish), pp 116-135

Salomons W (1988) Impact of metals from mining and industry on the hydrosphere. In: G Strigel (ed) Metals and metalloids in the hydrosphere; impact through mining and industry, and prevention technology. Proceedings of an IHP workshop, Bochum, Germany, 21-25 September 1987. Unesco, Paris, pp 1-41

Salomons W, Förstner U (eds)(1988) Environmental management of solid waste: dredged materials and mine tailings. Springer, Berlin Heidelberg New York, 396 pp

Siuta J (1976) Comment to the "Map of soils resistance to degradation" (in Polish). Institute of Environment Protection (ed) Warsaw, pp 15

Siuta J (1990) Structure of degradation and renovation of biologically active land in Poland (in Polish). Institute of Environment Protection (ed) Warsaw, pp 55

Sroczyński J (1989) The influence of environment pollution in Upper Silesia on health of inhabitants. (in Polish) Conference materials: functioning barriers of Upper Silesia, Regional Group for Protection and Management of Environment in Upper Silesia, Katowice, pp 43-64

Strzyszcz Z (1990) Acid precipitation research in Poland. In: Bresser AHM, Salomons W (eds) Acid precipitation, international overview and assessment, vol 5. Springer, Berlin Heidelberg New York, pp 215-228

Twarowski R, Gendolla T, Institute of Meteorology and Water Management in Wroclaw. Variation in surface loading of Karkonosze mountain area with rainfall pollutants (in Polish). Manuscript (with permission)

Twarowski R, Gendolla T (1993) Rainfall chemistry in Lower Silesia (in Polish). Materials of 6th Conference on Chemistry of rainfalls, surface and ground waters, Łódź, 16-18 Sept 1993.

Ure AM, Berrow ML (1982) The elemental constituents of soils. In: Bowen HJM (eds) Environmental chemistry, vol 2; a specialist Periodical Report. A review of the literature published up to mid-1980. The Royal Society of Chemistry, Burlington House, London, W1V OBN, pp 94-203

Wasilewski M, Twarowski R, Institute of Meteorology and Water Management in Wroclaw. The influence of rainfall pollution on the water quality in Karkonosze region (in Polish). Manuscript (with permission)

Wlodarczyk Z (1983) Acidity state of Polish soils. Proceedings of symposium: liming as a factor for increasing of soil fertility.(in Polish) Pulawy, pp 37-44

WSSE (Voivodeship Sanitary-Epidemiology Station in Katowice)(1991) Air pollution in Katowice voivodeship in 1988-1990.

WUS (Voivodeship Statistical Office) in Katowice (1993) Report on "environment protection". (With permission of Ecology Department of voivodeship office in Katowice)

# 10 Mercury and the New Gold Rush in the South

*L. D. Lacerda, O. Malm, J. R. D. Guimarães, W. Salomons, and R.-D. Wilken*

## 10.1 Introduction

Mercury contamination is considered one of the worst hazards among the anthropogenic impacts upon the environment. Mercury is one of the few metal pollutants which has already caused human deaths due to ingestion of contaminated food. It is estimated that globally more than 800 humans have died and more than 17 000 have been afflicted due to mercury poisoning (Lodenius 1985; D'Itri and D'Itri 1977, 1991, 1992). Also, the diverse uses of mercury have resulted in severe contamination of fish in many regions with significant economic impacts on populations depending on fisheries and other aquatic resources. It was in the fishing villages of Minamata and Niigata, where in the 1950's and 1960's more than 100 people died and several hundreds were handicapped for life due to consumption of mercury contaminated fish.

The seriousness of the situation has lead to intense control of Hg emissions to the environment in most countries. This has resulted in a significant decrease in the global Hg inputs to the environment. For instance, estimates of anthropogenic Hg emissions to the atmosphere from the earlier 1970's generally range from 10 000 to 30 000 ton $a^{-1}$ (Andren and Nriagu 1979; WHO 1976). By the late 1980's however, the estimates of range decreased to 1000 to 11 000 ton $a^{-1}$ (Fitzgerald et al. 1984; SCOPE 1985; Nriagu 1990). Chlor-alkali plants using Hg cells, Hg fungicides and other organic Hg compounds previously used in agriculture are virtually disappearing in most countries. For example, mercury consumption in chlor-alkali plants in Brazil dropped from 80 tonnes per year in 1979 to only 18 tonnes per year in 1985, as a direct result of the substitution of Hg cells, responsible for 90% of Hg utilization in chlor-alkali plants prior to 1980 (Bezerra

1991). The decrease in industrial uses of Hg and the banning of Hg fungicides in Brazil reduced the industrial and agricultural inputs of Hg to the environment to a few tons per year (Ferreira and Appel 1991). As a direct result of the decrease in Hg inputs to the environment from anthropogenic sources, a significant decrease in Hg content of fish in previously contaminated areas occurred throughout the world (Nuorteva et al. 1979; CETESB 1980; Boldrini et al. 1983; Eysink 1990).

Notwithstanding the sharp decrease in Hg emissions from industry and agriculture, Hg releases into the environment have recently aroused new concern regarding Hg utilization in gold and silver mining. Applied worldwide up to the beginning of the present century, this "technology" faced a decline to near cessation due to the invention of the cyanide leaching process. Recently however, a confluence of economic and social situations, mostly in developing countries located in the tropics, has resulted in a new rush for gold and silver by individual entrepreneurs, for whom Hg amalgamation is a cheap, reliable and simple-to-carry-out operation.

Environmental problems caused by the utilization of Hg in gold and silver mining have been reported since Roman times. Circa 77 A.D. Roman authorities imported 5000 kg of Hg from Spain to be used in gold amalgamation in Italy. Curiously less than 100 years later this was forbidden in mainland Italy (D'Itri and D'Itri 1977; Nriagu 1990). With the development, in 1554 in Spanish America of the "Patio" amalgamation process by Bartolomeu de Medina, Hg amalgamation reached its peak. In 1870 over 70% of Mexican silver was produced through this technology. During colonial times between 1570 and 1900, it is estimated that of Hg release into the environment from the extraction of silver and gold in central and northern South America reached 196 000 tons (Nriagu 1993). During the late 1800's North America experienced its gold rush and the Hg related problems associated with it. Tailings containing Hg of up to 500 $\mu g\ g^{-1}$ are still present in central Canada, while in the Carson River area over 100 000 tons of tailings containing Hg with concentrations up to 4900 $\mu g\ g^{-1}$ are being continuously reworked, potentially mobilizing over 7000 tons of Hg, from tailings (Lechler and Miller 1993).

After the development of cyanide leaching process, Hg amalgamation virtually disappeared as a significant mining technology up to the 1970's. A dramatic increase of gold prices, from US$ 58.16 /oz tr in 1972 to US$ 436.6 /oz tr in 1985, triggered a second gold rush in many tropical countries, particularly in Latin America. In Brazil, Hg amalgamation was responsible for the production of only 5.9 tons of gold in 1972. In 1989 this technology was responsible for the production of nearly 100 tons of gold, mostly from the Amazon region. This new gold rush rapidly spread through other Amazonian countries (in particular Venezuela, Peru and Colombia), and later to the Philippines, Thailand and Tanzania.

## 10.2 Mercury Use in Gold and Silver Mining and its Dispersal Mechanisms in Aquatic Systems

The different mining processes using Hg amalgamation in tropical America and Asia result in different wastes, Hg dispersal mechanisms, degrees of mobility, and biological availability.

The various procedures can be grouped in two large categories occurring in different gold-mining regions.

The first involves the recuperation of gold from soils and rock, with gold content ranging from 4.0 to 20.0 g per ton. This consists of digging large amounts of Au-rich material, generally soils containing quartz veins, passing it through grinding mills and centrifuges to produce a Au-rich gravimetric concentrate. The concentrate is moved to small amalgamation ponds, a few square metres large or to amalgamation drums, where it is mixed with liquid Hg and later separated in round pans. Then the amalgam-Hg mixture obtained is separated by pressing through leather or cloth into the highly viscous amalgam ($Au_2Hg$ and $Au_3Hg$) and liquid Hg. It is common to roast the amalgam in pans in open air (temperatures of 350° to 600°C). Thus releasing nearly all the Hg content of the amalgam directly to the atmosphere. The impure gold, still with about 1% Hg, is sold nearby.

Any residue of the concentrate is returned to the amalgamation pond and continued to be reworked until exhaustion of the gold in the area. After an area is exhausted from the gold, tailings are left with spots of high Hg concentrations in amalgamation ponds (CETEM 1989; Andrade et al. 1988; Lacerda et al. 1991 a, b; Ramos and Costa 1990).

The second typical process is carried on in most Amazonian rivers. Gold is extracted from bottom sediments by dredging. After removal of large stones the material is passed through carpeted riffles which retain heavier particles. This operation lasts for 20–30 h, then the dredging stops and the heavy fraction is collected in barrels for amalgamation. The amalgam is then separated in the same way described above.

Pfeiffer and Lacerda (1988) have estimated Hg losses to the environment in a mining site in the Madeira river, Rondonia State, NW Amazon (Fig. 10.1). They estimated that the major proportion of Hg loss is during the burning of the Au-Hg, reaching 50 to 60% of the total Hg loss, an extra 5% is vaporized during the various extraction steps. From 40 to 50% was lost to rivers during the amalgamation process as metallic Hg, and from 5 to 10% during the recuperation of the Hg used in the process also to rivers. The estimated emission factors (EF) from this gold-mining area in the Madeira river, is 1.32 (kg of Hg released to the environment for the production of 1.0 kg of gold); in the Philippines, Cramer (1990) estimated the EF to be 1.0, while CIMELCO (1991) estimated EFs ranging from 0.9 to 1.1 for the mining areas near the Peru-Ecuador border along the Puyango-Tumbes river. Müller (1993) reported EFs ranging from 1 to 1.5 for the Tapajós

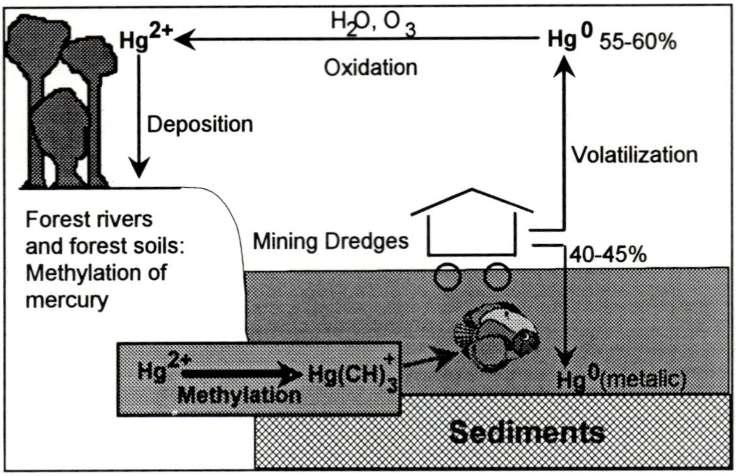


**Fig. 10.1.** Mercury cycle in gold-mining sites in large rivers of the Amazon. Based on Malm et al. (1990)

river in south-eastern Amazon. Although estimates from historical mining are very scarce, Nriagu (1993) reported EFs ranging from 0.85 to 1.5 for silver mining in colonial Peru.

Mercury lost to rivers as metallic Hg is preferentially accumulated in bottom sediments and generally presents very low mobility. In the Madeira river, a tributary of the Amazon river, with mean flow of 23 000 to 49 000 $m^{-3}s^{-1}$ during the dry and rainy seasons respectively, Hg concentrations in sediments close to operating dredges can reach values up to 157 $\mu g\ g^{-1}$ decreasing to background values ($\approx$ 0.04 $\mu g\ g^{-1}$) a few kilometres downstream (Pfeiffer et al. 1989a, b; Malm 1991). In smaller rivers Hg dispersion is even lower with background concentrations being found only a few hundred metres from the source (Andrade et al. 1988).

The low mobility displayed by Hg is mostly due to its chemical form. Mercury emitted to rivers is present as elemental Hg with its high density of 13.6 and low reactivity. Drops of metallic Hg have been frequently found in exposed river sediments during the dry season. However, Martinelli et al. (1988) working in the same Madeira river reported high Hg concentrations (up to 1.04 $\mu g\ g^{-1}$) in floating macrophytes from oxbow lakes 100 km away from the nearest Hg source. Müller (1993) studied the deposition of Hg in the lower section of the Tapajós river and found evidence of Hg transport over several hundred kilometres from mining sites associated with the suspended particles. This suggests a long range Hg transport mechanism associated with suspended particles. Nriagu et al. (1992) suggested that nearly 70% of the Hg released into the Madeira river is exported downstream as far as the Amazon river, 1000 km away. The dredging and consequently the resuspension of bottom sediments, facilitates this process. Sediment transport and remobilization may explain fish contamination in the Madeira river circa 200 km downstream from major mining sites (Malm et al. 1990).

Another example of long range transport of Hg from gold-mining areas was demonstrated for the Paraíba do Sul river, south-eastern Brazil. Gold production in this river is small (only between 20 to 50 kg per year). However, gold-mining is the major source of Hg to the lower basin of the river. It is estimated that from 150 to 300 kg of Hg has been released in the river since 1985 (Lacerda et al. 1993). Concentrations drop from a maximum of 550 $\mu g\ kg^{-1}$ in the fluvial sediments to 110 to 150 $\mu g\ kg^{-1}$ in the estuary and river plume. Only at a water depth of over 70 m, nearly 70 km offshore, are Hg concentrations down to background values, between 30 and 50 $\mu g\ kg^{-1}$ (Lacerda et al. 1992).

In areas where gold is mined from soils or rock, Hg is deposited and accumulated in mine tailings. An important question is: how is Hg eventually transported and spread to pristine adjacent environments. Although only few studies have dealt with this subject, a general pattern of Hg distribution in tailings can be stated:

- In the majority of the tailings Hg is present as metallic Hg (CETEM 1989; Lacerda 1990; Andrade et al. 1988; Ramos and Costa 1990; Lechler and Miller 1993).
- Mercury distribution in tailings is a function of the distribution of mining activities in the area rather than any geochemical process. In areas where old pre-concentration ponds existed, Hg concentration can reach a few micrograms per gram of tailing material, while mean Hg content in the whole tailings body are similar to local background values (CETEM 1989).
- Different mining techniques may result in different Hg concentrations in tailings. Typically these concentrations range from less than 0.5 $\mu g\ g^{-1}$ up to 4900 $\mu g\ g^{-1}$ dry weight (Ramos and Costa 1990; Andrade et al. 1988; Lechler and Miller 1993).

In small creeks draining mine tailings in central Brazil, Andrade et al., (1988) found that Hg contamination of the Crixás deposit, Goiás State, decreases from a maximum value of circa 16 $\mu g\ g^{-1}$ at the tailings to background levels of circa 0.7 $\mu g\ g^{-1}$ in the first 200 m along the drainage area. The same behaviour was found by Lacerda et al. (1991a) at the Tanque dos Padres tailings, Mato Grosso State. Mercury concentrations decrease from values up to 3.0 $\mu g\ g^{-1}$ in the tailing to background levels ($\approx 0.04\ \mu g\ g^{-1}$) in the first 100 m along the drainage area. Although Hg leaching from tailings seems a slow process, through the years the migration of tailings themselves may be the most important Hg dispersal mechanism. In the Carlson City mining district over the past 134 years, Hg-rich tailings were transported over 100 km from the Six Mile Canyon and tributaries to the Carlson River valley, forming a Hg-rich 1.5 $km^2$ fan area (Lechler and Miller 1993; Miller et al. 1993).

Brabo et al. (1991) found that pelitic bottom sediments are easily remobilized to the water column in south Amazonian rivers affected by gold-mining. This results in high Hg concentrations in suspended particles ranging from 0.2 to 1.1 µg $g^{-1}$ dr. wt., similar to the concentrations found in bottom sediments which ranged from 0.1 to 3.0 µg $g^{-1}$ dr. wt. This mechanism is reported as the major process controlling Hg dispersion in these rivers.

Changes in Hg concentrations in water and suspended particles, as well as in major physico-chemical parameters in water draining tailings in Mato Grosso State, occur during storm events. The water becomes progressively more acidic (pH= 6.8 to 5.1) and oxidized (Eh= 280 to 430 mV), and enriched in suspended particles which can reach values 10 times higher than during dry conditions (6.0 to 66 mg $l^{-1}$) These changes occur just a few minutes after the beginning of rain. Mercury concentrations in suspended particles increase from <0.02 to 0.61 µg $l^{-1}$, although dissolved Hg concentrations remain below 0.04 µg $l^{-1}$. These results show the capacity of rain to erode fine particles enriched with Hg from the tailings, suggesting that erosion, followed by Hg transport associated with suspended particles, is the main pathway of Hg contamination of the tailings drainage. The fact that dissolved Hg concentrations remain constant during rainstorms suggests that only small, if any, amounts of Hg are remobilized into solution and that Hg concentration in water is independent of Hg content in sediments (Andrade et al. 1988). Nriagu et al. (1992) studying Hg partitioning between dissolved and particulate forms in the Madeira river waters also showed that dissolved Hg concentrations are independent of their concentrations in suspended sediments.

Mudroch and Clair (1986) found the same Hg distribution pattern in waters draining tailings in Canada. They have also suggested that contamination of sediments was due to transport and deposition of suspended particles brought into the drainage water by erosion of mining wastes. Lechler and Miller (1993) found similar behaviour of Hg in waters draining tailings in the Carlson River Valley, United States.

The fact that none of these studies found significant changes in dissolved Hg concentrations, even during strong storm events, suggests that only small amounts of Hg are remobilized into solution and that Hg concentration in water is independent of Hg content in sediments (Andrade et al. 1988). Thus, controlling waste erosion and suspended particle transport from tailings, by means of dams, impermeable barriers and other protecting devices, should be quite efficient in preventing Hg contamination in surrounding areas. The correlationship between Hg dispersal and erosion and transport of suspended particles suggests that the dispersal mechanisms should vary with the season, either through intensive erosion during the rainy season or by a dilution effect upon existing Hg content in rivers. Few studies, however, have measured Hg concentrations in environmental samples through a yearly cycle. The existing results are therefore still preliminary.

Rodrigues et al. (1991), measured total Hg concentrations (dissolved and particulate) in water of the Baía river, southern Brazil, and found that during the dry season, Hg concentrations ranged between 0.05 µg $l^{-1}$ to 0.88 µg $l^{-1}$, while during the rainy season Hg concentrations ranged between 1.0 µg $l^{-1}$ to 12.0 µg $l^{-1}$. Also, the variability among data during the same sampling period was much larger during the rainy season, suggesting pulse contamination of Hg, as a consequence of leaching and transport of contaminated soil particles to the river. Tumpling et al. (1993) showed significant seasonal variation in the concentration of volatile Hg in waters draining tailings in Mato Grosso State, central Brazil, with concentrations ranging from 0.3 $\mu g\ l^{-1}$ during the dry season to 5.2 $\mu g\ l^{-1}$ during the rain season, suggesting stronger leaching of Hg from tailings. In the Carajás mining site, Fernandes et al. (1990 a,b) found contrary results. They measured highest total Hg concentrations in river water during the dry period, ranging from 0.10 to 0.74 $\mu g\ l^{-1}$ (when total suspended solids concentrations varied between 25 and 100 mg $l^{-1}$), while lower concentrations were found during the rainy season: <0.05 to 0.38 $\mu g\ l^{-1}$ (when total suspended solids concentrations varied from 50 to 400 mg $l^{-1}$). They explained their results as the effect of dilution and the fact that mining activities in the area are restricted to the dry season. The difference in mining procedure seems to be a key factor controlling Hg dispersal. In areas where gold is mined from soil or rocks, during the rainy season the leaching and erosion of contaminated particles results in higher total Hg concentrations in rivers, while in areas where Hg is mined from river bottom sediments Hg concentrations in water are lower during the rainy season due to dilution and the decreasing of the mining activity itself.

## 10.3 Mercury Distribution in Aquatic Sediments in Gold-mining Areas

The fate of Hg in water and sediments in gold-mining sites has been studied mostly in large surveys in a great number of mining sites; few however, have been studied in detail. A summary of Hg concentrations in sediments of various Amazon river systems (Fig. 10.2) is presented in Table 10.1. Also, mean values for non-contaminated areas in the Amazon and for northern hemisphere regions are presented for comparison. By comparing Hg concentrations in areas with different mining procedures, it is clear that Hg concentrations in water and sediments draining tailings are considerably lower than in other contaminated areas, in particular those in large rivers of the Amazon region where Hg is released directly into rivers.

In areas such as the Mutum Parana and Madeira rivers, a total of 25 tons of Hg is released annually to the environment (Lacerda et al. 1989), which is higher than the total amount of Hg released in most areas where gold is extracted from rocks and soils. However, Hg concentrations in these tailings drainages are still slightly higher than in non-contaminated Amazon rivers. The most contaminated rivers systems are those in the Amazon region itself where gold-mining is traditional and large scale, such as in the Madeira River basin, which includes the Mutum Parana river.

**Fig. 10.2.** Major gold-mining sites in Brazil

Most bottom sediment samples collected along Madeira river, one of the best studied gold-mining areas, (approximately 150 samples) show values close to local background (around 0.05 μg $g^{-1}$), and are always smaller than the global average value for bottom sediments of 0.3 μg $g^{-1}$ (Salomons and Förstner 1984). Where elevated levels were found, they could be explained by the proximity of operating dredging boats or special hydrodynamic conditions. In the confluence of the Madeira and Mutumparaná rivers (the latter is used for parking, maintenance, small mining activities as well as a waterway for gold-mining boats) regularly high Hg concentrations were observed. (1.64; 19.5; 157.3; and 6.53 μg $g^{-1}$) in four different periods (March/87; June/87; June/88 and October/89 respectively). This may be attributed to physically favourable conditions for $Hg^0$ retention. In October 1989 and 1991 metallic Hg was found in rock crevices at the upper border of Teotônio waterfall after lowering of the water level in the dry season. Highest Hg values (20 μg $g^{-1}$) in the Madeira river bed were also observed in newly exposed sediments of upstream waterfalls (Malm 1991). Strong gradients have also been frequently found in both sediments and water Hg concentrations from point sources of Hg, such as tailings. In general these gradients are very steep indicating the low dispersability of Hg from the releasing points.

**Table 10.1.** Mercury concentration (μg $g^{-1}$ dr.wt.) in bottom sediments from gold-mining areas in Brazil compared with non-contaminated sites in the Amazonian region and with world rivers.

| Locality | Hg in sediments | Author |
|---|---|---|
| Madeira river, Amazon | 0.05–19.72 | Lacerda et al.(1987); Malm (1991) |
| Mutum Parana, Amazon | 0.21–157.30 | Pfeiffer et al. (1989a) |
| Itacaiunas-Paraopebas, Carajas | 0.04–3.37 | Fernandes et al. (1990 a,b) |
| Tapajós river, Pará | <0.01–0.14 | Padberg (1990) |
| Venezuelan Amazon rivers | 0.12–4.81 | Shrestha et al. (1989) |
| Peru-Ecuador, Puyango river | 2.30 | CIMELCO (1991) |
| Paraíba do Sul river, SE Brazil | 0.30–0.90 | Pfeiffer et al. (1989b) |
| Tanque dos Padres, Mato Grosso | 0.05–0.18 | Lacerda et al. (1991a) |
| Crixás tailings, Goias | 0.27–12.8 | Andrade et al. (1988) |
| Crepori/Marupa rivers, Pará | 0.10–3.0 | Brabo et al. 1991 |
| Amazon rivers [a] | ≈0.04 | Malm (1991) |
| World rivers [a] | ≈0.30 | Salomons and Förstner (1984) |

[a] non-contaminated

Metallic Hg is stable in aquatic environments under reducing to slightly oxidizing conditions. If covered by sediments, $Hg^0$ may remain in this form for long periods. The fate of $Hg^0$ in those systems seems to depend mainly on physical and gravimetric parameters that determine its transport and accumulation. Adsorption or complexation of metals to particles should be expected to happen with mercuric ion but not with $Hg^0$. Available data of Hg concentrations in bottom sediments do not show correlation with organic matter content as would be expected in case of particulate transport. On other hand, physical processes controlled by gravimetric and hydrological parameters are apparently responsible for transport and preferential accumulation conditions for $Hg^0$ in specific localities. This was observed in a fluvial system in the United States (Thiboudeaux 1979), and now in Amazon rivers (Malm 1991). Amalgams of Hg were observed in definite locations at upstream waterfalls. Two amalgams collected in 1991 weighed 56 and 137 mg, and their Hg contents were 53% and 65% respectively (Malm et al. 1993).

Most surveys on Hg concentrations in fluvial sediments of the Amazon rivers were done in the clay–silt fraction of sediments (<63 m). Few studies considered the effect of grain size upon the Hg distribution. Since metallic Hg is the major form of Hg released into these rivers, grain size seems to play a minor role in determining Hg sediment content. Along the river there are various points of Hg release, which explain the multiple peak distribution found (and also confirms the

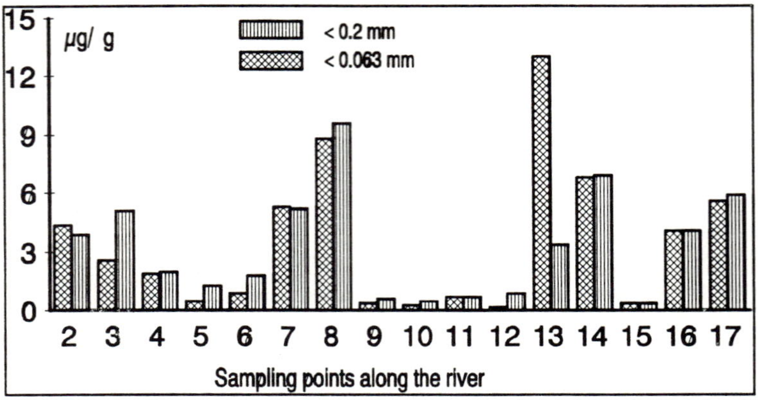


**Fig. 10.3.** Mercury concentrations in sediments of different grain size along the drainage of the Cumaru mining site, Para State, northern Amazon

low mobility of metallic Hg released) in two different grain size sediment fractions. It is clear that Hg distribution is independent of grain size. This is probably a result of the low $Hg^{2+}$ concentration in river water (≈ 0.02 µg $l^{-1}$), which would be preferentially adsorbed on fine particles.

Another aspect of Hg fate is the presence of methylated forms. Unfortunately only one survey has been completed on the concentration of this bio-available form of Hg in water and sediments of Amazonian rivers (Padberg 1990), although others have been recently started in the region. The study of Padberg (1990) was carried on in the Tapajós river, where one of the largest mining sites in the Amazon is located. Table 10.2 summarizes its major results. It is clear that methylation is presently occurring in this river system; notwithstanding the low concentrations found. Methylmercury concentrations reached nearly 10% of the total Hg content in water but only 2% of the total Hg content of sediments.

**Table 10.2.** Total mercury and methylmercury concentrations in water and sediments of the Tapajós river , Pará State, northern Amazon. Based on Padberg (1990)

| | Total mercury | Methylmercury | Percent of total |
|---|---|---|---|
| Water (ng $l^{-1}$) | | | |
| Itaituba | 3.2–5.3 | 0.2–0.6 | 8.8 |
| South Itaituba | 1.8–8.7 | 0.2–0.5 | 8.4 |
| Sediment (µg $kg^{-1}$) | | | |
| Itaituba | 144 | 0.8 | 0.6 |
| South Itaituba | 2.9–93 | 0.07–1.9 | 2.2 |

## 10.4 Mercury Distribution in Soils

Soils present special integration characteristics due to the low mobility and transport of chemicals in their domain. Thus, they are good indicators of atmospheric deposition, serving as a final or temporary sink but also as a source of volatile Hg species. The importance of studying this compartment is to understand the dispersion of $Hg^0$ in air and soil systems.

Analysis of urban soil samples, collected all on the same day in the surroundings of reburning areas in Porto Velho City, SW Amazon, indicates a dispersion circle of around 600 m as well as the effect of prevailing winds on Hg deposition. (Malm et al. 1991). Data showed that the proximity of gold shops is strongly correlated to the Hg soil concentrations, independently from origin (different cities, forest or urban soils).

No relation was found between Hg concentrations and organic matter content, either in surface soil or in core samples, reinforcing the idea that physical factors are more important than chemical ones for explaining dispersion and retention of Hg, especially when emitted in the metallic form.

Forest soils from the Madeira river watershed, north-western Amazon, were analyzed by Lacerda et al. (1987), and showed Hg concentrations ranging from 35 to 300 $\mu g\ kg^{-1}$ dr. wt. The highest values were found close to river sections where intense mining takes place. Malm et al. (1991) also studied forest soils in the Madeira River area. These authors found similar results, with Hg concentrations ranging from 30 to 340 $\mu g\ kg^{-1}$ dr. wt. in forest soils relatively far from mining sites. In forest soils close to amalgam burning areas, Hg concentrations were much higher, ranging from 420 to 9900 $\mu g\ kg^{-1}$ dr. wt. Also they found a 100% increase in Hg concentrations in surface soils when compared to deeper horizons in soil profiles. They associated high Hg content in the topsoil with atmospheric deposition and high organic matter content (up to 27%). Soils collected in urban areas of the same region where roasting of gold with Hg impurities occurs, exhibited much higher Hg concentrations, ranging from 460 to 64 000 $\mu g\ kg^{-1}$ close to gold dealer shops, to 30 to 1300 $\mu g\ kg^{-1}$ dr. wt. at least 300 m away from those shops (Malm et al. 1991).

In another study in the Poconé region, central Brazil, 70% of the analyzed soils samples from a 10 000 $km^2$ area under the influence of various mining sites showed very low Hg concentrations, being smaller than 30 $\mu g\ kg^{-1}$ dr. wt.(which is considered the local background Hg concentration). In circa 30% of the samples, Hg concentrations ranged from 30 and 100 $\mu g\ kg^{-1}$ dr. wt. Close to mining sites however, Hg concentrations reached 270 $\mu g\ kg^{-1}$ dr. wt. (Lacerda et al. 1991a). Compared to European and Canadian soils, with Hg concentrations between 70 and 300 $\mu g\ kg^{-1}$ dr. wt. (Mitra 1986), the values found in this region are relatively lower. However, if one considers the local background Hg concentrations in Amazon soils, Hg contamination is very clear, at least close to mining sites.

In general, all studies on soil contamination in the Amazon region showed that Hg concentrations dropped rapidly away from their sources. Extremely high Hg concentrations have been found in urban street dust. In Porto Velho town, Madeira River area, Malm et al. (1991) found Hg concentrations up to 36 000 μg $kg^{-1}$ of dry dust, while Thornton et al.(1992) found up to 250 000 μg $kg^{-1}$ of dry dust in the mining village of Itaituba, southern Amazon. In a study of the characterization of Hg species emitted by roasting of gold in gold dealer shops, Marins and Tonietto (1991), suggest that particulate Hg is the major species of Hg emitted rather than Hg vapour.

## 10.5 Atmospheric Dispersal

The roasting of the Au:Hg amalgam in gold-mining sites causes high Hg inputs into the atmosphere. The evaporated Hg is returned to land surfaces and waters with wet and dry deposition and accumulates in soils and sediments (Pfeiffer and Lacerda 1988; Lacerda et al. 1989; Marins et al. 1991). This process renders Hg a much broader distribution – reaching remote areas relative to mining sites – than when released into rivers as metallic Hg. In these areas, the absence of a direct Hg inputs from rivers and mining effluents themselves, allows a steady accumulation of Hg in lakes and reservoir sediments in equilibrium with atmospheric deposition rates over these areas. Therefore, sediment cores from remote lakes and reservoirs may provide an excellent way to determine the history of Hg contamination in gold-mining areas and estimates of atmospheric deposition.

The distribution of Hg in sediment profiles was studied in lakes of the Pantanal area, which receive atmospheric inputs of Hg from important gold-mining sites located along its northern border, but not from direct mining effluents. Mercury concentrations were higher in surface sediments of lakes ranging from 62 to 80 μg $kg^{-1}$, and decreased steadily therein to nearly constant values between 12 to 30 μg $kg^{-1}$, which are similar to background values reported for the area (Lacerda et al. 1991b). In these two lakes peaks of organic matter content and Hg concentration presented the same distribution throughout the profile, although elevated Hg concentrations were restricted to the first 2.0 cm of the cores.

The Hg concentrations found are very low when compared to remote lake sediments from the northern hemisphere. Evans (1986) reported background values for remote Canadian lakes to be circa 100 mg $kg^{-1}$, while surface sediments presented values up to 560 mg $kg^{-1}$. Andren and Nriagu (1979) considered background values for northern hemisphere lakes of circa 330 mg $kg^{-1}$. Rekolainen et al. (1986) found background Hg concentration in remote lakes of Finland to range between 50 to 250 mg $kg^{-1}$ but found values as high as 550 mg $kg^{-1}$ in the surface layers of these lakes. These data are considerably higher than the concentrations found in the Pantanal lakes, and are due to both the presence of Hg-rich base rocks in most Canadian and northern European areas and anthropogenic Hg

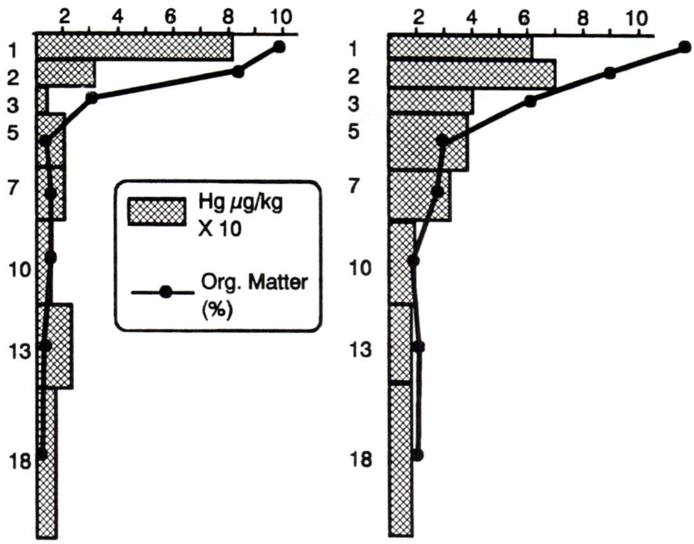


**Fig. 10.4.** Mercury (µg $kg^{-1}$) and organic matter (%) concentrations in sediment cores (depth in cm) from remote lakes in the Pantanal swamps, Mato Grosso State, central Brazil

emissions in these areas.

As gold-mining activities, and therefore Hg emissions to the atmosphere, started in the area in 1981, we can use Hg distribution throughout the sediment profiles to estimate sedimentation rates and Hg accumulation for the region. Assuming Hg background concentrations to be between 10 to 30 µg $kg^{-1}$, the data of profiles showed a steady increase in Hg concentration above around 2.0 cm of depth. Using the Hg concentrations found in the top 2 cm (60 to 80 µg $kg^{-1}$) we can estimate a Hg accumulation rate ranging from 90 to 120 µg Hg $m^{-2}$ per year. Although Hg concentrations in surface and deep sediments of the Pantanal lakes are lower than those values reported for industrialized areas, the Hg accumulation rate is very similar to such areas. This suggests the importance of the Hg contribution from gold-mining to the Pantanal area, even considering that the remote lakes studied are relatively far from direct Hg inputs from gold-mining (Lacerda et al. 1991 b).

## 10.6 Mercury Methylation

In 1970 Jensen and Jernelov (1969) demonstrated that the addition of $Hg^{2+}$ to surface sediments of a Swedish lake caused the formation of methylmercury (MeHg), in a process mediated by microbial activity. This offered a possible explanation for the, until then intriguing evidence that most Hg in fish and other aquatic organisms is present as MeHg, while Hg in industrial effluents and fallout is mainly inorganic. This prompted many investigations about the nature of Hg methylation in the environment, its preferential sites and controlling factors. A great variety of microorganisms are able to methylate Hg, including strepto and staphylococci, lactobacilli, yeasts and different fungal cultures. Mercury can be methylated by aerobic and anaerobic bacteria but only a small percentage of the obligate anaerobes can form MeHg. Additionally, the latter produce less MeHg than facultative anaerobes under the same conditions (Robinson and Tuovinen 1984). Mercury methylating microorganisms therefore appear to occur on most environmental substrates and even in fish and human intestines (Furutani, Rudd and Turner 1980).

The mechanism of Hg methylation is still not well understood, despite its evident toxicological relevance. It appears to involve methylcobalamin (vitamin $B_{12}$) as the donor of methyl groups to $Hg^{2+}$ in a non-enzymatic transfer, as suggested by the fact that this compound is a methyl donor for the synthesis of methionine via homocysteine methylation in bacteria (Landner 1971). $Hg^{2+}$ is believed to be the likely direct precursor or MeHg, though the formation of dimethylmercury from phenylmercury acetate by some microorganisms in soil and water has already been reported (Matsumura et al. 1971). Mercury can be methylated by abiotic or photochemical processes but their significance remains unclear. Weber et al. (1985) demonstrated that fulvic acids could mediate MeHg formation. Jernelov et al. (1972) found a steep increase in MeHg concentrations in the Saint Clair river near the release point of effluents containing methyllead, and concluded that this was the result of an abiotic transalkylation between inorganic Hg and alkyllead. Such a process may be relevant in limited areas subject to methyllead and methyltin containing effluents. Bisogni (1979) suggests that MeHg formation may not result from direct biological methylation, but rather from a transalkylation reaction with biologically formed methylation compounds.

Once formed, MeHg is able to resist many environmental stresses except specific biochemical degradation processes, again mediated by microbial activity. These demethylation mechanisms are much better understood than methylation itself. Bacterial resistance to organic and inorganic Hg compounds is determined by plasmids that often also encode the resistance to other heavy metals and to antibiotics (Robinson and Tuovinen 1984). The resistance to MeHg involves the detoxification of this compound by cleavage of the C-Hg bond by the organomercurial lyase, enzyme followed by the elimination of the metal from the medium by conversion to volatile $Hg^0$ through the action of a mercury reductase enzyme. Thus, it may be stated that mercury resistance and volatilization are essentially synonymous. The production of both enzymes is induced by exposure to Hg, and resistance to this metal has been shown to be governed by at least four genes whose genetic and physical structure has already been mapped (Robinson and Tuovinen 1984). This has led some authors to suggest that native bacteria from contaminated sites could be genetically manipulated to enhance their ability to demethylate and volatilize MeHg (Winfrey and Rudd 1990). Abiotic processes can also reduce $Hg^{2+}$ ions, as shown by Alberts et al. (1974) who found that humic acids could stimulate $Hg^0$ evolution.

Despite the better knowledge on demethylation mechanisms, few authors have been able to estimate demethylation rates in environmental samples, perhaps because of the relatively complex procedures involved (Ramlal et al. 1985, 1986). Usually, only net methylation, the balance between MeHg production and decomposition, is measured. These studies have shown, however, that methylation and demethylation react in a complex and variable way to parameters such as $Hg^{2+}$ and MeHg concentration and availability, pH, redox potential, composition of the microbial population and other synergistic or antagonistic effects of environmental and biological processes (Korthals and Winfrey 1987). The differences in the optimal conditions for these activities may render them mutually exclusive when present in the same organism and defined microenvironments in which one activity predominates (Robinson and Tuovinen 1984).

Compeau and Bartha (1985) and Gilmour and Henry (1991) presented experimental evidences suggesting that sulphate reducing bacteria are the main methylating organisms in anoxic estuarine and lake sediments, respectively. More attention has been paid to Hg methylation in bottom sediments, but this process also occurs in the water column, though apparently at slower rates (Xun et al. 1987).

In-lake MeHg formation has long been considered the main source of MeHg to fish but Hg methylation in soils has already been demonstrated (Rogers 1976, 1977), and MeHg has been measured in run-off and precipitation (Lindqvist et al. 1984; Lee and Hultberg 1990). These findings, together with the observation of high Hg concentrations in fish from coloured lakes and the correlation between DOC and MeHg in humic waters suggest that the terrestrial ecosystems may be relevant MeHg contributors to lakes and rivers, possibly due to abiotic methylation by fulvic and humic acids in the watersheds (Lee et al. 1985). The relative importance of terrestrial and in-lake MeHg production would, therefore, vary with factors such as the drainage area and the lake water residence time (Winfrey and Rudd 1990).

Fish in reservoirs and artificial lakes tend to exhibit marked increases in Hg content after impoundment and it may take many years for concentrations to return to the pre-impoundment values (Gilmour and Henry 1991). It has been hypothesized that the flooding of large (and generally forested ) areas may cause a sudden mobilization of Hg present in soils and vegetation, the methylation of which will be favoured by the increased microbiological activity and reducing conditions associated to the decomposition of the flooded vegetation (Jackson 1986, 1988). Recently, average total Hg concentrations of up to 2.8 $\mu g^{-1}$ were measured on carnivorous fish samples (piranha, *Serrasalmus* sp.) of the Tucuruí reservoir on the Tocantins river, eastern Amazon (Aula et al. 1994).

Factors controlling net methylation rates depend on specific environments. This complicates the construction of general predictive models of Hg methylation and bioaccumulation applicable to a wide variety of environments (Winfrey and Rudd 1990). Most research on Hg behaviour has been conducted in temperate or boreal aquatic systems; very little is known about Hg biogeochemistry in tropical waters. In the particular case of the Amazon, the diversity and complexity of the aquatic ecosystems, and the nomadic and largely unpredictable nature of gold-mining activities create additional difficulties.

The data available to date on MeHg concentrations in the environment and in human populations in the Amazon are scarce, but point to some trends that are already familiar (Malm et al. 1990). Malm et al. (1994) showed that the form of Hg in fish of the Madeira and Tapajós rivers, as well as in hair of fish-eating dwellers in these areas was essentially MeHg.

Much of the present knowledge about Hg methylation and demethylation has been gleaned from radiochemical techniques. The advantages and limitations of this approach have been reviewed in detail by Furutani and Rudd (1980) and Gilmour and Henry, (1991). Incubation of environmental samples spiked with $^{203}$Hg, followed by extraction and measurement of Me$^{203}$Hg formed during incubation are especially useful for testing hypotheses on the influence of single variables on Hg methylation rates and on preferential methylation sites. The obtained rates are considered as estimates of maximum potential Hg methylation rates since the carrier Hg added along with $^{203}$Hg increases Hg concentration and availability in the sample.

Simplified versions of the radiochemical technique originally described by Furutani and Rudd (1980) have been developed and applied by Guimarães (1992) and Guimarães et al. (1994) for determining the net Hg methylation rates in sediment, water and soil samples of the Madeira and Tapajós river basins. Methylation was not detectable in unfiltered water samples of these rivers incubated with $^{203}HgCl_2$ at Hg concentrations equivalent to 5–20 $\mu g\ l^{-1}$ for up to 3 days. Similar results were obtained for forest soil–water suspension samples incubated under the same conditions. High methylation rates were, however, found in surface sediment samples of the Madeira river area, as shown in Table 10.3. Rates of $10^{-3}$ and $10^{-5}$ % $g^{-1}\ h^{-1}$ were found for the samples of the typical white-water Madeira river, where low methylation rates were anticipated due to its high pH, conductivity and suspended sediment load. In the Mutum-Paraná river and Novo creek, typical black water forest streams, the rates reached $10^{-2}$, which is already comparable or even higher than seasonal methylation peaks observed in temperate lakes (Furutani and Rudd 1980; Ramlal et al. 1986; Korthals and Winfrey 1987). Rates higher than this by one order of magnitude were however found in the highly organic flocculent surface sediment of the flooded forest area of the Samuel reservoir and in similar sediments collected just downstream from the dam. Both of these sites are characterized by mildly reducing conditions. These data suggest that intrinsic methylation rates are high in the Samuel reservoir,

**Table 10.3.** Potential net $^{203}Hg$ methylation rates in some surface sediment samples of the Madeira river and tributaries, including Samuel reservoir. Average and range (N = 2) for samples of water-sediment slurries of up to 100 ml incubated in situ for 15–24 h with 44 kBq of $^{203}Hg$ (approximately 2 μg Hg)

| Sampling site | Hg[a] | P[b] | Methylation rate (% $g^{-1}$ $h^{-1}$) |
|---|---|---|---|
| Madeira river, upstream Mutum-Parana river | 0.26 | 2 | 6.2 (5.1–7.3)· $10^{-5}$ |
| Madeira river, upstream Teotonio falls | 0.46 | 1–2 | 1.5 (1.4–1.6)· $10^{-3}$ [c] |
| Mutum-Parana river | 3.7 | < 1 | 1.0 (0.84–1.2)· $10^{-2}$ |
| Novo Creek, midstream | N.A.[d] | 2–3 | 3.0 (2.8–3.2)· $10^{-2}$ |
| Jamari river, aprox. 500 m downstream Samuel reservoir | 0.13 | < 1 | 6.6 (3.2–10)· $10^{-1}$ |
| Samuel reservoir, flooded forest | 0.08 | 1–2 | 6.9 (6.4–7.4)· $10^{-1}$ |
| Samuel reservoir, ca. 100 m from the dam | 0.05 | 20 | 2.9 (2.1–3.7)· $10^{-3}$ |

[a] Total Hg in sediment (<74 mesh fraction) expressed in μg $g^{-1}$ dr. wt.
[b] Approximate depth (metres)
[c] Incubation of samples kept at 4°C for 69 days
[d] Not analyzed

which may explain the finding of total Hg concentrations approaching or exceeding the 0.5 μg $g^{-1}$ safety limit in many carnivorous fish caught in the reservoir, despite the very low Hg concentrations in the bottom sediment (Table 10.3).

Petrick (1993) showed that approximately 80% of total Hg in sediments of the Madeira river are in the metallic form. Since $Hg^{2+}$ is believed to be the substrate for methylation, Hg oxidation will be a rate-limiting factor for Hg methylation. Gold-mining in the Madeira river is performed by mechanical dredges and their operation favours $Hg^0$ oxidation by continually bringing huge amounts of sediment to the surface. In the Tapajós river and tributaries gold is extracted from the bottom sediment by small rafts but most frequently co-alluvial gold is extracted from soils of the river banks or watersheds. Data on Hg speciation in the Tapajós river sediments are not yet available, but methylation experiments similar to those described above for the Madeira River area indicated that Hg finds favourable conditions for methylation from the point of its release in gold-mining areas down to the Tapajós river itself. Net methylation rates in the sediments of ponds in recently abandoned gold-mining fields were in the $10^{-3}$ range while in the sediments of streams that drain these mining areas and in the Tapajós river rates were one order of magnitude higher (Guimarães et al. 1993).

## 10.7 Mercury in Fishes

Fish is the main source of MeHg for humans. Hg concentrations in carnivorous (piscivorous) fish from several gold-mining areas (Fig. 10.5) are systematically higher than in fish from lower trophic levels. The carnivorous fishes must be monitored for public health reasons and because they are good indicators of the degree of contamination of the aquatic system. Thus the bioaccumulation of Hg by the fishes was confirmed in the Amazon ecosystems and in the case of carnivorous fishes, the occurrence of the biomagnification process was observed.

Fish from control areas (i.e. those with no direct effect of Hg contamination) usually have levels lower than 0.2 μg Hg $g^{-1}$ wet weight in rivers and 0.15 μg $g^{-1}$ in oceans. In contaminated areas levels can reach 2 μg $g^{-1}$ or more, and, in carnivorous fish, up to 20 μg $g^{-1}$, a value already considered lethal to fish (Meili 1991).

Ingestion of contaminated fish by humans may pose serious risks when fish is a staple of the diet, because the methylated form has very high intestinal absorption rates (>95%) and high chemical stability.

Each river basin in the Amazon has its own characteristics. As an example we will discuss the Madeira river. Fish sampling was conducted from June 1987 to October 1991 from Guajará-Mirim to Manicoré including some tributaries and the reservoir of Samuel Hydroelectric Power Plant. Fish were separated in three main alimentary habit classes, according to the Goulding classification (Goulding 1979): (1) carnivorous, eating predominantly other fish; (2) microfagous, ingesting organic matter in decomposition, algae and slimy clay with some sediment;

**Table 10.4.** Mercury concentrations in fish from Amazon gold-mining areas (μg $g^{-1}$ wet weight)

| | Average | Range | N |
|---|---|---|---|
| **Madeira river**[a] | | | |
| Carnivorous | 0.70 | 0.07–2.89 | 73 |
| Microfagous | 0.18 | 0.03–0.65 | 28 |
| Omnivorous | 0.13 | 0.02–0.50 | 24 |
| **Tapajós river**[b] | | | |
| Carnivorous (all river long) | 0.55 | 0.04–3.77 | 60 |
| –Upstream (Alta Floresta, Rato river, Itaituba, and Brasília Legal) | 0.69 | 0.18–3.77 | 43 |
| –Downstream (Ponta de Pedras and Santarém) | 0.19 | 0.07–0.55 | 17 |
| Other (npn carnivorous) | 0.10 | 0.01–0.31 | 16 |

[a] Malm (1991)
[b] Malm et al. (1994)

and (3) omnivorous fishes eating mainly fruits and seeds. The total number of species studied was over 25; the majority being carnivorous since they pose the highest human health risks. Statistical analysis for fish collected at the Madeira River basin showed an average value of 0.70 $\mu g\ g^{-1}$(w. wt.) for the carnivorous group, 0.18 $\mu g\ g^{-1}$ for the microfagous and 0.13 $\mu g\ g^{-1}$ for the omnivorous (Table 10.4). When considering all carnivorous fish in Madeira River area it is important to remark that nearly 25% of samples are from the Samuel reservoir and Jamari river. These have witnessed no gold-mining activity, which for this class of fish, an average value of 0.37 $\mu g\ g^{-1}$ was observed. If only Madeira River fish are considered the average value would be close to 0.9 $\mu g\ g^{-1}$.

Probably the information of greatest relevance is that though gold-mining occurs almost exclusively between Guajará-Mirim and Porto Velho, fish collected from 1987 to 1991 at Humaitá and Manicoré (180 and 500 km downstream from mining areas, respectively) exhibited values close to those from fish gathered at Porto Velho or closer to *Garimpos* (Malm 1991). This suggests Hg is being transported over long distances in inorganic form, associated with organic matter or suspended particles, and then methylated further downstream. Biota must play an important role in the transport and dilution of MeHg. Nevertheless, fish collected in August 1992 showed lower levels downstream in Humaitá. This can be related to the strong decrease in mining activities upstream of Porto Velho.

The Tapajós river basin has been prospected for a longer time than the Madeira (30 and 15 years respectively), but Hg concentrations in fish of the same species and weight range are higher in the latter basin. Elevated concentrations are found in the Tapajós river basin only in its upper reaches (Teles Pires river; Malm et al.

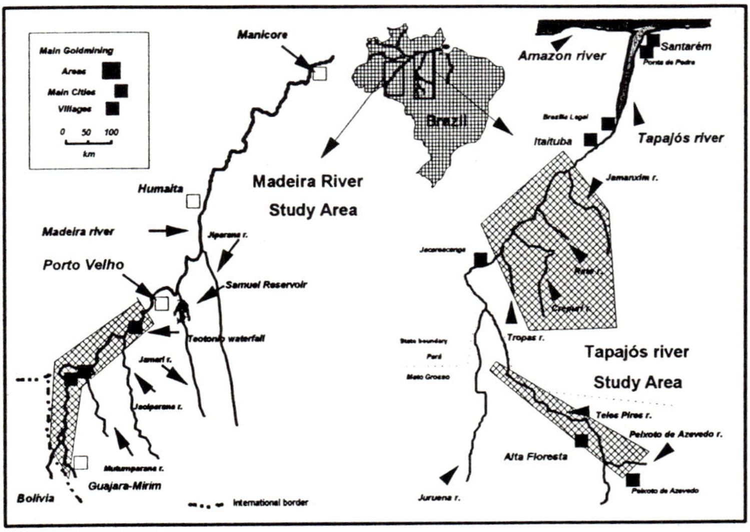


**Fig. 10.5.** Major riverine sites of mercury studies in fish

1994a,b). Probably long distance transport of $Hg^0$ in the Tapajós river basin is less efficient than in the Madeira. Fish from some other areas close to Hg sources like the Tapajós river basin (Ramos and Costa 1990; Padberg 1990; Thornton 1992; Malm 1991, 1994) and the Tocantins river basin (Aula et al. 1994) show the same accumulation pattern. That signifies a prevalence of the biomagnification process in carnivorous fish. In some carnivorous fish species, a correlation between Hg and fish weight was observed. Data confirm methylation and biomagnification processes in carnivorous groups with an average value in the entire Madeira river basin of 0.70 $\mu g\ g^{-1}$ and in the Tapajós basin of 0.55 $\mu g\ g^{-1}$ both of which are above the 0.5 $\mu g\ g^{-1}$ limit established by Brazilian legislation (Brasil 1975). Methylmercury values observed in 26 fish showed an average percentage of around 90% of the total Hg content. This value can be considered low if compared with our preliminary data from the Madeira river (98%) or with Swedish or Minamata area fishes (values often higher than 95%; Akagi et al. 1979).

## 10.8 Contamination in Humans

Human contamination by Hg has been reported from various Amazonian gold-mining sites. Surveys showed that contamination is mostly reflected by higher mercury concentrations in body indicators (such as hair, urine and blood). Nearly all surveys have shown these concentrations to be higher than expected background levels. However, only a few studies have actually detected symptoms or clinical signs of mercury poisoning. Most health authorities agree that lack of information on Hg levels among the Amazonian population is probably due to difficult logistics and poor health conditions which may mask Hg poisoning symptoms.

These information gaps regarding most remote Amazonian populations have hampered a better understanding the epidemiology of Hg contamination in humans from the Amazon region. Therefore in this review we will restrict ourselves to the presentation and discussion of the data on Hg concentrations in humans where it exists.

Human exposure to Hg from gold-mining in the Amazon region occurs via two main pathways. The first pathway involves occupational exposure, and affects miners who burn the amalgam in the field, and workers exposed to Hg at gold dealer shops. The latter group includes workers who are directly involved in the roasting of the ore for purification, and others who spend at least 8 h a day in these shops involved in various administrative tasks. The second main pathway involves environmental exposure, and affects urban populations living close to gold dealer shops, and rural populations ingesting potentially contaminated food, including various Indian populations.

### 10.8.1 Metallic Mercury

Emissions in indoor areas in cities or villages with poor air circulation can cause high human exposures to $Hg^0$ mainly via inhalation. Surrounding environments can also be contaminated. However, in those situations it is possible to implement procedures for Hg recovery, using retorts in the gold-mining areas and efficient exhaustion systems to collect $Hg^0$ vapour in gold dealer shops. Despite the relatively small amounts of Hg reburned in the shops compared to the total Hg used in the whole process (around 5% of total losses), this source represents a potential risk of contamination for people occupationally exposed, as well as for those living in the vicinities of the shop.

According to literature, inhaled $Hg^0$ is high absorbed through the lungs (more than 85%) and after some time to the blood stream, part is oxidized and accumulated in the kidneys. Mercuric ion is excreted in urine; urine analysis is the best indicator of Hg exposure and inorganic Hg body burden. In cities and villages reburning of amalgam (bullion) in gold dealer shops contaminates indoor areas and the vicinities, providing the critical occupational exposure pathway for $Hg^0$ to people. Urine samples have been used to determine Hg exposure to workers in shops and to *garimpeiros* (gold-miners). The highest average values were observed in persons working indoors with little air renovation or in reburning rooms without conditioned air. Mercury concentrations in urine from the people exposed in shops were usually high, ranging from 8.45 to 1168 $\mu g\ ml^{-1}$, with an average value of 228 $\mu g\ ml^{-1}$ in a total of 75 urine samples analyzed. Workers or people exposed during burning in open areas in *garimpos*, show much lower urine Hg values with a range from 0.27 to 74 $\mu g\ ml^{-1}$, and an average of 12.4 $\mu g\ ml^{-1}$ over 37 samples analyzed, despite the much bigger amounts of Hg manipulated here than in the shops. Maximum acceptable concentration in urine is 50 $\mu g\ ml^{-1}$ (WHO 1991). In fact, concentrations in the shop workers are quite high and this group is the critical one with regard to $Hg^0$ inhalation.

Branches et al. (1992), surveyed workers in gold dealer shops in various areas of the Tapajós mining site, Pará State, eastern Amazon. Out of 45 individuals, they found Hg concentration in urine to range from 10 to 1168 $\mu g\ l^{-1}$, with an average of 250 $\mu g\ l^{-1}$. Mercury concentrations in the ambient air of these shops varied from 7.18 to 107.8 $\mu g\ m^{-3}$. However no correlation has been found between Hg concentrations in humans and in the ambient air of the shops.

Although occupational exposure surely results in the most serious cases of human poisoning in the Amazon, it seems to be very restricted to a specific critical group. Environmental exposure on the other hand, seems to affect a much larger population, including those not necessarily involved in gold-mining.

### 10.8.2 Methylmercury

Human hair is well accepted as the best indicator for assessment of contamination in populations exposed to MeHg. Mercury concentration in hair reflects blood concentration at the moment hair is being formed. In a survey of Hg concentrations in hair of the population working in mining sites along the Madeira river, Malm (1992) showed a median value of 4,55 $\mu g\ g^{-1}$, with concentrations ranging from 0.2 to 24.1 $\mu g\ g^{-1}$. On the other hand, among the population not directly involved in gold-mining, the author found much higher values, ranging from 0.5 to 71.3 $\mu g\ g^{-1}$ and with a median value of 7.9 $\mu g\ g^{-1}$ (Table 10.5). Malm et al. (1992) analyzed Hg concentrations in 56 persons living in the Tapajós river mining area, who had a diet of high fish consumption but were not directly involved in gold-mining. Data for these people were compared to Hg in hair samples from the urban population of Santarem City, eastern Amazon, who also were not directly involved with gold-mining. Fish eaten by the two groups was also analyzed. Mercury concentrations in fish consumed by the riverine population were higher than those measured in the Santarem area. Corresponding Hg concentrations in hair of the former group ranged from 0.74 to 17.7 $\mu g\ g^{-1}$, and were up to 5 times higher than those measured in the latter group. Therefore, a direct relationship was established between Hg concentrations in fish and in humans.

A more complete survey of populations not directly involved with gold-mining, was conducted among the Yanomami Indians in Roraima State (Castro et al. 1991). Mercury concentrations in 162 individuals ranged from 1.42 to 8.14 $\mu g\ g^{-1}$ in females and from 1.40 to 7.90 $\mu g\ g^{-1}$ in males. The overall result showed that for certain groups, over 40% of the samples had Hg concentration in hair higher than 6.0 $\mu g\ g^{-1}$ (Table 10.6). Methylmercury concentrations measured in these samples showed concentrations ranging from 0.96 to 3.02 $\mu g\ g^{-1}$, with a percentage variation ranging from 40 to 90% of the total Hg concentrations.

During the past 5 years, the territory of the Yanomami Indians has been invaded by *garimpeiros*, who have caused widespread pollution of the environment.

**Table 10.5.** Mercury concentrations in human urine and hair samples from different mining sites in Brazilian Amazon

| Author | Hg concentration | Observations |
|---|---|---|
| Branches et al.(1992) | 10 – 1168 $\mu g\ l^{-1}$ urine (median = 250 $\mu g\ l^{-1}$) | Workers of gold commercialization shops in Itaituba, PA |
| Malm (1992) | 0.2 – 24.1 $\mu g\ g^{-1}$ hair (median = 4.55 $\mu g\ g^{-1}$) | Workers in mining sites along the Madeira river, RO |
| Malm (1992) | 0.5 – 71.3 $\mu g\ g^{-1}$ hair (median 7.9 $\mu g\ g^{-1}$) | Riverine population consuming fish as the main diet along Madeira river |
| Malm et al. (1992) | 0.74 – 17.7 $\mu g\ g^{-1}$ hair | Riverine population consuming fish as the main diet along Tapajós river |
| Castro et al. (1991) | 1.42 – 8.14 $\mu g\ g^{-1}$ hair (40% – 90% metal-Hg) | Indian groups living close to mining sites in the Amazon |

Fortunately, the Yanomamis' diet is not based on fish but rather on hunting animals and gathering fruits and vegetables. Analysis of 162 hair samples collected in February and March 1990 showed Hg concentration values from 1.40 and 8.14 $\mu g\ g^{-1}$ (Castro et al. 1991). No significant differences were found between male and female, adults or children. The general average value was 3.61 $\mu g\ g^{-1}$, well below the established limit of 6 $\mu g\ g^{-1}$.

**Table 10.6.** Mercury concentrations ($\mu g\ g^{-1}$ dr. wt.) in hair samples of different groups of Yanomami Indians and percentage of samples with concentrations higher than 6.0 $\mu g\ g^{-1}$. After Castro et al. (1991)

| Indian Group | Hg concentration | % > 6.0 $\mu g\ g^{-1}$ dr. wt. |
|---|---|---|
| Koremutheri | 5.01 +/- 1.04 | 20 |
| Maisibiu | 5.03 +/- 1.67 | 40 |
| Kanau | 3.40 +/- 0.73 | 0 |
| Xahoxi | 3.99 +/- 1.46 | 10 |
| Maharun | 2.64 +/- 0.86 | 0 |
| Irobrerebe | 2.91 +/- 1.83 | 12 |
| Wakahusibiu | 3.21 +/- 1.42 | 7 |
| Heron | 3.72 +/- 0.97 | 2 |
| Mucajai | 4.13 +/- 1.22 | 7 |

Along Madeira river (Fig. 10.5), dwellers from cities as Humaitá and Manicoré show higher Hg values in hair than in Porto Velho. This is remarkable since PortoVelho is closer to the mining sites. This difference is probably caused by differences in diet. The citizens from Porto Velho, with better social and economical conditions, have a more diversified protein supply. The fish ingestion rate among individuals surveyed in Porto Velho is less than two meals a week while at Humaitá and Manicoré it is between three and four meals per week. Smaller villages surely will present much higher fish ingestion frequency, up to several meals a day. So, in downstream villages fish is a more obligatory food.

Gold-miners' fish consumption is lower, eating on average close to one fish meal per week with a maximum of four fish meals per week. In fact, because of their better economical conditions gold-miners eat more meat and in the case of fish, they prefer the non-carnivorous. The Hg concentrations in hair samples from gold-miners reflect their low consumption of fish (Table 10.7).

A detailed study on Hg levels in hair was conducted in the upper Tapajós river basin. The observed differences in Hg levels could be explained by differences in fish consumption and by the amount of Hg in the fish.

**Table 10.7.** Mercury concentrations in hair samples (μg $g^{-1}$) of riverine populations and gold-miners along the Madeira and Tapajós rivers

| | Average | Range | N |
|---|---|---|---|
| **Madeira river** | | | |
| Riverine population (females) | 9.22 | 0.58 – 37.18 | 22 |
| Riverine population (males) | 11.77 | 0.52 – 71.35 | 25 |
| Gold-miners (males) | 5.29 | 0.22 – 24.17 | 30 |
| **Tapajós river** | | | |
| Upstream | | | |
| Jacareacanga | 25.46 | 5.74 – 51.61 | 10 |
| Brasília Legal | 26.03 | 4.79 – 151.22 | 55 |
| Downstream | | | |
| Ponta de Pedras | 11.37 | 6.92 – 22.25 | 46 |
| Santarém | 2.66 | 0.74 – 06.54 | 11 |

In summary, the existing data on Hg contamination in humans in the Amazon gold-mining region shows that environmental exposure, in particular through eating of Hg contaminated fish, is the major pathway of human contamination. Also, there already seems to be widespread contamination in Amazonian human populations, including some Indian tribes.

## 10.9 Technologies for Cleaning Up Mercury Contaminated Areas

Mercury contaminated sites are a problem worldwide. Not only Brazil, but especially Germany has a lot of problems with it. Mostly it is a monocontamination with no other contaminants involved. So a remediation is principally easier than a decontamination from different heavy metals and/or organic pollutants too. There are two different solutions of Hg contaminations: the fixation of Hg by sulphides to an insoluble compound as HgS or the separation of Hg from the matrix.

**Fixation of Mercury.** HgS is a well known example for an insoluble compound. As cinnabar it is known as a mercury ore, stable over millions of years. Its solubility is known to be $10^{-54}$ $mol^2$ $l^{-2}$ in water. From this fact the idea arose, to transform mercury and its compounds into HgS by means of sulphides. With this method it is possible also to oxidize elemental Hg at least on its surface. But it is a problem to transfer metallic Hg inside a ball to HgS. This is a surface problem – the bigger the surface the better the oxidation.

However, the insolubility is not guaranteed. In a surplus of alkalipolysulphide the complex $[HgS_2]^{2-}$ is formed, which is better soluble in water. The solubility of HgS is enlarged by a factor of $10^7$ by a 1 M Chloride solution. Even a 0.1 mM $Cl^-$ solution results in an increase of solubility by a factor of 400. Microorganisms are able to oxidize HgS to the more soluble compound $HgSO_4$. The alkali- or ammoniapolysulphides, used to transfer the Hg-compounds to HgS, are harmful and toxic to the environment, so fixation should only be used when the product is stored safely under undisturbable and closed conditions.

**Main Possibilities of Demercurization.** The main pathways of demercurization are:

- Extraction methods for cleaning up of Hg contaminated soils
- Use of certain bacteria to reduce $Hg^{2+}$ to $Hg^0$
- Electrochemical separation of Hg
- Thermal volatilization of Hg and Hg compounds

Extraction methods depend on the species of Hg to be extracted. They must be, when only one extraction step should be used, of nearly the same behaviour. In a contaminated soil we have the compounds: metallic Hg, HgS, and soluble mercury salts. From the Hg budget the organic Hg compounds as methylmercury don't contribute to the extraction yield. Before extraction all compounds should be transformed to $Hg^{2+}$, than extractions by sulphur containing and chelating agents can take place. The chemicals with an optimum for this purpose are still under development.

Certain bacteria are able to reduce $Hg^{2+}$ to $Hg^0$, which can be stripped off by gas. The Hg-species must be available to the microorganisms, and that means, that $Hg^{2+}$ must be in solution, in a specific concentration, and in an environment suitable for them. The technology for decontamination of solid material is still in the laboratory phase.

$Hg^{2+}$ can also be separated by electrochemical methods. If it is in solution, it can be adsorbed on an electrode with about 1 A/dm$^2$ and a pH of 6–7.5, as described by de Souza. It seems also possible to desorb Hg into solution and oxidize $Hg^0$ by use of NaOCl, produced directly in the electrochemical step from NaOH and $Cl^2$. This method has not been tested under field conditions yet and may have a disadvantage in the chlorinated compounds originated by the process from the organic substances in the soil.

It is well known, that Hg evaporates from a given matrix by heating. Under natural conditions the evaporation of Hg corresponds to its vapour pressure which is at 20°C $1.22 \cdot 10^{-3}$ mm Hg. The boiling point is 356.6°C, but $Hg^0$ evaporation already takes place about 100°C, when natural samples are heated constantly with increasing temperature. Other compounds evaporate at higher temperatures up to 450°C.

In Germany a reprocessing plant for Hg is in operation. It cleans the soil with a grain size <100 µm of a Hg factory site in Marktredwitz, where Hg was processed for over 200 years. The working principle is a thermal desorption at 310°C under slight vacuum, 100 hPa. The processed material is guaranteed to be under 50 mg Hg/kg soil, dry weight, and is stored after the process in a mono landfill deposit.

A general question arises about the aim and the quality objectives for a decontamination. The German lists for Hg contamination are still under discussion and are given in Table 10.8.

**Table 10.8.** Reference values for Hg-contaminated soil

| Check values | A: reference value | B: test value for analysis | C: test value for sanitation |
|---|---|---|---|
| | 0.5 | 2 | 10 mg Hg/kg |

As can be seen, sites with a concentration under 2 mg Hg/kg have not to be decontaminated. In Germany and other countries soil up to 50 mg Hg/kg can be used in industrial sites, so also up to this value a decontamination is often not necessary.

With the technical possibilities nowadays a decontamination of soil is reasonable from 50 mg Hg/kg to higher values. A decontamination of lower concentrations gives a lot of new contaminated materials by the chemicals used and if thermal desorption is performed, the soil fertility can be destroyed.

So many of the tailings in Brazil are not cleanable with available technologies under reasonable circumstances. This is based on the concentrations reported and the volumes involved. Only hot spots such as the amalgamation sites can be cleaned. Further research can perhaps be helpful to find mild processes for decontamination.

## 10.10 Conclusions and Outlook

Mercury amalgamation is typically a low-efficiency technology. However it is cheap, reliable and easy to operate at the individual scale. Therefore, notwithstanding the recent making the acquisition and use of Hg more difficult in many countries, this type of technology will only be limited either by the exhaustion of gold and silver deposits or by a drastic decrease in gold prices. Both alternatives are unlikely to occur in the next 50 years, and the extension of new prospecting areas can be expected particularly in the northern part of South America and Southeast Asia. This technology has been responsible for the emission of over 3000 tons of Hg to the environment in the Brazilian Amazon alone during the last 10 years. In 50 years we estimate that over 50 000 tons of Hg will be spread in the environment worldwide due to Hg amalgamation for the exploitation of gold and silver ores.

Mercury is being released in the metallic form. In order to become substantially soluble, $Hg^0$ needs to be oxidized, it may then be transported as inorganic salts or soluble complexes, biologically incorporated, methylated and finally accumulated in the food web. The balance between these different processes determines the fates of Hg in the environment. Data has shown that in river systems Hg is occurring mainly in the metallic form. So, the dynamics of resuspension of sediments and Hg oxidation must be a critical factor regulating the availability of Hg in those systems.

Little is known about the long-term fate of Hg in tropical systems. Based on a comparison of recycling of elements in tropical and in temperate climates, it is expected that the dispersion of Hg is more wide (Lacerda and Salomons 1991). However, much more research has to be carried out to determine the long-term risk to the human population and the ecosystem.

Regardless of all the experience acquired over the past 25 years on Hg mobility and methylation, little is known about simple and efficacious measures to control methylation processes (Lindqvist et al. 1984; D'Itri 1991). Correction of pH with additions of $CaCO_3$ or CaO, a measure that should enhance volatile dimethylmercury formation, did not produce the expected results. It did not reduce significantly methylation and or bioaccumulation. Chemical immobilization of Hg with selenium (Hg-Se has a very high stability constant as is the case with sulphide), can reduce Hg availability and its toxicity to the biota. On the other hand, at the Se amendment levels required to toxicological consequences of Hg, may cause a toxic response of selenium (Pelletier 1985).

It seems therefore, that in the present state-of-the-art, any human attempts of manipulation, aiming to reduce the potential toxic consequences of Hg once it is dispersed in the dynamic tropical aquatic ecosystems, will be at least as unfeasible as it has been in temperate areas.

Since it is expected that the use of Hg will continue in the foreseeable future one can only make recommendations to decrease exposure to Hg:

1. The release of Hg to the atmosphere can be reduced by installing treatment systems in the gold dealer shops.
2. The use of retorts by *garimpeiros* should be encouraged. Their use will make it possible to recycle Hg and decrease losses to the environment.
3. The consumption of carnivorous fish should be discouraged.

In a number of areas "hot spots" of Hg exists. These are those areas where the gold was recovered from soils and Hg is now slowly released from the tailings deposits including the former panning ponds. Technologies have been developed to treat these contaminated areas and offer an solution for either the recovery of the existing Hg, or at least for the immobilization of Hg under non-bioavailable chemical species.

## References

Akagi H, Mortimer DC, Miller DR (1979) Mercury methylation and partition in aquatic systems. Bull Environm Contam Toxicol 23:372-376

Alberts JJ, Schindler JE, Miller RW (1974) Elemental evolution mediated by humic acid. Science 184, 895-897

Andrade JC, Bueno MIMS, Soares PV, Choudhuri A (1988) The fate of mercury released from prospecting areas (Garimpos) near Guarinus and Pilar, Goiás (Brazil). An Acad Brasil Ciênc 60:293-303

Andren AW, Nriagu JO (1979) The global cycle of mercury. In: Nriagu JO (ed) The biogeochemistry of mercury in the environment. Elsevier North Holland, Amsterdam, pp 1-22

Aula I, Braunsweiller H, Leino T, Malin I, Porvari P, Hatanaka T, Lodenius M, Juras A (1994) Levels of mercury in the Tucuruí reservoir and its surrounding area in Pará, Brazil. In: Watras CJ, Huckabee JW (eds) Mercury pollution: integration and synthesis. Lewis Publ, Boca Raton, pp 21-40

Bezerra JFM (1991) Estimativas de cargas de mercúrio para o meio ambiente por atividades industriais. Caracterização de fontes. In: Hacon S, Lacerda LD, Carvalho D, Pfeiffer WC (eds) Riscos e Consequências do Uso do Mercúrio UFRJ, Rio de Janeiro, pp 91-109

Bisogni JJ (1979) Kinetics of methylmercury formation and decomposition in aquatic environments. pp 211-227 In: Nriagu JO (ed) The biogeochemistry of mercury in the environment. Elsevier North-Holland Biomedical Press, Amsterdam, 696 pp

Boldrini CV, Padua HB, Pereira DN, Rezende EK, Juras AA (1983) Contaminacao por mercurio nos rios Moji-guacu e Pardo. Rev DAE 135:106-117

Brabo ES, Costa MQ, Ramos JFF (1991) Avaliacao preliminar da contaminacao por mercurio nos rios Crepari e Marupe. In: III Simp Geol Amazônia 1:331-336

Branches FJP, Malm O, Bastos WR, Pfeiffer WC (1992) Clinical findings, its relation with air mercury concentrations and urinary Hg levels among gold shop workers, Amazon, Brazil. Abs 4th Ann Meet Intern Soc Environm Epidemiol, Mexico. Session: 03.1

Branches FJP, Erickson T, Aks SE, Hryhorczuk DO (1993) The price of gold: mercury exposure in the Amazonian rain forest. The Journal of Clinical Toxicology 31:295-306

Brasil (1975) Ministério da Saude, Resolução No 18/75 da Comissão Nacional de Normas e Padrões para Alimentos. Resolução de 18 de agosto de 1975. Diário oficial da União, Brasília, 19 de Dezembro de 1975. Seção 1, 16378 pp

Castro MB, Albert B, Pfeiffer WC (1991) Mercury levels in Yanomami Indians hair from Roraima, Brazil. Proc 8th Intern Conf Heavy Metals in the Environment (ed) Cep Consultants, Edinburgh, UK, 1:367-370

CETEM (1989) Relatório Anual do Projeto Poconé. Centr Tecnol Mineral, Rio de Janeiro, 287 pp

CETESB (1980) Avaliação da contaminação atual dos rios Moji-guacú e Pardo e seus reflexos sobre comunidades biológicas. Comp Est Tecnol San Bas, São Paulo, 3 vols 637 pp

CIMELCO (1991) Proyecto Binacional Puyango-Tumbes. Consorcio Cimelco Construtores, Tumbes, Peru

Clearly D (1990) Anatomy of the Amazon gold rush. The Macmillan Press Ltd, London, 245 pp

Compeau GC, Bartha R (1985) Sulfate reducer bacteria: principal methylators of mercury in anoxic estuarine sediments. Appl Environ Microbiol 50, 498-452

Cramer SW (1990) Problems facing the Philippines. International Mining -July 1990: pp 29-31

D'Itri FM (1991) What we have learned since Minamata. Environment Monitoring Assessment 19:165-182

D'Itri FM (1992) Mercury pollution and cycling in aquatic systems. In: Morges, CEP consultants (eds) 5th Inter Conf Environmental contamination, Edinburgh, pp 64-74

D'Itri PA, D'Itri FA (1977) Mercury contamination: a human tragedy. John Wiley, New York, 85 pp

Drummond RA, Olson GF, Batterman AR (1974) Cough response and uptake of mercury by brook trout *Salvelinus fontinalis,* exposed to mercuric compounds at different hydrogen ion concentrations. Trans Am Fish Soc 103:244-249

Evans RD (1986) Sources of mercury contamination in the sediments of small headwater lakes in south-central Ontario, Canada. Arch Environm Contam Toxicol 15:505-512

Eysink GGJ (1990) A presença de mercúrio nos ecossistemas aquáticos do Estado de São Paulo. In: Hacon S, Lacerda LD, Pfeiffer WC, Carvalho D (eds) Riscos e Consequências do Uso de Mercúrio. UFRJ Rio de Janeiro pp 12-28

Fernandes RS, Guimarães AF, Bidone ED (1990a) Monitoramento do mercurio na area do Projeto Carajás. Saneamento Ambiental 6:34-41

Fernandes RS, Guimarães AF, Bidone ED (1990b) Monitoramento do mercúrio na area do Projeto Carajás. Bios 2:37-44

Ferreira RCH, Appel LE (1991) Fontes e usos de mercurio no Brasil. Proc I Intern Symp Environmental Studies of Tropical Rain Forests. 1:211-217

Fitzgerald WF, Gill GA, Kim JP (1984) An equatorial Pacific Ocean source of atmospheric mercury. Science 224:597-599

Frischmuth A, Weppen P, Deckwer W-D (1993) Microbial transformation of mercury (II), I. Isolation of microbes and characterization of their transformation capabilities. J Biotechn 29:39-55

Furutani A, Rudd JWM (1980) Measurement of mercury methylation in lake water and sediment samples. Appl Environ Microbiol 40:770-776

Furutani A, Rudd JWM, Turner MA (1980) Mercury methylation by fish intestinal contents. Appl Environm Microbiol 40:777-782

Gilmour CC, Henry E (1991) Mercury methylation in aquatic systems affected by acid deposition. Environ Pollut 71:131-169

Gilmour CC, Henry EA, Mitchell R (1992) Sulfate stimulation of mercury methylation in freshwater sediments. Environ Sci Technol 26:2281-2287

Goulding M (1979) Ecologia da pesca do Rio Madeira. CNPq - Inpa, Manaus - Amazonas, 172 pp

Guimarães JRD (1992) Padronização de técnicas radioquímicas visando estudos de metilação e volatilização do Hg em sistemas aquáticos de áreas de garimpo de ouro na região amazonica. PhD Thesis, Federal University of Rio de Janeiro, 109 pp

Guimarães JRD, Malm O, Pfeiffer WC (1993) Radiochemical measurements of net mercury methylation rates in sediments and soils near goldmining fields in the Tapajós river region, Brazilian Amazon. Proceedings of the International Conference on Heavy Metals in the Environment, Toronto, September 1993. 2:100-103

Guimarães JRD, Malm O, Pfeiffer WC (1994) A simplified radiochemical technique for measurements of net mercury methylation rates in aquatic systems near goldmining areas, Amazon, Brazil. Sci Tot Envir, in press

Hahne HCH, Kroontje B (1973) Significance of pH and chloride concentrations on behaviour of heavy metal pollutants: mercury(II), cadmium(II), zinc(II) and lead(II). J Environ Qual 2:44-450

Henning R (1993) Reinigung von quecksilberkontaminierten Böden mit Hilfe eines kombinierten Wasch- und Destillationsverfahrens. In: Arendt F, Annokkée GJ, Bosman R, Brink WJ van den (Hrsg) Altlastensanierung '93. Kluwer, Dordrecht, pp 1335-1344

Jackson TA (1986) Methylmercury levels in a polluted prairie river-lake system: seasonal and site-specific variations and the dominant influence of trophic conditions. Can J Fish Aq Sci 43 1873-1887

Jackson TA (1988) The mercury problem in recently formed reservoirs of northern Manitoba (Canada): effects of impoundment and other factors on the production of methylmercury by microorganisms in sediments. Can J Fish Aquat Sci 45:97-121

Jensen Sand, Jernelov A (1969) Biological methylation of mercury in aquatic organisms. Nature 223 753-754

Jernelov A, Lann G, Wennergren T, Fagerstrom B, Andersson R (1972) Analysis of methyl mercury concentrations in sediments from the St. Clair system. Unpubl Report Swed Water Air Poll Res Inst

Korthals ET, Winfrey MR (1987) Seasonal and spatial variation in mercury methylation and demethylation in an oligotrophic lake. Appl Environ Microbiol 53:2397-2404

Lacerda LD (1990) Ciclo biogeoquímico do mercúrio na Amazônia. In: Hacon S, Lacerda LD, Pfeiffer WC, Carvalho D (eds) Riscos e Consequências do Uso do Mercúrio. UFRJ, Rio de Janeiro, pp 80-90

Lacerda LD, Solomons W (1991) Mercury in the Amozon: a chemical time bomb? Duth Ministry of Housing, Physical Planning and the Environment, Haren, 61 pp

Lacerda LD, Pfeiffer WC, Silveira EG, Bastos WR, Souza CMM (1987) Contaminação por mercúrio na Amazônia: análise preliminar do Rio Madeira, RO. An I Congr Bras Geoquím 2:295-299

Lacerda LD, Pfeiffer WC, Ott AT, Silveira EG (1989) Mercury contamination in the Madeira River, Amazon: mercury inputs to the environment. Biotropica 21:91-93

Lacerda LD, Marins RV, Souza CMM, Rodrigues S, Pfeiffer WC, Bastos WR (1991a) Mercury dispersal in water, sediments and aquatic biota of a gold mining tailings drainage in Pocone, Brazil. Water Air Soil Pollut 55:283-294

Lacerda LD, Salomons W, Pfeiffer WC, Bastos WR (1991b) Mercury distribution in sediment profiles of remote high Pantanal lakes, central Brasil. Biogeochemistry 14:71-77

Lacerda LD, Carvalho CEV, Rezende CE, Pfeiffer WC (1993) Mercury in sediments from the Paraiha do Sul River Continental shelf, SE-Brazil. Mar Pollut Bull 26:220-222

Landner L (1971) Biochemical models of the biological methylation of mercury as suggested from methylation studies in-vivo in *Neurospora crassa.* Nature 230:452-453

Lechler PJ, Miller JR (1993) The dispersion of mercury, gold, and silver from contaminated mill tailings at the Carson River mercury superfund site, west-central Nevada, USA. In: Abrão JJ, Wasserman JC, Silva-Filho EV (eds) Proc Inter Symp Perspectives for Environmental Geochemistry in Tropical Countries, Univ Fed Fluminense, Niteroi, Brazil, pp 433-436

Lee YH, Hultberg H (1990) Methylmercury in some swedish surface waters. Environm Toxicol and Chem 9:833-841

Lee YH, Hultberg H, Andersson I (1985) Catalytic effect of various metal ions on the methylation of mercury in the presence of humic substances. Water Air Soil Pollut 25:391-400

Lindqvist O, Jernelov A, Johanson K, Rodhe H (1984) Mercury in the Swedish environment. Global and local sources. Report PM 1816, Nat Swed Environ Protec Board, Solna, Sweden

Lodenius M (1985) The mercury problem and fishing in Finland. In: DO Hall, N Myers, NS Margaris (eds) Economics of ecosystem management. Dr. W Junk Publs, Dordrecht, Netherlands, pp 99-103

Malm O (1991) Estudo da contaminação ambiental e humana por Hg na região garimpeira de ouro do Rio Madeira, Amazônia. PhD Thesis Instituto de Biofísica CCF$^{o}$, UFRJ, Rio de Janeiro, Brazil, 113 pp

Malm O, Pfeiffer WC, Souza CMM, Reuther R (1990) Mercury pollution due to gold mining in the Madeira river basin, Brazil. Ambio 19:11- 15

Malm O, Pfeiffer WC, Souza CMM (1991) Main pathways of mercury in the Madeira river area, Rondonia, Brazil. In: Procc 8th Intern Conf Heavy Metals in the Environment, Edinburgh 1:515-518

Malm O, Branches FJP, Castro MB, Pfeiffer WC (1992) Mercury contamination in riverine population through ingestion of fish in the Tapajós Basin, Amazon, Brazil. Abs 4th Ann Meeting Inter Soc Environment Epidemiology, México. Session: 03.1

Malm O, Guimarães JRD, Pfeiffer WC (1993) Accumulation of metallic mercury and natural amalgams findings in Madeira river basin bottom sediments, Amazon. Proc Inter Symp Perspectives for Environmemtal Geochemistry in Tropical Countries, Univ Fed Fluminense, Niteroi, Brazil, pp 391-393

Malm O, Castro MB, Bastos WR, Branches FJP, Guimarães JRD, Zuffo CE, Pfeiffer WC (1994a) An assessment of Hg pollution in different goldmining areas, Amazon, Brazil. Accepted in Science of Total Environment.

Malm O, Branches FJP, Akagi H, Castro MB, Pfeiffer WC, Harada M, Bastos WR, Kato H (1994b) Mercury and methylmercury in fish and human hair from Tapajós river basin, Brazil. Accepted in Sci Total Environ

Marins RV, Tonietto GB (1991) An evaluation of the sampling and measurements techniques of mercury in the air. Ann 3rd Intern Congr Environ Chem in Brazil, pp 103

Marins RV, Imbassay JA, Pfeiffer WC, Bastos WR (1991) Contaminação atmosférica de mercúrio em area produtora de ouro no distrito de Poconé, Mato Grosso, MT Proc I Inter Symp Environm Studies Tropical Forest Manaus, 1:199-203

Martinelli LA, Ferreira JR, Forsberg BR, Victoria (1988) Mercury contamination in the Amazon: a gold rush consequence. Ambio 17:252-254

Matsumura F,Gotoh Y, Boush GM (1971) Phenylmercuric acetate: metabolic convertion by microorganisms. Science 173:49-51

Meili M (1991) Mercury in boreal forest lake ecosystems. PhD Thesis, Acta Univ Ups, Comprehensive Summaries of Uppsala Dissertations from the Faculty of Science 336. Uppsala University, Sweden 36 pp

Miller J R, Lechler PJ, Rowland J (1993) Heavy metal transport by physical processes in the Carson river valley, west-central Nevada, USA: implications to the distribution and storage of metal pollutants in tropical environments. In: Abrão JJ, Wasserman JC, Silva-Filho EV (eds) Proc Inter Symp Perspectives for Environmental Geochemistry in Tropical Countries, pp 131-136

Mitra S (1986) Mercury in the ecosystem. Transtech Publ Columbus, Ohio, 327 pp

Mudroch A, Clair TA (1986) Transport of arsenic and mercury from gold mining activities through an aquatic system. Sci Total Environm 57:205-216

Müller G (1993) Long distance mercury transport in Rio Tapajós, Pará, Brazil. In: Abrão JJ, Wasserman JC, Silva-Filho EV (eds) Proc Inter Symp Perspectives for Environmental Geochemistry in Tropical Countries, Univ Fed Fluminense, Niteroi, Brazil, pp 363-364

Nriagu JO (1990) Global metal pollution poisoning the biosphere. Environment 32(7):7-33

Nriagu JO (1993) Mercury pollution from silver mining in colonial South America. In: Abrão JJ, Wasserman JC, Silva-Filho EV (eds) Proc Inter Symp Perspectives for Environmental Geochemistry in Tropical Countries, pp 365-368

Nriagu JO, Pacyna JM (1988) Quantitative assessment of worldwide contamination of air, water and soils by trace metals. Nature 333:134-139

Nriagu JO, Pfeiffer WC, Malm O, Souza CMM, Mierle G (1992) Mercury pollution in Brazil. Nature 356:389

Nuorteva P, Lodenius M, Nuorteva SL (1979) Decrease in Mercury levels of *Esox lucius (L.)* and *Abramis farenus (L.)* (Teleostei) in the Haneenkyro water course after the phenylmercury ban in Finland. Aquilo Ser 19:97-100

Padberg S (1990) Mercury determinations in samples from Tapajós (Itaituba). Unpubl manuscript, Inst Angewandte Physikalische Chemie, Jülich, Germany, 13 pp

Pelletier E (1985) Mercury-selenium interactions in organisms: a review. Mar Environm Res 18:111-132

Petrick FR (1993) Bindung und Akkumulation von Quecksilber in den vom Goldabbau kontaminierten Flussedimenten des rio Madeira, Rondônia, Brasilien. PhD Thesis, Ludwig Maximilians Universität, München 112 pp

Pfeiffer WC, Lacerda LD (1988) Mercury inputs into the Amazon region, Brazil. Environm Technol Lett 9:325-330

Pfeiffer WC, Malm O, Souza CMM, Bastos WR, Torres JP (1989a) Mercury contamination in goldmining areas of Rio de Janeiro state, Brazil. Proc Inter Conf Heavy Metals Environ, Geneve, vol 1:222-225

Pfeiffer WC, Lacerda LD, Malm O, Souza CMM, Silveira EG, Bastos WR (1989b) Mercury concentrations in inland waters of Rondônia, Amazon, Brazil. Sci Tot Environm 87/88:233-240

Pfeiffer WC, Lacerda LD, Malm O, Souza CMM, Silveira EG, Bastos WR (1991) Mercury in the madeira River ecosystem, Rondonia, Brazil. Forest Ecol Manegt J 38:239-245

Ramlal PS, Rudd JMW, Furutani A, Xun L (1985) The effect of pH on methylmercury production and decomposition in lake sediment. Can J Fish Aquat Sci 42:685-692

Ramlal PS, Rudd JWM, Hecky RE (1986) Methods for measuring specific rates of mercury methylation and degradation and their uses in determining factors controlling net rates of mercury methylation. Appl Environ Microbiol 51:110-114

Ramos JFF, Costa MQ (1990) Distribuição de mercúrio em dois garimpos do Estado do Pará. In: Hacon S, Lacerda LD, Pfeiffer WC, Carvalho D (eds) Riscos e Consequências do uso do Mercúrio. UFRJ, Rio de Janeiro, pp 70-79

Rekolainen S, Verta M, Liehu A (1986) The effect of airborne mercury contents in some Finnish forest lakes. Publ Water Res Inst Finland 65:11- 20

Robinson JB, Tuovinen OH (1984) Mechanisms of microbial resistance and detoxification of mercury and organomercury compounds: physiological, biochemical and genetic analyses. Microbiol Rev 48:95-124

Rodrigues BA, Leuzi E, Luchese EB, Rauber T (1991) Níveis de concentração de mercúrio nas águas dos Rios Paraná e Baía, Região de Porto Rico Acta Limnol Bras (in press)

Rogers RD (1976) Methylation of mercury in agricultural soils. J Environ Qual 5:454-458

Rogers RD (1977) Abiological methylation of mercury in soil. J Environ Qual 6:463-467

Salomons W, Förstner U (1984) Metals in the hydrocycle. Springer, Berlin Heidelberg New York, 349 pp

SCOPE (1985) Rapporteur's report – mercury. Scient Comm Problems Environm Toronto, Canada, 16 pp

Shrestha KP, Quilarque XR (1989) A preliminary study of mercury contamination in the surface soil and river sediment of the Roscio district, Bolivar state, Venezuela. Sci Tot Environm 79:233-239

Souza V P de (1991) Descontaminacao de rejeito de garimpo contendo mercurio pelo processo de eletrooxidacao. In: Poconé: um campo de estudos do impacto ambiental do garimpo / Org. por Marcello M Veiga e Francisco Rego C Fernandes, Rio de Janiero: CETEM/CNPq

Steffan RJ, Korthals ET, Winfrey MR (1988) Effects of acidification on mercury methylation, demethylation and volatilization in sediment of an acid-susceptible lake. Appl Environ Microbiol 54:2003-2009

Thibodeaux LJ (1979) Chemodynamics - Environmental movement of chemicals in air, water and soil. John Wiley and Sons, New York, 491 pp

Thornton I, Cleary Dand Worthington S (1992) Mercury contamination in the Brazilian Amazon. A cooperative research study conducted by Gedebam, Brazil e Sol 3 (Switzerland and Luxemburg), 45 pp

Tumpling WV, Wilken RD, Einax J (1993) Pathways and contaminations of mercury in the Pantanal area, Brazil. In: Abrão JJ, Wasserman JC, Silva-Filho EV (eds) Proc Inter Symp Perspectives for Environmental Geochemistry in Tropical Countries, Univ Fed Fluminense, Niteroi, Brazil, pp 403-406

Weber JH, Reisinger H, Stoeppler M (1985) Methylation of mercury(II) by fulvic acid. Env Technol Lett, 6:203-208

WHO (1976) Environmental health criteria n° 1 – mercury. World Health Organization, Geneva, 131 pp

WHO (1991) IPCS-inorganic mercury – environmental health criteria 118 Geneva, Switzerland, 168 pp

Wilken R-D (1993) Mercury problems in Germany and the analytical approach. Intern Conference on Heavy Metals in the Environment Toronto, September (1993. CEP Consultants Ltd, Edinburgh, 1:372-375

Winfrey MR, Rudd JWM (1990) Environmental factors affecting the formation of methylmercury in low pH lakes. Environm Toxicol and Chem 9:853-869

Xun L, Campbell NER, Rudd JWM (1987) Measurement of specific rates of net methylmercury production in the water column and surface sediments of acidified and circumneutral lakes. Can J Fish Aquat Sci 44:750-757

# 11 Non-linear Release of Metals from Aquatic Sediments

*U. Förstner*

## 11.1 Introduction

Modern research on particle-bound contaminants quite probably originated with the idea, that sediments reflect the biological, chemical and physical conditions in a water body (Züllig 1956). Based on this concept, the historical evolution of limnological parameters could be traced back from the study of vertical sediment cores. In fact, already early in this century, Nipkow (1920) suggested that the alternating sequence of layers in a sediment core from Lake Zürich might be related to variations in the trophic status of the lake system. Geochemical investigations of stream sediment has become standard practice in mineral exploration since the beginning of the sixties (Hawkes and Webb 1962). Similarly, lake sediment geochemistry has been used as a guide to mineralization, particularly intensive on lakes of the Canadian shield. This approach attracted even more attention when mineral exploration was followed by large-scale mining and processing activities: "Both the exploration and environmental geochemist can be looking to the same type of areas, those with high metal concentrations, but obviously from a different motivation" (Allan 1974). During the past 20 years, research on metal-contaminated sediments evolved to four aspects, which in an overlapping succession also reflect the development of knowledge on particle-bound pollutants:

- The evaluation of *solid/solution relations* of metals in surface water;
- The study of in situ *processes and mechanisms* of metal transfer in various compartments of the aquatic ecosystems;
- The assessment of the *environmental impact* of particle-bound metals, i.e. development of sediment quality criteria;
- The development of *remedial measures,* in particular, of dredged materials, by integrated, multidisciplinary research

For all four aspects, in particular for the assessment of the environmental impact of particle-bound metals and for the selection of remedial options for metal-rich dredged materials, new experience with non-linear and time-delayed responses has significant implications on both scientific and technological strategies.

## 11.2 Geochemical Concepts for Metallic Pollutants in Sediments

Three geochemical concepts demonstrate the advantage of a long-term strategy compared to the traditional approaches in research on contaminated sediments: (i) the mobility concept, (ii) the concept of capacity controlling properties, and (iii) the concept of final storage quality; the first and second being relevant for process and effect studies, the third for problem solutions.

### 11.2.1 Mobility Concept

For systems involving solution/solid interactions, "mobility" typically reflects the flux of metal species in a certain medium, which contains both accelerating and inhibiting factors and processes. Accelerating factors comprise effects of pH-lowering, redox changes, inorganic and organic complexation, and microbially mediated species transformations such as biomethylation. Within the spectrum of "barriers", physical processes include adsorption, sedimentation, and filtration; chemical barriers comprise mechanisms such as complexation and precipitation; and biological barriers are often associated with membrane processes, which can limit translocation of metals (e.g. from plant roots to the shoots and fruits). "Complexation" in its various forms can both inhibit and accelerate metal fluxes, particularly in biological systems consisting of different types of membranes. Three major factors affect the distribution of metals between solution and particulate matter: (1) the chemical form of dissolved metals originating both from natural and human-induced sources; (2) the type of interactive processes, such as sorption/desorption-controlled or precipitation-controlled mechanisms; and (3) concentration and composition of particulate matter, mainly with respect to surface-active phases.

Prediction of short-, middle-, and long-term effects of metal contaminated particulates, in the framework of sediment quality objectives and criteria development has to consider the environmental "speciation" of critical metals – in the meaning of "describing the distribution and transformation of metal species and various media" (Bernhard et al. 1986). Long-term prognosis, in particular, of the behaviour of metals at critical sites requires both knowledge of interactions of element species in solid matter and solution, and an estimation of future borderline (particularly "worst case") conditions in a dynamically evolving medium (Förstner 1993). In sediments, typical driving forces for intensified matrix/element-interactions are strong chemical gradients of redox conditions, pH-values and organic ligand concentrations, all three factors mainly being induced by degradation of organic matter.

### 11.2.2 Storage Capacity Controlling Properties

The conceptual approach of "storage capacity controlling properties" as the link between the geochemical cycles comprising driving forces such as organic matter/redox-sensitive elements and cycles of mobilizable micropollutants forms the key aspect in the framework of the "chemical time bomb" concept (Stigliani 1991). To make the scientific objectives clearer, it is useful to distinguish between two different mechanisms: The first is direct saturation, by which the capacity of a sediment for toxic chemicals becomes exhausted. The second way to "trigger" a time bomb is through a fundamental change in a chemical property of the solid matrix that reduces its capacity to adsorb (or keep adsorbed) toxic materials.

Typical "macro-environments", where the assessment of metal mobility is of particular relevance for estimating both the actual and future effects of critical components, are (Förstner et al. 1990): (a) Excessive treatment of soil with organic-rich agricultural waste materials; (b) transformation of metal-rich components in mining residues; (c) long-term development of municipal and industrial waste material; and (d) behaviour of polluted sediments in the estuarine mixing zone. In these dynamic environments characteristic factors and processes interact with the storage capacity controlling properties:

- High-energy mechanical processes, such as resuspension of particles by wave action, bioturbation, and dredging activities
- Strong chemical gradients, e.g. of redox conditions, pH-values, and organic ligand concentrations
- Distinct temporal variations and developments, such as seasonal changes of biological activity and typical successions in metabolic transformations, e.g. reductive processes in sediments and waste materials

Regarding the interactions between metals and sediment components, pH-conditions are of prime importance. Typical curves for the adsorption of metals onto inorganic substrates, such as iron oxyhydroxides, increase from almost nothing to near 100% as pH increases through a critical range 1-2 units wide (Benjamin et al. 1982). It is important to note, that the location of the pH-adsorption-"edge" depends on adsorbent concentration. This effect is due to the existence of a range of specific site-binding energies; high-energy adsorption sites, since they are fewer in number, become limiting first. For systems rich in organic matter, metal adsorption curves cover a wider pH-range than is observed for inorganic substrata.

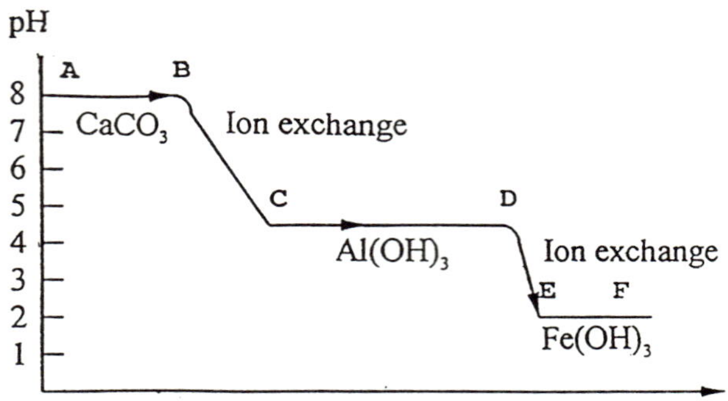


**Fig. 11.1.** Acid buffering processes in soil/water systems (after Prenzel 1985)

Acid consuming processes in terrestrial and aquatic ecosystems include (Swift 1977): (1) mineralization of cations, (2) assimilation of anions, (3) protonation of anions, (4) reductions, (5) weathering of metal oxide components, (6) reverse weathering of anions, and, in particular, (7) dissolution of carbonate minerals. A schematic curve of different buffering levels in soil–water systems (Fig. 11.1 after Prenzel 1985) combines two factors: the horizontal axis, which usually refers to the "capacity" factor, indicating the amount of strong acid added, while the vertical axis indicates the "intensity" factor, i.e. the pH or the degree of base saturation of the exchange complex. As long as $CaCO_3$ is present, added protons will be consumed by dissolution reactions, and the pH of the system will remain constant (A → B in Fig. 11.1). Between B and C the exchangeable sites of clay minerals are displaced by hydrogen ions. Ultimately, this process leads to the dissolution of Al, e.g. clay minerals (C → D); due to the abundance of Al in clay minerals, the buffering capacity in this process is high. When the accessible reservoir of Al is exhausted, the dissolution of ferric oxide will consume hydrogen ions and resist the pH decrease (E → F). With regard to long-term effects, major processes consuming hydrogen ions in ecosystems include release of basic cations by decomposition of organic matter, specific anion adsorption, mineral weathering, and unbalanced reduction of oxidized compounds (Binkley et al. 1989). The process of specific anion adsorption in sediments mainly refers to interactions of sulphate and phosphate with aluminium and iron sesquioxides (Tabatabei 1987).

### 11.2.3 Extension of Salomons' Scheme of "Coupled Cycles": Geo- and Bioaccumulation

With respect to the modelling metal partitioning between dissolved and particulate phases in a natural system, e.g. for estuarine sediments, the following requirements have been listed by Luoma and Davis (1983):

- The determination of binding intensities and capacities for important sediment components
- The determination of relative abundance of these components
- The assessment of the effect of particle coatings and of multi-component aggregation on binding capacity of each substrate
- The consideration of the effect of major competitors ($Ca^{2+}$, $Mg^{2+}$, $Na^+$, $Cl^-$)
- The evaluation of kinetics of metal redistribution among sediment components

It seems that thermodynamic models are still restricted because of various reasons: (1) adsorption characteristics are related not only to the system conditions (i.e. solid types, concentrations and adsorbing species), but also to changes in the net system surface properties resulting from particle/particle interactions such as coagulation; (2) influences of organic ligands in the aqueous phase can rarely be predicted as yet; (3) effects of competition between various sorption sites, and (4) reaction kinetics of the individual constituents cannot be evaluated in a mixture of sedimentary components.

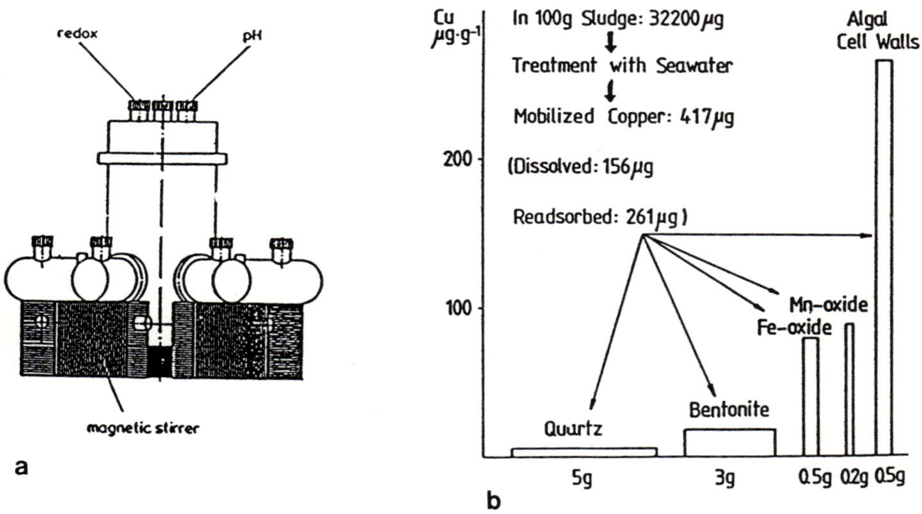


**Fig. 11.2a–b.** Metal transfer between sedimentary components (Calmano et al. 1988)

At present, experimental studies on the dissolved/solid interactions in such complex systems seem to be more promising. One approach uses a six-chambered device, where the individual components are separated by membranes, which still permit phase interactions via solute transport of the elements (Calmano et al. 1988); in this way, exchange reactions and biological uptake can be studied for individual phases under the influence of pH, redox, ionic strength, solid and solute concentration, and other parameters (Fig. 11.2a). In an experimental series on the effect of salinity (i.e. disposal of anoxic dredged mud into sea water) quantities of model components were chosen in analogy to an average sediment composition: 0.5 g algal cell walls (=5%), 3 g bentonite (=30%), 0.2 g manganese oxide (=2%), 0.5 g goethite (=5%), and 5 g quartz powder (=50%).

The effect of salinity on metal remobilization from contaminated sediments is different for the individual elements. While approximately 16% and 9% of cadmium and zinc, respectively, in the dredged mud from Hamburg harbour is released, for metals such as copper the factor salinity increase seems to be less important in the transfer both among sediment substrates and to aquatic biota. This is, however, untrue as can be demonstrated from a mass balance for the element copper in Fig. 11.2b. It is indicated that only 1.3% of the inventory of Cu of the sludge sample is released when treating with sea water. Only one-third stays in solution, equivalent to approx. 40 $\mu g\ l^{-1}$, and there is no significant difference to the conditions before salt addition. Two-thirds of the released Cu is readsorbed at different affinities to the model substrates. Slight enrichment of Cu is observed in the iron hydroxide (approx. 80 ppm) and manganese oxide (100 ppm), whereas the cell walls – a minor component in the model sediment – have accumulated nearly 300 ppm of Cu.

The dominant role of organic substrates in the binding of metals such as Cd and Cu is of particular relevance for the transfer of these elements into biological systems. It can be expected that even at relatively small percentages of organic substrates these materials are primarily involved in metabolic processes and thus may constitute the major carriers by which metals are transferred within the food chain. Including these mechanisms of metal enrichment on mineral surfaces ("geoaccumulation"; Müller 1979), organic matrices and, in particular, by organisms, the concept of coupled geochemical cycles (Salomons 1993) can be extended to the following scheme:

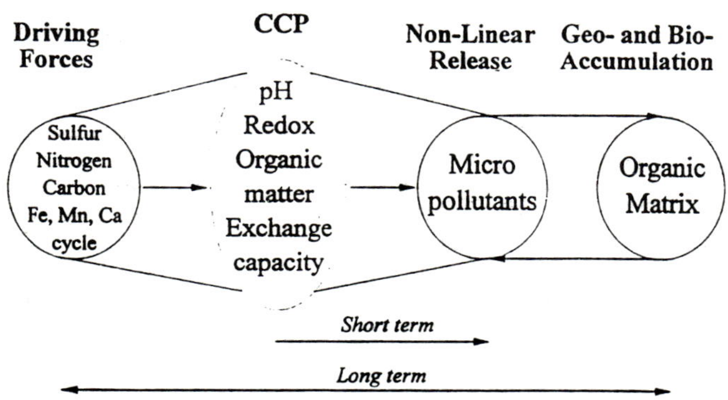


**Fig. 11.3.** Extension of Salomons' (1993) concept of "coupled geochemical cycles"

### 11.2.4 Final Storage Quality

The final storage approach is one way to develop and control landfills on a conceptual basis. It has been defined by the Swiss Federal Government in 1986 and received wider attention by the book edited by Peter Baccini (1989) on *Landfills-Reactor and Final Storage*: "Landfills with solids of final storage quality need no further treatment of emissions into air and water". Including new experience from impact evaluations related to capacity controlling properties, the mobility concept of environmental geochemistry can be implemented into waste management practice by different ways of optimizing barrier systems. As shown from the examples of large-mass wastes – dredged material, mining residues and municipal solid waste, long-term immobilization of critical pollutants can be achieved by promoting less soluble chemical phases, i.e. by thermal and chemical treatment, or by providing respective milieu conditions. A common feature of geochemically-designed deposits is their tendency to increase stability in time, due to the formation of more stable minerals and closure of pores, thereby reducing water permeation.

## 11.3 Redox/Ph-Reactions: Variations in Time and Space

### 11.3.1 Redox Reactions – Temporal Developments (Meyer et al. 1994)

In aquatic environments, a predictable sequence of electron acceptors is used during the microbial degradation of organic matter. These reactions determine the redox potential and may alter the pH; they also determine, directly or indirectly, the bioavailability of many inorganic and organic chemicals. Figure 11.4 shows the set of sequential redox reactions and dissociation equilibria of the classical biogeochemical paradigm.

A net release of $CO_2$ occurs as a result of most oxidations of $C_{org}$, thus providing a driving force to decrease the pH in systems that are not at equilibrium with the atmosphere. However, the natural buffer systems and the presence of acid-soluble minerals containing anions of weak acids (e.g. $CaCO_3$) will at least partly offset the associated increase in $H^+$ concentration when $H_2CO_3$ dissociates.

Besides the direct detrimental effect of anoxia on most aquatic fauna, a major concern in the classical biogeochemical paradigm is the release of trace metals (symbolized by *Me* in Fig. 11.4). Trace metals adsorb to manganese oxides and iron oxyhydroxides, which exist in particulate and colloidal form in oxic water–sediment systems. Because the reduced forms are relatively soluble in water, the $MnO_2$-Me and FeOOH-Me complexes dissolve and the metal ions may become

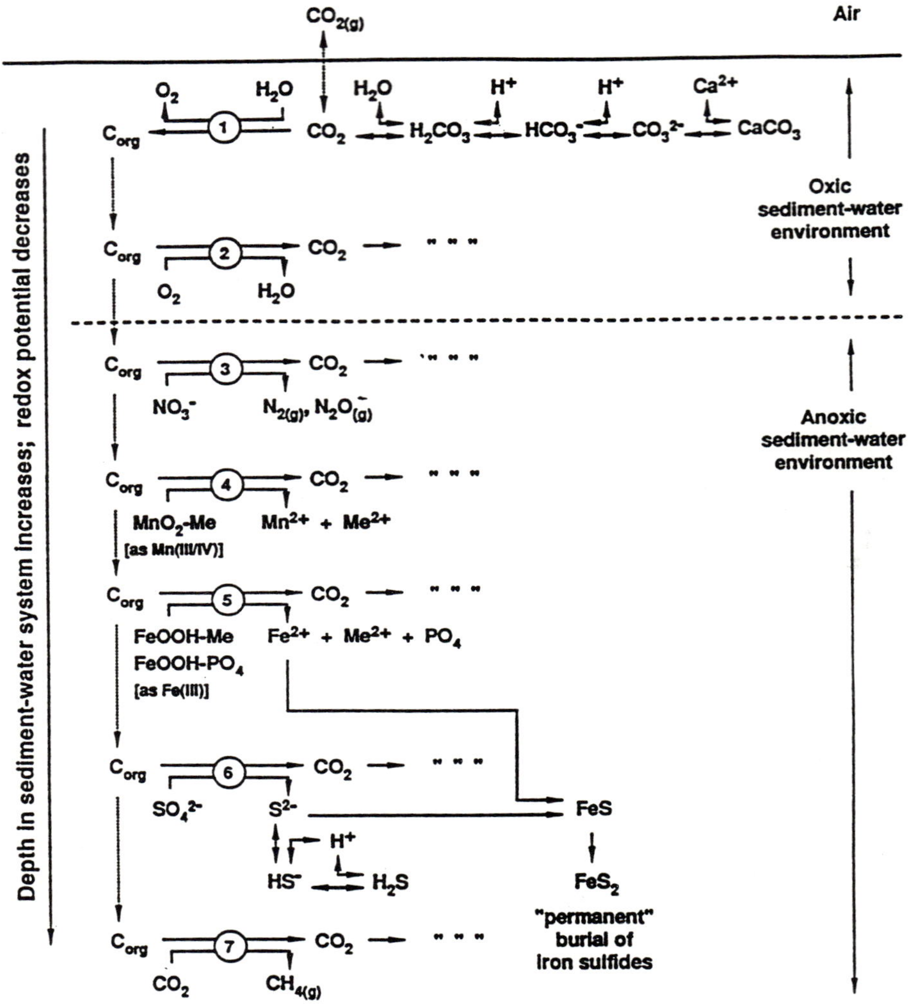


**Fig. 11.4.** Schematic diagram of the biogeochemical processes in water–sediment systems (Meyer et al. 1993). *1* - photosynthesis, *2* - oxidative metabolism, *3* - denitrification, *4* - manganese reduction, *5* - iron reduction, *6* - sulphate reduction, and *7* - methanogenesis

bioavailable.

If the system is not disturbed, each redox pair will buffer the redox potential until the reservoir of the electron acceptor is exhausted (Fig. 11.5a). However, if a mixing event transports oxygenated water down to the sediment surface before the $O_2$ reservoir is depleted, the use of other inorganic electron acceptors will not be required; Fig. 11.5b shows conceptually the stepwise sequence of decreasing redox potentials that might be observed between two disturbances.

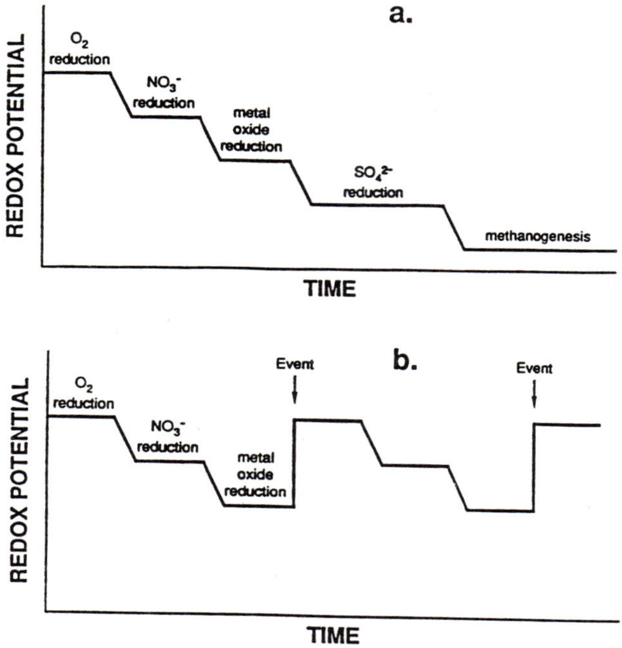


**Fig. 11.5a–b.** Stepwise decreases in redox potential over time in a sediment, as progressively less energetically favourable electron acceptors are used to oxidize organic matter (**a**) in a closed system that proceeds to $SO_4^{2-}$ reduction without disturbance, and (**b**) in a system that is reoxygenated periodically during the phase when a metal oxide is being used as the electron acceptor (Meyer et al. 1994)

Periodical physical, chemical and biological phenomena linked to redox processes – with time-scales of minutes to years – are listed in Table 11.1, in terms of driving forces, resulting processes, and effects on bioavailability (Meyer et al. 1993).

### 11.3.2 Spatial Patterns of Redox Effects: Sediment Diagenesis and Estuarine Environment

The sedimentary environments corresponding to the processes given in Fig. 11.5a are, respectively: "oxic", "post-oxic", "sulphidic" and "methanic" (Berner 1981). The sequence of events affecting these transformations can be evaluated both from pore water analyses and from mineral concentrates of typical "*diagenetic*" (spectrum of mechanisms and effects by which sedimentary constituents after deposition from the water column are changed in their chemical form, involving new equilibrium between solids and dissolved species) phases in short sediment cores.

The succession of typical pore water constituents has been depicted by Froelich et al. (1979) from the study of deep-sea sediment cores from the eastern equatorial

**Table 11.1.** Periodic Disturbances that affect redox-related processes and bioavailability of chemicals (Meyer et al. 1994)

| Time scale | Phenomenon | Driving Force | Resulting process | Effect on Bioavailability |
|---|---|---|---|---|
| 1 year | seasonal production | $Co_2$ consumption on photic zone | pH increase (up to pH 10) in photic zone during summer | Release of $PO_4$ from shallow-water sediments |
| 1 year | seasonal production | Increased supply of $_{org}C$ | Trace-metal uptake by growing phytoplankton and/or their exudates | Decrease of available trace-metal concentration in photic zone |
| 1 year | seasonal production | Increased supply of $_{org}C$ | Increased supply of metals in biota, and increased reducing intensity in surface sediment (0–5 cm) | Chain of events iniated:<br>1. Initial release of trace metals, due to oxidative decomposition of biota<br>2. Subsequent binding of trace metals to sulfides, when $SO_4^{2-}$ is reduced |
| 1 year | seasonal production | Increased supply of $_{org}C$ | pH decrease (-0.5 pH unit) in hypolimnion and oxic upper layer of sediment, due to decomposition | Release of trace metals to dissolution and exchange reactions [a] |
| 1 year | Seasonal stratification of lakes | Diminished supply of $O_2$ | Release of $Fe^{2+}$, $Mn^{2+}$ and $S^{2-}$ in hypolimnion (min. at several m deep) | Possible release of CO and N but no effect on Cu Pb, and Zn |
| 1 month | Lunar tidal cycle (max. current velocities are lunar-dependent | Gravitational attraction of moon | Periodic inundation, flushing and aeration of extreme inter-tidal zone sediment | Resuspension and oxidation reducing sediment; can include longer-term (years) change in sediment character [a] |
| 1 day | Photosynthesis | Diurnal $CO_2$ consumption | Diurnal pH shift (usually <0.5 pH unit) | Small diurnal release of $PO_4$ from shallow-water sediments? [a]<br>Diurnal hydrolysis/adsorption of metals? [a] |
| | Photosynthesis | Diurnal $_{org}C$ production | Diurnal uptake of metals by phytoplankton (probably undetectable) | Diurnal changes in trace-metal availability (probably undetectable [a] |
| 1 day | Photoreduction | Light in photic zone | Reduction of Fe-, Mn- and other metal oxides | Fe (essential nutrient) available for uptake by organisms as $Fe^{2+}$; more general availability of trace metals (e.g. Zn) ? [a] |
| 1 day | Photooxidation | Light in photic zone | Partial or complete oxidation of organic ligands (e.g. humics) | Release or uptake of inorganic and organic ions [a] |
| 12 hours | Semi-diurnal tidal cycle (prime determinant of instantaneous current velocities | Gravitational attraction of moon | Periodic inundation, flushing and aeration of inter-tidal zone sediment | Resuspension and oxidation of reducing sediment; can include longer-term (years) change in sediment character [a] |
| Minutes | Advection in rippled, sandy sediment | Surface gravity waves | Advection of water through surface sediment (0–5 cm) | Exchange flux of material at most active aquatic site – the sediment–water interface |
| Minutes | Pore-water flushing | Physical motion of organisms | Exchange of water between sediment environment and overlying water | Perturbation of redox environment within sediment |

[a] A reasonable hypothesis, rather than an established fact

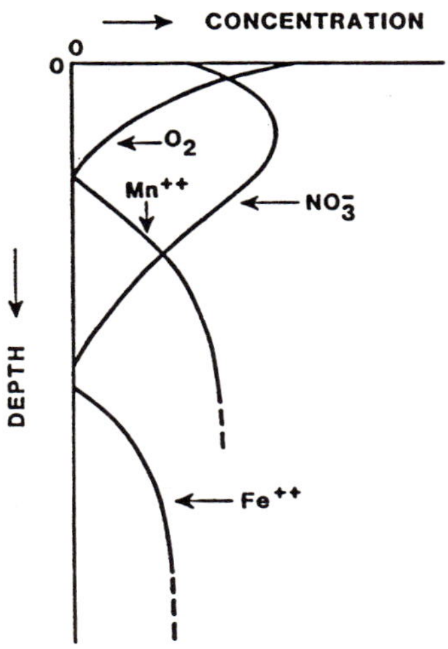


**Fig. 11.6.** Schematic succession of concentration profiles of oxygen, nitrate, manganese and iron in pore waters from deep-sea sediment cores (Froelich et al. 1979)

Atlantic (Fig. 11.6). Oxygen is consumed in the uppermost deposits during decomposition of organic matter. This process, which involves dissolution of carbonate, will release ammonia, which is subsequently oxidized to nitrate; $NO_3^{2-}$ diffuses down to be denitrified in the lower part of the core. Not far below the nitrate maximum organic carbon starts to become oxidized by simultaneous reduction of $MnO_2$ (upward diffusing $Mn^{2+}$ will be precipitated in the stronger oxidizing part of the sediment sequence); at greater sediment depth, the process of iron reduction to $Fe^{2+}$ begins, thus indicating the persistence of iron oxides, but not manganese oxides, under anoxic conditions (Berner 1981). In stronger reducing environments the metal distribution should be controlled primarily by the formation of (highly insoluble) sulphides; at lower $H_2S$ concentrations, it should be controlled by the stability of carbonate and phosphate phases.

Typical spatial distribution patterns of master variables such as $O_2$ and pH are found in estuaries. The circulation pattern of water determines the residence time of water, transfer mechanisms and accumulation of particles, as well as the interactions between dissolved and solid element species (Duinker 1986).

In Fig. 11.7, characteristic parameters are presented for a longitudinal section of the Elbe River estuary. The "salinity gradient" is usually located between river km 670 and 700, depending primarily on water discharge from the upstream catchment area. Stronger spatial and temporal variabilities are observed for the concentration and composition of suspended matter and dissolved organic carbon. These parameters are typically affected by inputs from upstream and the seasonal variations in biological production. The strongest variations in both time and space are found for oxygen content. In spring, with increasing bioactivity and decreasing waterflow, the oxygen minimum is primarily caused by intensive nitrification. In summer, the "oxygen hole", sometimes with critical values below 2 mg $l^{-1}$, shifts upstream in the direction to Hamburg harbour.

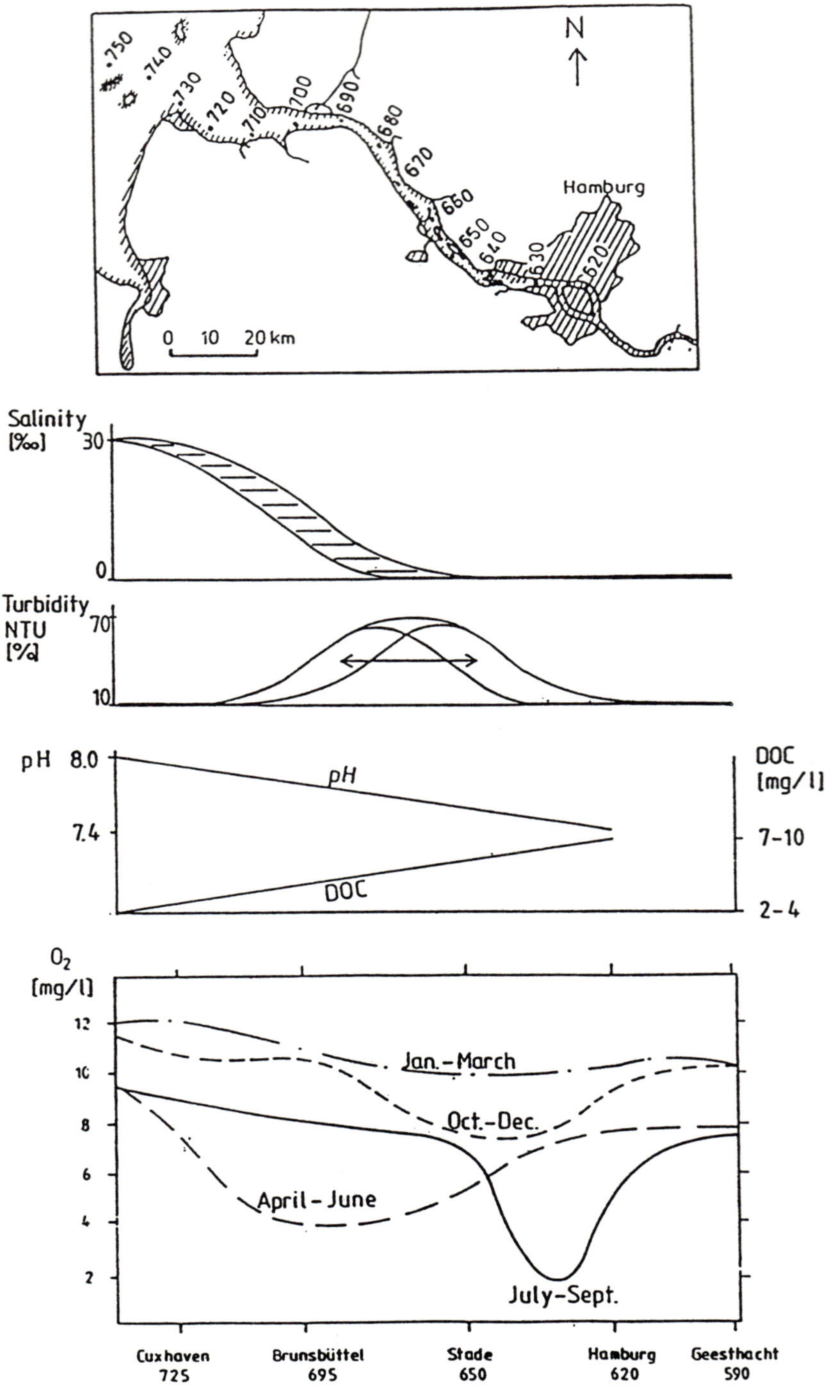


**Fig. 11.7.** Variations of typical parameters in a longitudinal section of the tidal Elbe river (Tent 1987; after ARGE Elbe Reports 1978-1985)

### 11.3.3 Oxidation of Sulphides and Proton Release

Acidity is perhaps the most serious long-term threat from metal-bearing wastes. For decades, water seeping from mine refuse has delivered increased metal concentrations into receiving waters. The threat is especially great in waters with little buffer capacity (i.e. in carbonate-poor areas where dissolved metal pollution can be spread over great distances). The acidity production can develop many years after disposal, once the neutralizing or buffering capacity in a pyrite-containing waste is exceeded.

The major process (see Eqs. 1–3) affecting the lowering of pH-values (down to pH 2 to 3) is the exposure of pyrite ($FeS_2$) and other sulphide minerals to atmospheric oxygen and moisture, whereby the sulphidic component is oxidized to sulphate and acidity ($H^+$ ions) is generated. Bacterial action can assist the oxidation of $Fe^{2+}(aq)$ in the presence of dissolved oxygen.

$$4\ FeS + 9\ O_2 + 10\ H_2O = 4\ Fe(OH)_3 + 4\ SO_4^{2-} + 8\ H^+, \text{ or} \quad (1)$$

$$4\ FeS_2 + 15\ O_2 + 14\ H_2O = 4\ Fe(OH)_3 + 8\ SO_4^{2-} + 16\ H^+ \quad (2)$$

*Pyrite*

$$2\ NH_4^+ + 3\ O_2 = 2\ NO_{2-} + 2\ H_2O + 4\ H^+ \quad (3)$$

The acidification of a sediment/water system begins after hydrogen ions are generated during the oxidation (e.g. during dredging or resuspension of mainly fine grained material containing less carbonate than needed for long-term neutralization). Primary emissions containing high metal concentration issue from waste rocks and tailings, while tailing ponds are primarily responsible for secondary effects on groundwater. An important and long-term source of metals are the sediments reworked from the flood plain, mainly by repeated oxidation and reduction processes. High concentration factors were found in inland waters affected by acidic mine effluents.

**The Concept of Acid-Producing Potential.** The concept of acid-producing potential (APP) was initially developed in the prediction and calculation of acid mine drainage and waste tailings management (Anonymous 1979; as summarized by Ferguson and Erickson 1988). Our findings on the effects of periodical redox processes on both APP and metal mobility in estuarine sediments (Kersten et al. 1985; Kersten and Förstner 1991) have further enhanced research interest in this field.

Periodical redox processes can spur an increase or decrease in APP or pH in a sediment/water system. In a closed system, periodical redox processes can lead to the change or transfer between APP(s) and APP(aq) but the total APP of the system does not change. The processes are reversible. The hydrogen ions produced in the oxidation will be consumed by the following reduction. Contrarily, in an open system, the total APP of the system will change depending on the properties of the system and the reaction processes. Under certain conditions total APP in the system increases, while under other conditions total APP in the system decreases. Some processes are irreversible. The components producing or consuming $H^+$ leave the system and cause the change in APP(s), APP(aq) and permanent ANC (acid neutralizing capacity). The following chemical changes affecting increases or decreasing of the ANC in aerobic and anaerobic samples have been found (Fig. 11.8):

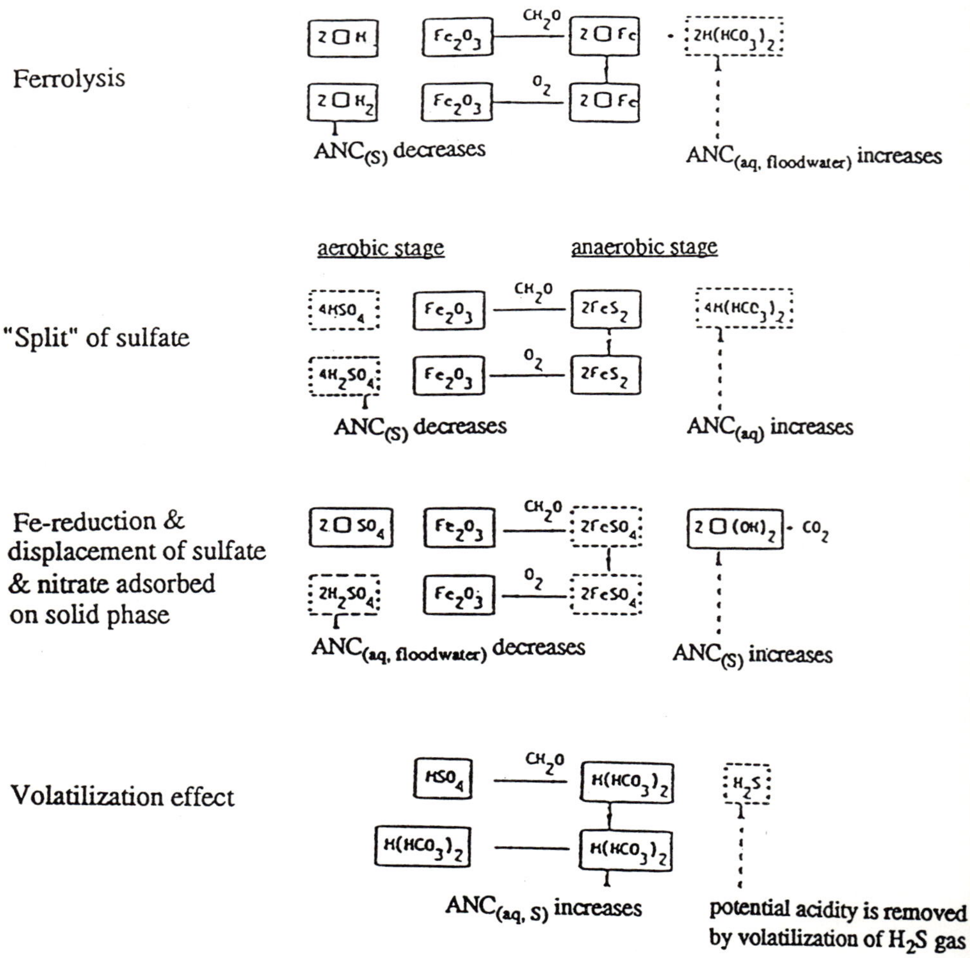


**Fig. 11.8.** Chemical changes in periodical aerobic (*left*) and anaerobic (*right*) samples. Components in *broken-line rectangles* are mobile. *Small squares* denote exchangeable sites. *M* stands for a divalent cation ($Ca^{2+}$ or $Mg^{2+}$) and $CH_2O$ refers to oxidizable organic matter (Van Breemen 1987)

**Change of APP by the "Split" of Sulphate.** Permanent acidification in alternating aerobic and anaerobic systems (Kittrick et al. 1982) is proceeding in two stages (Van Breemen et al. 1983): The first stage is characterized by the reduction of sulphates, in particular in tidal flats or sea-bottom sediments. Most of the sulphide formed is fixed in the sediment as FeS or $FeS_2$ (pyrite). It leads to the increase in APP(s) while ANC(aq) ($HCO_3^-$) formed during sulphate reduction (Eqs. 4 and 5) is removed by tidal turbulence or by diffusion into the overlying water. As a result mobile ANC(aq) ($HCO_3^-$) and immobile potential acidity ($FeS_2$) are separated or "split". The increase in APP(s) leads to a permanent decrease in the ANC(aq) after next aeration and oxidation and results in extreme acidification of the system.

$$SO_4^{2-} + 2\,H^+ = H_2S + 2\,O_2 \tag{4}$$

$$CO_2 + H_2O = HCO_3^- + H^+ \tag{5}$$

If during the reduced stage $M(HCO_3)_2$ were retained in the system (e.g. as precipitated carbonate) the acidity formed by sulphide oxidation would be neutralized exactly by the carbonate formed without permanent change in ANC and APP. A similar process involving formation of FeS in young non-acid marine clay sediments may also lead to a rapid acidification of the system (Van Breemen 1975).

**Change in APP by Ferric Iron Reduction and Displacement Reactions.** Alkalinization by ferric reduction and desorption or displacement of $SO_4^{2-}$ from the solid surface has been reported (Van Breemen et al. 1983). In the reduction process (Fig. 11.1) the sulphates adsorbed on positive colloids, basic iron sulphates or aluminium sulphates (e.g. jarosite and jurbanite) serve as proton acceptors in the reduction of ferric iron:

$$Fe_2O_3 + 1/2\,CH_2O + 4\,H^+ = 2\,Fe^{2+} + 1/2\,CO_2 + 5/2\,H_2O \tag{6}$$

The product ferrous sulphate, not ferrous hydrogen carbonate (Fig. 11.8), appears in solution (Van Breemen 1987). In fact $OH^-$ adsorbed as anion displaces the sulphate group adsorbed on the surface of the positive colloids. On the other hand part of the ferrous sulphate is oxidized at the interface between the solids and the water, producing a ferric oxide coating on the solid surface and releasing sulphuric acid in the water. The ferrous sulphate concentration gradient caused by these processes promotes further transport of ferrous sulphate to the sediment or soil surface. These processes will lead to a decrease in APP(s) and a permanent increase in ANC(s) by removal of $SO_3$ ($H_2SO_4$).

**Change in APP by "Ferrolysis" in Periodical Redox Processes.** Ferrolysis ("dissolution by iron"; Brinkman 1970, 1979; Eaqub and Blume 1982) in redox processes proceeds in two stages (Fig. 11.8). When the system becomes reduced, part of the $Fe^{2+}$ becomes exchangeable and displaces other cations such as $Ca^{2+}$ and $Mg^{2+}$. The displaced cations together with the anions that appear simultaneously with dissolved $Fe^{2+}$ (mainly bicarbonate and some organic anions (Van Breemen 1988a) can be removed by percolation or by diffusion into the surface water followed by lateral flow. If the supply of cations by flood water, groundwater, or mineral weathering is negligible, the surface complex may eventually be depleted of bases. The depletion is not immediately apparent in the reduced stage when the pH is high. During aeration of the system, however, exchangeable $Fe^{2+}$ is oxidized to essentially insoluble $Fe^{3+}$ oxide and $H^+$ takes the place of adsorbed $Fe^{2+}$ to the extent that formerly adsorbed base cations are leached (Van Breemen 1988b). It is suggested (Van Breemen 1987) that ferrolysis is typical for sediments or soils of older river or marine terraces in monsoon climates which have a seasonal perched water-table caused by submergence with rainwater. In those conditions, hydrology favours either lateral or vertical drainage and removal of bases liberated from the exchange sites by ferrous iron.

**Change in APP by Volatilization of $H_2S$.** A sediment/water system containing sulphate may become alkaline after volatilization of $H_2S$ from periodic sulphate reduction and retention of $M(HCO_3)_2$ in a saline alkaline lake or marsh (Fig. 11.8). If the system reoxidizes, it will only re-acidify partly, i.e. to such extent that reduced sulphur has been retained in the system (e.g. as FeS) and is available for the formation of sulphuric acid. The volatilization of $H_2S$ from reduced sediments causes a part of acidity leave to the system. The APP of the system will decrease. The escape of $H_2S$ from a sediment/water system depends on temperature, atmospheric pressure, vapour pressure of $H_2S$, and the state of the system (e.g. the disturbance of the water phase). Although the increase of ANC in the system is permanent due to volatilizaton of gaseous $H_2S$, alkalinization due to sulphate reduction would be temporary if $H_2S$ sulphur is fixed as iron sulphide and becomes oxidized later (Van Breemen 1987).

**Freshwater vs. Marine Systems.** A major difference between many freshwaters and marine environments is that sea water has a high capacity to buffer pH. This high buffering capacity tends to maintain constant conditions in sea water for reactions involving acid-base dissociations, ion exchanges involving protons, and pH-dependent dissolutions and precipitation (Meyer et al. 1994). In the marine environment with low sedimentation rates, there is continuous supply of sulphate (by diffusion) from the surface waters to the sediments, and all reducible iron is transformed to iron sulphides; the methanogenic state is reached only at very great depths. In freshwater environment, the sulphate concentrations are much lower, resulting in a partial transformation of the iron hydroxides to iron sulphides; the methanogenic state is reached relatively fast and the available reducible iron is converted to siderite (iron carbonate). As a result the reducible iron acts as a buffer against a decrease of the pH by decreasing the alkalinity (Salomons et al. 1987).

It has often been suggested that the pyrite content of marine sediments is higher than the pyrite content of freshwater sediments, because of the greater supply of $S^{2-}$ in marine sediments (Berner and Rainwell 1984). However, the high pyrite content in many freshwater sediments caused Davison et al. (1985) to question that generality. Formation of iron sulphides in sediments may depend as much on the rate of supply of electrons in the form of reactive $C_{org}$ as on the rate of supply of S (Davison 1988). Thus, it is not surprising that DiToro et al. (1990) found that the ranges of reported concentrations of acid volatile sulphide (AVS, a measure of mobilizable sulphides that mainly comprises iron monosulphides) were similar in freshwater and marine sediments. Sulphide can be supplied by the decomposition of organic matter as well as by reduction of sulphate; there should be not much difference between the two systems (Meyer et al. 1994).

### 11.3.4 Changes of pH-Conditions in River Sediments

Direct assessment of the pH-changes resulting from the oxidation of anoxic sediment constituents can be performed by ventilation of sediment suspensions with air or oxygen and subsequent determination of the pH-difference between the original sample and oxidized material. As greater is this difference, as higher is the

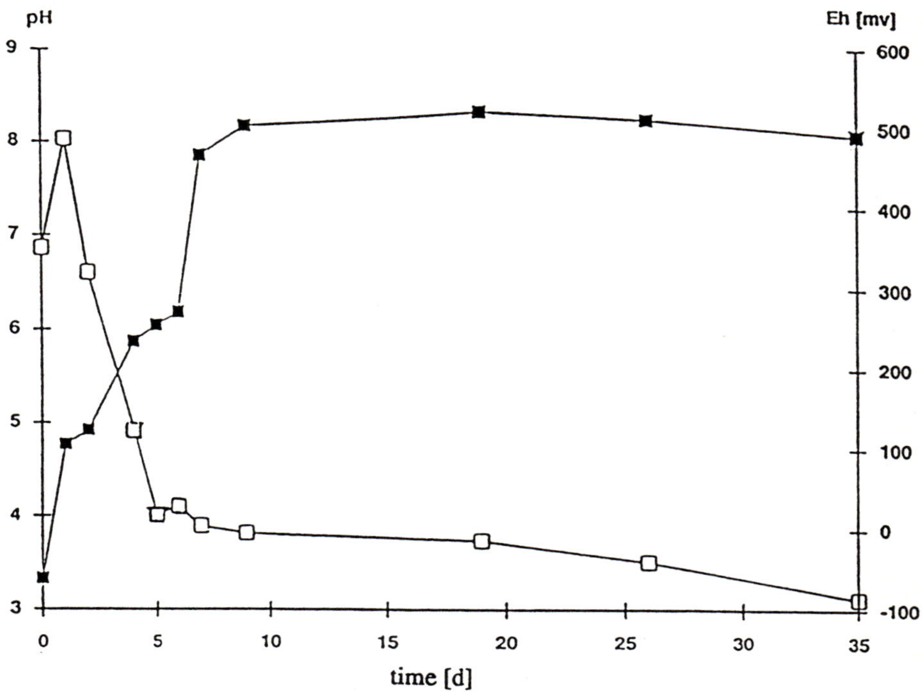


**Fig. 11.9.** Influence of redox potential on pH during oxidation of a sludge suspension from Hamburg Harbour (Calmano et al. 1992)

short-term mobilization potential of metals, e.g. during dredging, resuspension and other processes, by which anoxic sediments get into contact with oxygenated water or – following land deposition of dredged material – with atmospheric oxygen. A typical example demonstrating the temporal development of redox and pH-values in a sludge suspension from Hamburg harbour is presented in Fig. 11.9.

Results from titration experiments using 1 molar nitric acid on sediment suspensions of 100 g $l^{-1}$ are presented in Fig. 11.10. The titration curve of the Rhine river sediment exhibits a small plateau in the pH-range of 5.5 and 6, probably due to a certain fraction of carbonate, which is consumed by addition of 80 mmol of acidity. On contrary, the titration curves of both Elbe river sediments are continuously decreasing as a result of the low contents of carbonate in these samples. The sediment from the inland harbour basin of Harburg, originally sulphide-rich material which has been stored for 1 year in a closed bottle, has already reached an initial pH of 4.3; this is probably due to the consumption of the low residual buffer capacity by oxidation of parts of the sulphide fraction.

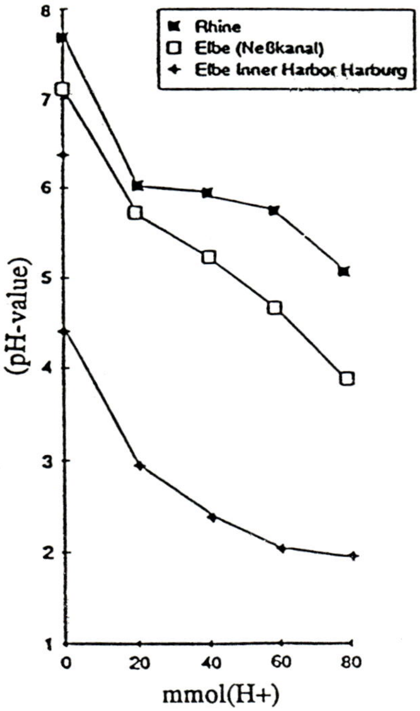


**Fig. 11.10.** Variation of pH-values from titration curves of suspensions (100 g $l^{-1}$) of sediment samples from Rhine and Elbe rivers after addition of 1 M nitric acid (Förstner et al. 1990)

## 11.4 Influence of Redox-Variations on Metal Mobility in Sediments

It can be expected that changes from reducing to oxidizing conditions, which involve transformations of sulphides and a shift to more acid conditions, increase the mobility of typical "B-" or "chalcophilic" elements, such as Hg, Zn, Pb, Cu, and Cd. On the other hand, the mobility is characteristically lowered for Mn and Fe under oxidizing conditions (Table 11.2).

**Table 11.2.** Relative mobilities of elements as a function of Eh and pH (Plant and Raiswell 1983)

| Relative mobility | Electron activity | | Proton activity | |
|---|---|---|---|---|
| | Reducing | Oxidizing | Neutral-alkaline | Acid |
| Very low | Al, Cr, Mo, V, U, Se. S. B, Hg, Cu, Cd, Pb | Al, Cr, Fe, Mn | Al, Cr, Hg, Cu | Si |
| Low | Si, K, P, Ni, Zn, Co, Fe | Si, K, P, Pb | Si, K, P, Pb, Fe, Zn, Cd | K, Fe(III) |
| Medium | Mn | Co, Ni, Hg, Cu | Mn | Al, Pb, Cu, Cr, V |
| High | Ca, Na, Mg, Sr | Ca, Na, Mg, Sr, Mo, V, U, Se | Ca, Na, Mg, Cr, Zn, Cd, Hg | Ca, Na, Mg, Co, (Mn) |
| Very high | Cl, I, Br | Cl, I, Br, B | Cl, I, Br, S, B, Mo, V, U, Se | Cl, I, Br, B |

Processes affecting metal mobilization from sulphide oxidation are highly complex. Factors involved are not only the extent of protonation, but also exchange processes involving interactions with $Fe^{3+}$ and earth alkalies as well as the buffering effect of organic substances (Fig. 11.11; Peiffer 1989). Part of the $Fe^{3+}$ formed according to Eq. 1 (see Sect. 11.3.3) is hydrolyzed, another portion is kept in solution due to the formation of organic complexes. The dominant process depends upon the oxygen potential as well as from the concentration and reactivity of organic matter (Stumm and Morgan 1981). The organic substance inhibits, among others, the oxidation of Fe(II). Exchange reactions (e.g. for adsorbed manganese and trace elements) can proceed in different ways:

$$Fe^{3+} + \text{Ads-Me}^{+} \Leftrightarrow \text{Ads-Fe}^{2+} + Me^{2+} \quad (7a)$$

$$2\ H_3O^{+} + \text{Ads-Me}^{+} \Leftrightarrow \text{Ads-H}^{2+} + Me^{2+} + H_2O \quad (7b)$$

$$Ca^{2+} + \text{Ads-Me}^{+} \Leftrightarrow \text{Ads-Ca}^{2+} + Me^{2+} \quad (7c)$$

Consider the possible precipitation of iron hydroxide,

$$Fe^{3+} + 6\ H_2O \Leftrightarrow Fe(OH)_3 + 3\ H_3O^+ \tag{8}$$

the oxidation reaction of FeS can, hypothetically, be completed by two reaction pathways:

$$FeS + 2.25\ O_2 + 4.5\ H_2O \Leftrightarrow Fe(OH)_3 + SO_4^{2-} + 2\ H_3O^+ \tag{9a}$$

$$FeS + 2.25\ O_2 + H_3O^+ + \text{Ads-}Me^+ \Leftrightarrow \text{Ads-}Fe^{2+} + SO_4^{2-} + Me^{2+} + 1.5\ H_2O \tag{9b}$$

Reaction (9a) is an acid-producing process, in contrast to reaction (9b). (Whether the decrease of the pH-values during oxidation is caused by this reaction or is due to the initialization of microbial dissimilative activity and production of $CO_2$ is still controversial; Peiffer 1989.)

Organic matter obviously plays a major role as a buffer for acidity. Either acid producing $Fe^{3+}$ is directly taken up or there is a permanent titration of the organic matter with protons from the hydroxide precipitation of $Fe(OH)_3$. This titration again leads to the release of metals according to Eq. (7b) and the gross reactions can be given as:

$$FeS + 2.25\ O_2 + 3.5\ H_2O + \text{Ads-}Me^{2+} \Leftrightarrow Fe(OH)_3 + SO_4^{2-} + \text{Ads-}H^{2+} + Me^{2+} \tag{10}$$

According to reaction (9b) and reaction (10), the concentration of released metals should be be equal to the sulphate concentration.

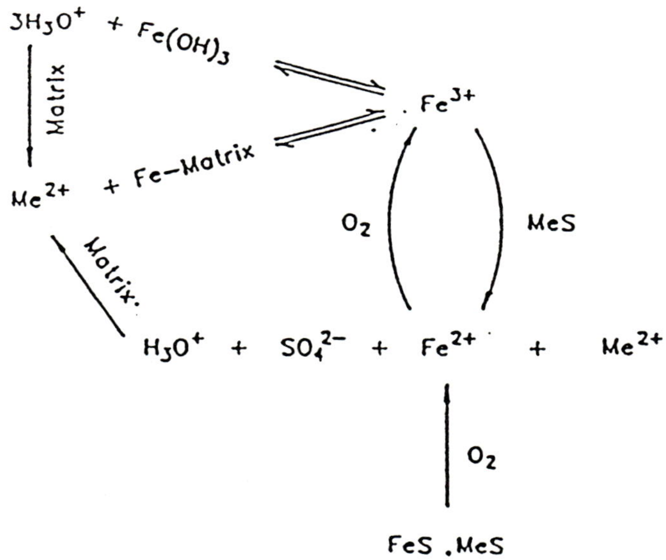


**Fig. 11.11.** Model conception on the oxidation of metal sulphides and the interactions between matrix and the newly formed reaction products (Peiffer 1989)

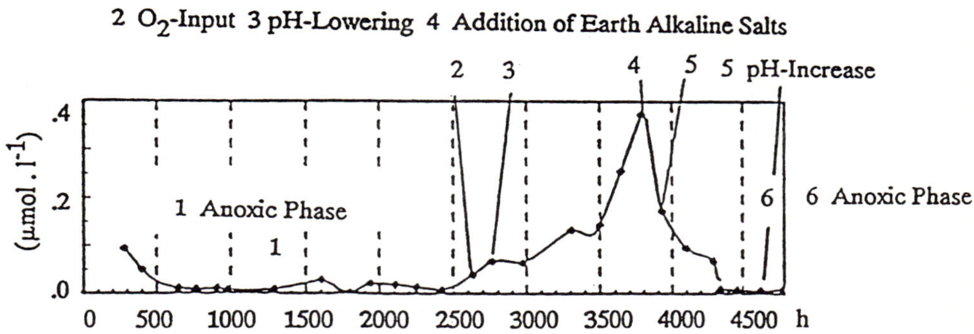


**Fig. 11.12.** Experiments on cadmium release from solid waste material (Peiffer 1989)

Experimental investigations by Peiffer (1989) on the long-term development of sewage sludge materials provide detailed insight into the sequence of processes taking place in the post-methanogenic stage of such deposits (Fig. 11.12). During an initial phase, anoxic cadmium is bound to sulphides, resulting in very low metal concentrations in solution (phase 1). Aeration by addition of dissolved oxygen initiates release of cadmium from the solid substrate (phase 2); this process is enhanced by the production of acidity, which lowers the pH from 6.7 to approximately 6.4 (phase 3). An even stronger effect is observed from the addition of alkaline earth ions (phase 4). The pH increase, which may be induced from buffering components within the system, leads to a reduction of dissolved cadmium (phase 5). Formation of new sulphide ions from the degradation of organic matter brings the concentrations of dissolved cadmium back to its original, extreme low level (phase 6). The observed pH decrease seems to indicate that zinc and cadmium are being exchanged for protons, whereas lead and copper, because of their much stronger bonding to the solid substrate, do not.

### 11.4.1 Metal Mobilization from Aquatic Sediments (Examples)

Examples on the release of potentially toxic metals from contaminated sediments are given by various authors, indicating the major factors, processes and rates of metal mobilization:

- Field evidence for changing cadmium mobilities was reported by Holmes et al. (1974) from Corpus Christi Bay Harbour : during the summer period when the harbour water was stagnant, cadmium precipitated as CdS at the sediment–water interface. In the winter months, however, the increased flow of oxygen-rich water into the bay resulted in a release of the precipitated metal.

- In the St. Lawrence estuary, Gendron et al. (1986) found evidence for different release mechanisms near the sediment–water interface – the profiles for cobalt resemble those for manganese and iron with increased levels downwards, suggesting a mobilization of these elements in the reducing zone and a reprecipitation at the surface of the sediment profile. On the other hand, cadmium appears to be released at the surface, probably as a result of the aerobic remobilization of organically-bound cadmium.
- Biological activities are typically involved in these processes – remobilization of trace metals has been explained by the removal of sulphide from pore waters via ventilation of the upper sediment layer with oxic overlying water, allowing the enrichment of dissolved cadmium that would otherwise exhibit very low concentrations due to the of formation of insoluble sulphides in reduced, $H_2S$-containing sediments. Emerson et al. (1984) suggest a significant enhancement of metal fluxes to the bottom waters by these mechanisms. It was evidenced by Hines et al. (1984) from tracer experiments that biological activity in surface sediments greatly enhances remobilization of metals by input of oxidized water. These processes are more effective during the spring and summer than during the winter months.
- From enclosure experiments in Narragansett Bay, Hunt and Smith (1983) estimated that by mechanisms such as oxidation of organic and sulphidic material, the anthropogenic proportion of cadmium in marine sediments is released to the water within approximately 3 years. For remobilization of copper and lead, approximately 40 and 400 years, respectively, is needed, according to these extrapolations.
- Prause et al. (1985) studied the release of Pb and Cd from contaminated dredged material after dumping in a harbour environment. During an observation period up to 24 h no significant Cd or Pb release could be found from the dredged sludge, but during the long term experiments extensive Cd remobilization was recognized. The contact of the polluted freshwater sediments with sea water finally cause the release of 1-2 mg Cd per kg solids. It is suggested that the reaction kinetics, mainly with respect to the initial mobilization of metals from solids, are controlled by microbial activity.

Metal release from tidal Elbe river sediments by a process of "oxidative remobilization" has been described by Kersten (1989; Fig. 11.13). Short (30 cm) sediment cores were taken from a site, where diurnal inundation of the fine-grained fluvial deposits takes place. In the upper part of the sediment column, total particulate cadmium content was approximately 10 mg $kg^{-1}$, whereas in the deeper anoxic zone the total particulate concentration of Cd was 20 mg $kg^{-1}$. Sequential extractions indicate that in the anoxic zone 60-80% of the Cd was associated with the sulphidic/organic fraction. In the upper – oxic and transition – zone the association of Cd in the carbonatic and exchangeable fractions simultaneously increase up to 40% of total Cd. This distribution suggests that the release of metals from particulate phases into the pore water and further transfer into biota is controlled by the frequent downward flux of oxygenated surface water. From the observed concentrations, it would be expected that long-term transfer of up to 50% of the Cd from the sediment subsurface would take place either into the anoxic zone located further below the sediment–water interface or released into the open water.

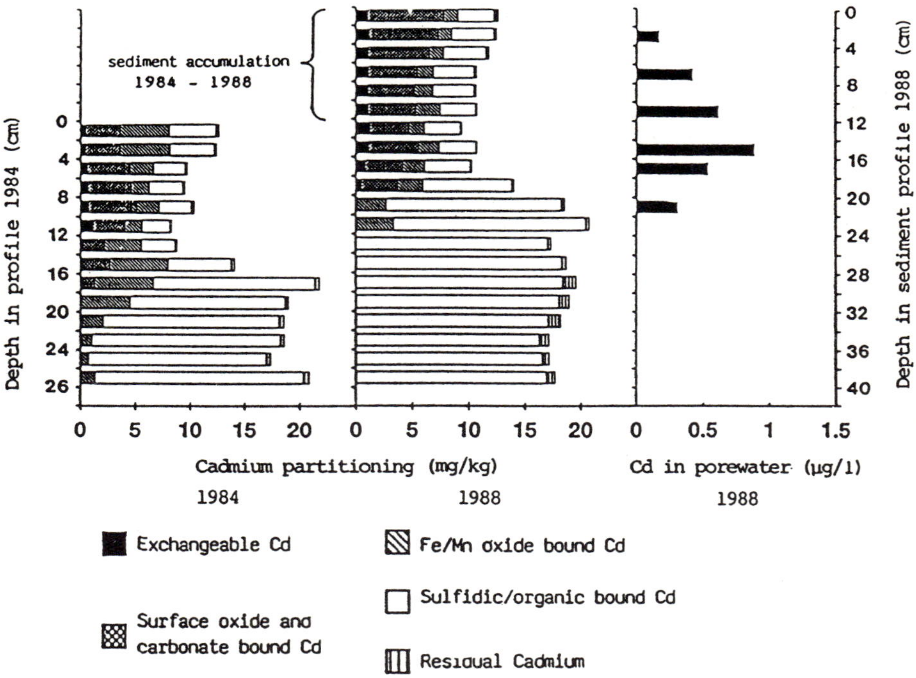


**Fig. 11.13.** Total concentrations and partitioning of Cd in a tidal flat sediment profile in the Heuckenlock area sampled in 1984 and 1988. Sedimentation rates were determined by the $^{137}$Cs-method. Cadmium pore water profile was determined at low tide (Kersten 1989)

Typical early diagenetic geochemical changes and subsequent element mobilization via the pore water result from dredging activities. A study performed by Darby et al. (1986) in an artificially created estuarine marsh demonstrates characteristic effects of oxidation (Table 11.3).

Compared to the river water concentration, the channel sediment pore water is enriched by a factor of 200 for iron and manganese, 30-50 for nickel and lead, approx. 10 for cadmium and mercury, and 2-3 for copper and zinc. When the expected concentration of metals following hydraulic dredging, which were calculated from a rate of pore water to river water of about 1:4, were compared with the actual measurements at the pipe exiting the dredging device, negative deviations were found for iron and manganese, suggesting reprecipitation of Fe/Mn-oxide minerals; the positive deviations of zinc (factor 80), copper, lead and cadmium (factors 7-8) indicate that during dumping of the sludge-water mixture significant proportions of these elements were mobilized and transferred into the effluent water.

**Table 11.3.** Mobilization of metals during dredging (after Darby et al., 1986). Concentrations in mg $l^{-1}$

| Metal | Channel sediment pore water (a) | River water conc. (b) | a/b | Effluent at an artificially created marsh | | |
|---|---|---|---|---|---|---|
| | | | | Expected conc. | Measured conc. | % Change |
| Mn | 6.94 | 0.03 | 230 | 1.34 | 1.19 | - 11 |
| Fe | 57.3 | 0.26 | 220 | 11.12 | 6.01 | - 46 |
| Ni | 0.054 | 0.001 | 54 | 0.011 | 0.035 | + 218 |
| Pb | 0.077 | 0.002 | 38 | 0.016 | 0.142 | + 788 |
| Hg (µg $l^{-1}$) | 3.2 | 0.26 | 12 | 0.82 | 2.0 | + 144 |
| Cd | 0.009 | 0.001 | 9 | 0.0025 | 0.019 | + 660 |
| Cu | 0.012 | 0.004 | 3 | 0.0055 | 0.051 | + 827 |
| Zn | 0.12 | 0.052 | 2 | 0.065 | 5.30 | + 8069 |

Pore water data from dredged material from Hamburg Harbour indicate typical differences in the kinetics of proton release from organic and sulphidic sources (Table 11.4). Recent deposits are characterized by low concentrations of nitrate, cadmium and zinc; when these low-buffered sediments are oxidized during a time period of a few months to years, the concentrations of ammonia and iron in the pore water typically decrease, whereas those of cadmium and zinc increase (with the result that these metals are easily transferred into agricultural crops!).

**Table 11.4.** Mobilization of metals and nitrogen compounds from dredged material after land deposition (Maaß et al. 1985 and other authors)

| Element or compound | Reduced water | Oxidized water |
|---|---|---|
| Ammonia | 125 mg $l^{-1}$ | < 3 mg $l^{-1}$ |
| Iron | 80 mg $l^{-1}$ | < 3 mg $l^{-1}$ |
| Nitrate | < 3 mg $l^{-1}$ | 120 mg $l^{-1}$ |
| Zinc | < 10 µg $l^{-1}$ | 5000 µg $l^{-1}$ |
| Cadmium | < 0.5 µg $l^{-1}$ | 80 µg $l^{-1}$ |

**Table 11.5.** Acid-producing potential (APP) of organic matter (Swift, 1977)

| Parameter | Mean content | (APP) | [H+][a] | pH[a] |
|---|---|---|---|---|
| Organic Carbon | 3.6% | | | |
| Organic Nitrogen | 0.27% | 0.19 mmol $g^{-1}$ | 0.019 | 1.71 |
| Organic Sulphur | 0.05% | 0.03 mmol $g^{-1}$ | 0.003 | 2.82 |
| $FeS_2$ | 1.0% | 0.33 mmol $g^{-1}$ | 0.033 | 1.50 |
| Total | | 0.55 mmol $g^{-1}$ | 0.055 | 1.28 |

[a] In 1:10 suspension

The acid-producing potential not only is related to the oxidation of sulphides, but oxidation of organic matter must be considered as well. Table 11.5 indicates that the contribution of protons from organic-N and organic-S in a sample containing approximately 5% organic carbon is equivalent to the acid-producing potential of 1% $FeS_2$. Studies of the long-term evolution and diagenesis in sewage sludge landfills and similar natural sediments (peat, organic soils) by Lichtensteiger et al. (1988) suggest that the transformation of organic material will last for geological time-scales ($10^3$ to $10^7$ years).

The different steps of short- and long-term transformations of chemically active constituents in dredged material from the Elbe river/Hamburg harbour and their effects on the mobility of typical trace metals are schematically given in Fig. 11.14. Oxidation of sulphides during stage B strongly increases the concentrations of cadmium and zinc in a relative short time. When acidity is consumed by buffer reactions (phase C), cadmium and zinc concentrations drop,

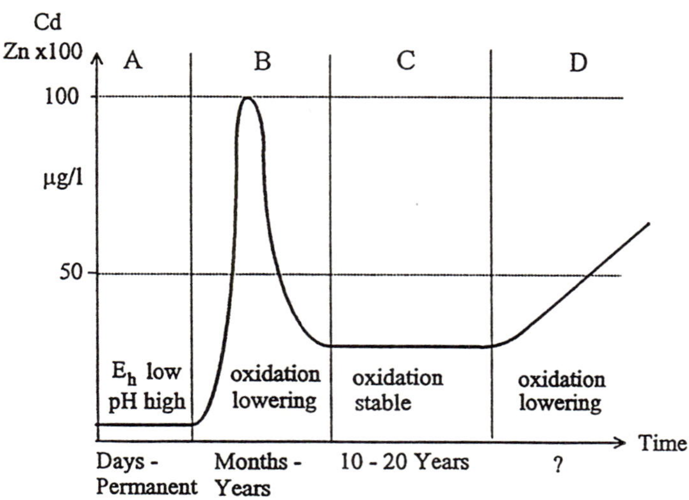


**Fig. 11.14.** Schematic diagram illustrating different phases of metal release from land-disposed dredged material (Maaß and Miehlich 1988)

but are still higher than in the original sulphidic system. In phase D, oxidation of organic matter again may lower pH-values and can thereby induce long-term mobilization of Zn and Cd.

Of the two major release processes, the first – oxidation of sulphides – can be predicted to some extent (see Sect. 11.5), whereas the implications of long-term degradation of organic matter on the release of less mobile elements such as copper and lead as yet cannot satisfactorily be described. Here, as with the effects of these interactions on the cycles of anionic elements such as arsenic, further research is needed.

### 11.4.2 Redox Variations and Metal Release in Estuaries

During the end of the seventies and the eighties most of the studies have performed, which led to an recognition of the typical estuarine processes related to metals. Such processes are suspected when a deviation is found from the theoretical mixing curve of salinity or chlorinity:

- Change of composition of the end member was studied by Duinker and Nolting (1978).
- Flocculation processes and their effects on the removal of trace metals have been investigated by Sholkovitz and co-workers (1976 to 1981).
- Studies on the release and/or uptake of dissolved metals by bottom sediments were undertaken, among others, by Salomons (1980).
- Biogeochemical implications on metals with the degradation of organic matter and biological uptake have been reviewed by Wangersky (1986).

There is a wide spectrum of investigation on changes in adsorption/desorption equilibria due to changes in pH, salinity, turbidity and formation of new particulate matter. Of the field studies, the working groups of Duinker (Duinker et al. 1982), Salomons (Salomons and Eysink 1981), Van Eck (Zwolsman and Van Eck 1993) and Wollast (1988) at the Scheldt River estuary; of Windom (Windom et al. 1983) and Church (1986) at eastern United States estuaries; and of Martin (Elbaz-Poulichet et al. 1987) at French and Chinese estuaries should be mentioned.

Initial data on the different behaviour of individual metals in a macrotidal estuary, that are examples with far-reaching tidal influences and in most cases typical variations in oxygen- and pH-conditions, has been given by Salomons and Eysink (1981) from the Scheldt River estuary. Zinc, copper and cadmium are typical examples for metals, which are increased at higher salinities; on the other side, nickel, manganese and – in particular – iron indicate highest dissolved concentrations at lower salinities. Data of Duinker and colleagues (1982) showed typical enrichment of particulate manganese at higher salinity; obviously, reprecipitation of dissolved manganese may take place in this area of higher pH-values and oxygen saturation.

Detailed studies have been performed on the biogeochemical factors influencing the residence time of metals by Church (1986) in Delaware Bay, one of the large tidal estuaries on the United States east coast. The Delaware retains some of its estuarine trace elements as sedimented estuarine particles, while others are more conservative and largely exported. Those retained include trace elements in primarily geochemical, so-called particle reactive chemistries – examples are on the left column: iron, manganese, cobalt. Those exported appear recycled by biochemical processes, such as zinc, copper and cadmium. Often, the behaviour of trace elements and nutrients are biogeochemically linked, for example cadmium and phosphates.

These field data agree with the findings of Santschi et al. (1982) from mesocosm experiments; 23 elements were added in radioactive from to fibreglass tanks, which contained 13 $m^3$ of water overlying 30 cm of silty-clay sediments. Removal of these elements was measured in summer and winter. (1) Particle-reactive elements were removed fast; distribution coefficients between sediment and water were higher, mainly in winter. (2) The recyclable element, e.g. manganese and cobalt are those, which are typically controlled by redox processes. (3) Biologically active elements are easily taken up by organisms; half lives are somewhat longer than those of the recyclable elements. (4) These elements exhibit long residence times in solution and are less strongly adsorbed to solids than the previous groups of elements. With biogeochemically active elements, prediction about their behaviour may become difficult. One the one hand, association with dissolved organic matter and colloids may increase their residence time in the water column; on the other hand, there are typical effects of filtering and aggregation by organisms, which make residence time of these elements substantially shorter than the average flushing time of the larger estuaries.

In the Elbe River estuary the findings by the GKSS Research Center (Irmer et al. 1986, 1988; Knauth et al. 1986, 1989) together with Delft Hydraulics Laboratory clearly demonstrate a significant transport of highly contaminated material from locations upstream of the Hamburg area into the estuary at high freshwater discharge rates, combined with an effective mixing of highly contaminated material from the upper estuary with less polluted sediments from downstream sections.

Compared to the "suspended solids effect" on the distribution of man-induced heavy metals in the Elbe estuary, chemical interactions seem to be relatively insignificant. From field investigations, indication on metal release has only been given from the increase of cadmium concentrations near the salinity gradient (Ahlf 1983). This effect can be explained by the formation of relative stable soluble chlorocomplexes with Cd and competition of earth alkali ions Mg and Ca for sorption sites on solid substrates. However, there are still conflicting results with regard to the mechanisms and kinetics of cadmium release from solids. Both Salomons (1980) and Prause et al. (1985) observed in their batch experiments with dredged sediments suspended in estuarine water that a considerable release of cadmium is found only after relative long periods of treatment. Prause et al. (1985), therefore, stressed that simple cation exchange mechanisms obviously have no importance because a new equilibrium at the surface of the sorbents will normally be established in short time; from studies applying antibiotics the authors suggested, that bacterial activity can stimulate cadmium remobilization by a complex web of substrate decomposition and potential organic ligand liberation or formation. It can be concluded, that a high particle residence time likely causes a stronger remobilization of the metal from the suspended matter and it is important to note that observations on the quantitative behaviour of cadmium as determined for a distinct estuary cannot be extrapolated to other estuaries, since hydrodynamic processes may differ to a large extent.

**Table 11.6.** Variability and potential effects of major geochemical parameters on physico-chemistry of metals in the Elbe River estuary

| Parameter | Determining variable | Variability | Physico-chemistry |
|---|---|---|---|
| Salinity | Water flow, weather | Small | Desorption |
| Suspended matter concentration | Water discharge, tidal phase, biological production | Medium | Adsorption |
| SPM composition | Water discharge, seasonable effects | Medium | Adsorption, bonding forms |
| pH | Salinity (buffer) | Small | Precipitation |
| Dissolved organic carbon | Season, salinity | Medium | Complexation, desorption |
| Redox potential | Season, temperature, residence time | High | Dissolution |

Two other parameters on potential mobilization of heavy metals, redox changes and the effect of DOC concentration, have not yet studied in the Elbe estuary on a larger scale. There are initial indications given by the Jülich research group (unpublished results), that lead concentrations in solution are higher in summer compared to winter samples, and this effect could be caused by higher content of dissolved organic matter from biological production, which influences solid/solution partition equilibria by complexation. Regarding the effect of redox variations, solid phases partitioning by means of sequential extraction procedures on suspended matter samples (which have been recovered during helicopter sampling in April and June 1986) indicate characteristic reprecipitation processes for manganese; this element, which is particularly sensitive to redox changes, shows elevated concentrations of easily reducible proportions in several samples between river kilometres 650 and 700 during the summer campaign (Fig. 11.7). From these data and from a view to the variations of oxygen in the Elbe estuary (Fig. 11.7), it seems that information on the effects of redox changes is difficult to be gained on a whole-system basis, e.g. from longitudinal profiles of dissolved metal concentrations and that sampling designs should be related to the temporal

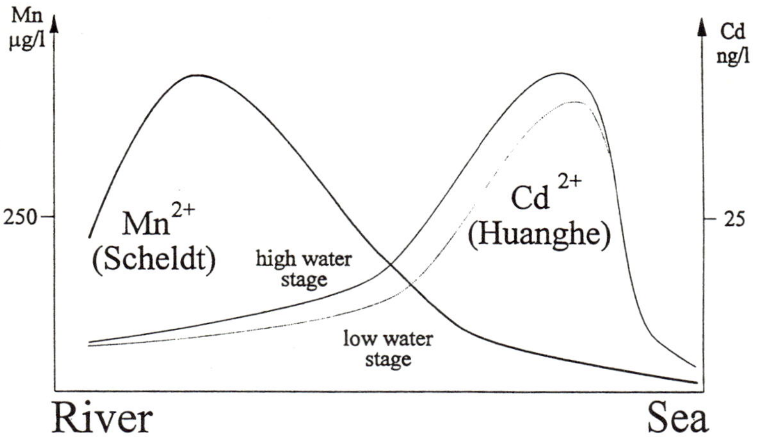


**Fig. 11.15.** Distribution of dissolved Manganese and Cadmium in the mixing zones of macrotidal estuaries of the rivers Scheldt (Wollast 1988) and Huanghe (Elbaz-Poulichet et al. 1987)

and spatial particularities of this parameter.

The full complex situation of metal transfer by redox variations can best be demonstrated from macrotidal estuaries (see above). In Fig. 11.15, a combination of data on dissolved manganese and cadmium and from the Scheldt and Huanghe estuaries is given. In many of these estuaries, tens or hundred of kilometres long, the upper parts show typical reducing conditions, and this is indicated by the manganese release from sediments and suspended matter. In the lower estuary, manganese may be reprecipitated as shown be Duinker et al. (1982) from the Scheldt estuarine. Lowest dissolved cadmium concentrations in Chinese estuaries (Elbaz-Poulichet et al. 1987) are at about 4 ng $l^{-1}$, comparable to the data observed in the Amazon river. In the Huanghe river, the second most important source of sediment discharge in the world, maximum concentrations of cadmium – up to 55 ng $l^{-1}$ – are observed in the marine waters of the Bohai Gulf, not in the mixing zone between river and sea water. In the adjacent Yellow Sea, there is no significant decrease of the values, e.g. to the concentrations found in the East China Sea, i.e. approximately 13 ng $l^{-1}$. Bohai Gulf is a shallow sea area – 1 to 20 m deep – with frequent storms favouring sediment resuspension. At higher water stage mobilization of cadmium is increased. A preliminary conclusion is, that remobilization of cadmium first needs oxidation of sulphidic forms, than complexation by chloride ions can take place.

The transformation of reactive components in solid matter, so-called early diagenesis, involves changes of element partition between solid and dissolved phases. The sequence of factors and events, which affect the release of biogeochemically active elements in macrotidal estuaries can be described as follows:

- A dynamic environment, where strong chemical gradients occur in the pore water of fine-grained sediments rich in organic matter and sulphides
- Frequent changes of river morphology, involving resuspension of these sediments, which leads to an oxidation of sulphide minerals, whereby solubility of typical heavy metals is increased
- The intensity of metal mobilization (sometimes also readsorption) depends on stability of either inorganic or organic complexes. Inorganic complexation by chloride ions is typical for cadmium; organic complexation generally follows the Irving-Williams rule. Intensity of readsorption is mainly controlled by the concentration of suspended particulates
- Compared to the relative strong effects during the oxidation of sulphide minerals, in particular iron sulphide, in anoxic sediments, suspended matter usually is varied between oxic and suboxic conditions, the latter characterized by Fe-carbonate minerals, whereas trace metals are still present in part as sulphides

For coastal marine environments, such as the Wadden Sea coast, the possible implications on metal concentrations in sediments have schematically been designed by Salomons. Figure 11.16 suggests, that after deposition and during storage in reduced coastal sediments, adsorbed metals are transferred to sulphidic bonding forms. During erosion processes, oxidation can take place involving remobilization of a certain proportion of the sediment-associated metals, which may subsequently be readsorbed in part to particulate matter. Under marine conditions, the extent of adsorption will mainly depend on the intensity of both complexation by chloride and competition with cations like $Mg^{2+}$, $Ca^{2+}$, and $Na^{+}$. Such oxidative events are relatively short compared to the longer periods of reduced conditions

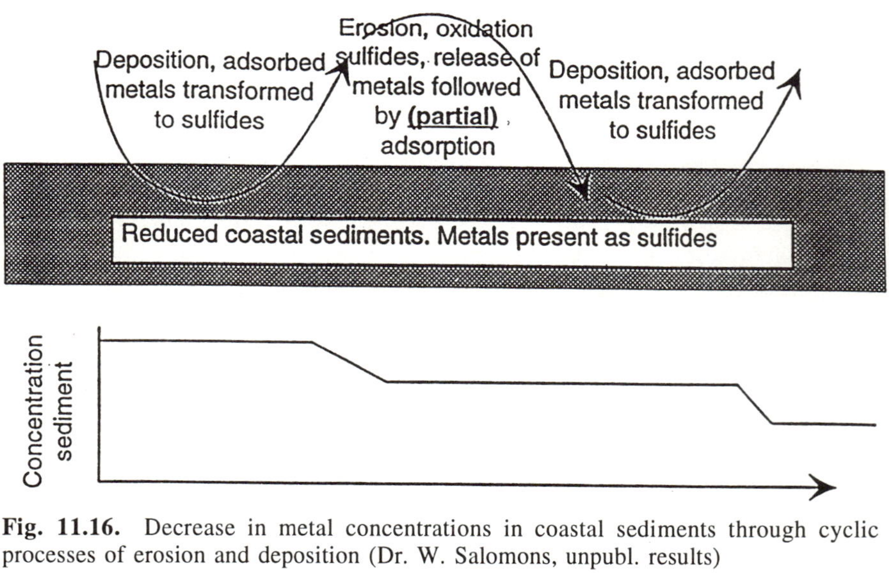


**Fig. 11.16.** Decrease in metal concentrations in coastal sediments through cyclic processes of erosion and deposition (Dr. W. Salomons, unpubl. results)

during deposition. However, it can be expected that through repeated but irregular (in contrast to the tidal interactions mentioned) cyclic processes of erosion and deposition a significant release of metals from coastal sediments will take place.

### 11.4.3 Experimental Mobilization of Metals Following Resuspension of Estuarine Sediments

Resuspension of sediments from the Elbe river can significantly decrease pH values due to high acid producing potential and low neutralizing capacity. Metals can be released into the dissolved phase, but may subsequently be readsorbed or precipitated in part to solid phases. To evaluate a regional and long-term perspective experiments were undertaken at stable neutral pH-values, and both total release and scavenging rates were extrapolated from time series of measured net release values over 630 hours (Calmano et al. 1994).

In laboratory simulation experiments, wet sediment material was transferred to a 180 ml reaction cell; 135 ml of artificial river water (2 m<u>M</u> $CaCl_2$ and 5 m<u>M</u> NaCl) was added to give a solid/liquid ratio of 1:64. The suspension was sampled at regular intervals and then centrifuged and filtered through a 0.2 μm membrane. The same volume of artificial river water used to wash the centrifugation tube was added into the reaction cell in order to keep the suspension volume constant. The experiments were carried out under exclusion of light.

The example in Fig. 11.17 shows that the initial lead release was high but

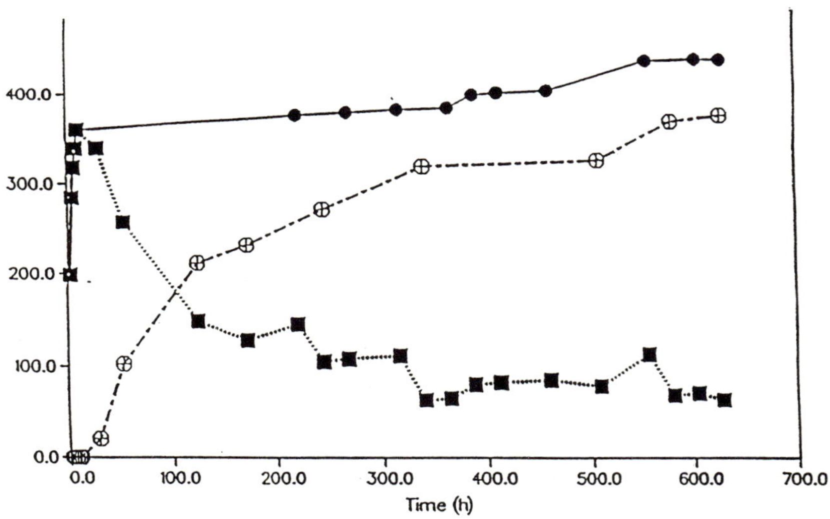


**Fig. 11.17.** Lead release from Elbe sediments. *Full quadrangles*: net release; *full circles*: total release; *open circles*: scavenging (adsorption/coprecipitation). Laboratory simulation experiments on wet sediments in a reaction cell indicate typical time evolution of interactive release and scavenging processes (Calmano et al. 1994)

then decreased continuously due to readsorption. After about 100 h the relative equilibrium between release and readsorption was reached. Curves for copper indicate continuous release until about 350 h, when scavenging rate became faster than the release rate. With respect to cadmium and zinc, release was found to be higher than the scavenging rate during the whole experiment.

The order of total release from the sediment has been found to be Cd (5%) > Zn (1.5%) > Cu (1%) > Pb (0.3%). The percentage scavenging of released metals was in the order of Pb (85%) > Cu (53%) > Zn (35%) > Cd (30%). The order of net release percentage of total metals from the sediment is Cd > Zn > Cu > Pb. For the release amount, however, the order is Zn > Cu > Pb > Cd. At the end of the experiment this order and the magnitude of released metals, were similar to that observed in the natural waters of the river Elbe at Hamburg.

Based on experimental results and relevant literature data, a four stage interaction model has been developed for metals in anoxic sediments to describe the behaviour of heavy metals in the system subsequent to oxidation (Fig. 11.18):

- *Release Stage:* includes amorphous sulphide oxidation, some pyrite oxidation and some organic detritus decomposition. At the beginning, amorphous sulphide oxidation is a dominant chemical process; the metals, which are bound to acid volatile sulphides will be released at the early stage. The second contribution is characterized by mineralization of ammonium and pyrite oxidation. At the end of the experiment, about 50% of the pyrite content was oxidized.

- *Transition Stage of Release and Scavenging:* Acid-volatile sulphide compounds have been exhausted; pyrite oxidation and organic detritus decomposition continues. On the other hand, readsorption becomes stronger. Phosphate, derived from the decomposed organic detritus, accumulates and then precipitates with some metals.

- *Scavenging Stage:* At this stage all processes including pyrite oxidation, organic detritus decomposition, and release from solid organic and inorganic compound decline. Iron and manganese oxide formation continues and scavenging of trace metals becomes a dominant process.

- *Eqilibrium Stage:* Release and scavenging processes have basically reached a new equilibrium, i.e. the release rate is equated to the scavenging rate. The metal concentration in solution will remain constant. The formation of new solid metal species and their redistribution still proceed during this period of "early diagenesis".

The time at which the highest concentration (peak value) of metal release is found can be called "characteristic value" of metal release and represented by an index $Omega_{Me}$ in hours. When the oxidation time is longer than this value, the dominating process is scavenging or relative equilibrium; when the oxidation time is less than $Omega_{Me}$, the dominating process is release of metals (Fig. 11.19). It has been suggested that the latter situation in particular will pose long-term problems to the aquatic ecosystem.

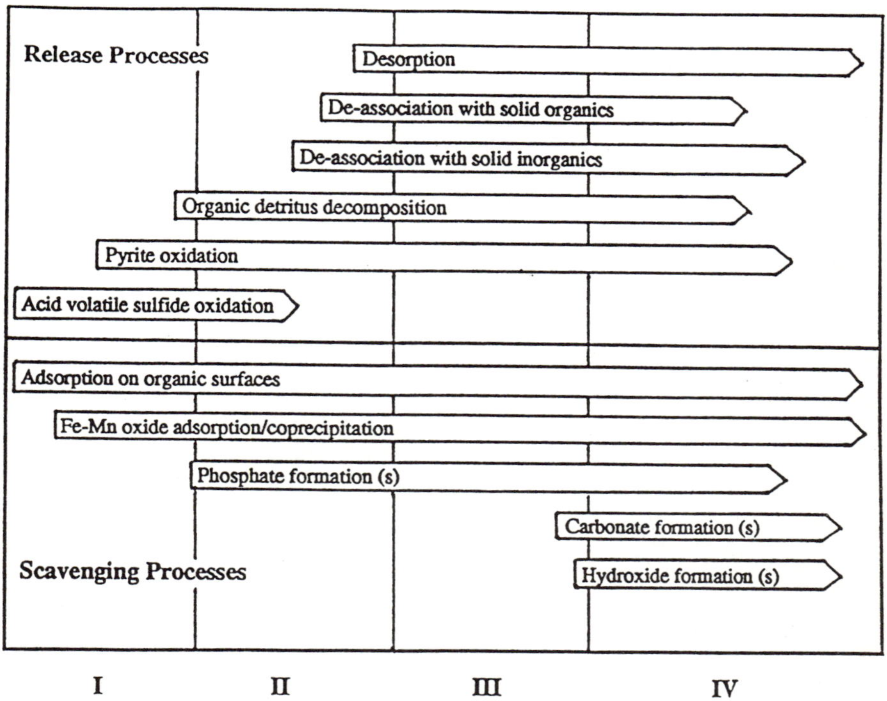


**Fig. 11.18.** Sequence of chemical and biochemical reactions involved in metal release and scavenging in suspension/oxidation processes (Calmano et al. 1994)

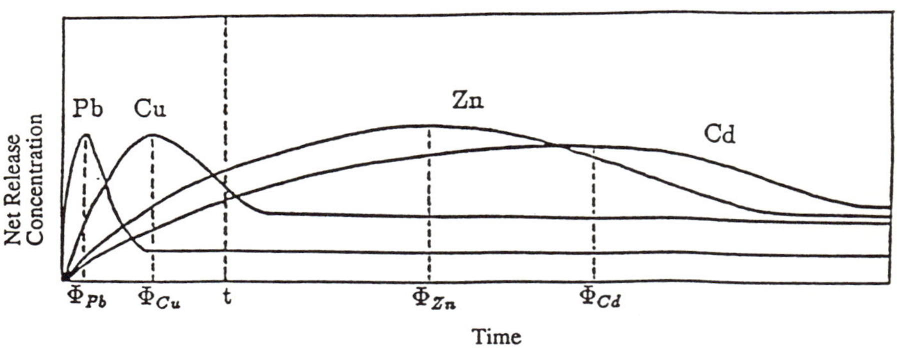


**Fig. 11.19.** Changes in relative concentrations of metals during resuspension/oxidation (Calmano et al. 1994)

## 11.5 Implications on Sediment Quality Criteria Assessment

Reasons for the development of sediment quality criteria have been in the mid-eighties that (1) sediments integrate contaminant concentrations over time; (2) long-term perspectives in water management need *"integrated strategies"*, in which sediment-associated pollutants have to be considered as well (generally, the contaminant level in the sediment have greater impact on the survival of benthic organisms than do aqueous concentrations); (3) management plans have increasingly been based on the *assimilative capacity* of a certain receiving system, and this requires knowledge of the properties of sedimentary components as the major sink; (4) permission for *dredging activities* and *deposition of dredged material* have to be based on standardized sampling protocols and test procedures. In this context, the conventions for the marine environment – Oslo and London – should be mentioned.

Efforts have first been undertaken by the United States Environmental Protection Agency to develop standard procedures for the assessment of environmental impact of sediment-bound pollutants. Further discussions led to the differentiation of biological and chemical-numerical approaches. Biological criteria integrate sediment characteristics and pollutant loads, but they do generally not indicate the cause of effects. With respect to chemical-numerical criteria, there is no immediate indication on biological effects; their major advantages lie in their easy application and amendment to modelling approaches.

Biological approaches on development and application of sediment quality criteria exhibit a common basis in the study of damaging impacts from contaminated sediments on organisms. The biological parameters *bioaccumulation*, *toxicity*, and *mutagenity* have to be considered separately in any case. Bioassays as well as field surveys are empirical considerations which cannot provide numerical criteria to be transferred on different situations.

Generally, it is difficult to establish clear cause-and-effect relationships between acute or chronic toxic effects on biota and the occurrence of specific pollutants in sediments. One major limitation is that not all sediment-associated chemicals can presently be identified; thus, unidentified compounds cannot be ruled out as principal aetiological factors.

Relatively simple and implementable liquid, suspended particulate and solid-phase bioassays have been carried out for assessing the short-term impact of dredging and disposal operations on aquatic organisms (Ahlf and Munawar 1988). Standardized tests are characterized by their lack of variability, but essential information (e.g. lethality, alterations of growth rate) can only be obtained with such single species test. The influence of the main environmental variables on the interaction of suspended particulates or in situ sediment contaminants and organisms should also be determined under simulated field conditions. In particular, benthic bioassay procedures, due to recent developments, are important in evaluating the relationship between laboratory and field impacts (Reynoldson 1987).

With restriction to the effects on benthic communities, the sediment quality "triad" by Chapman (1986) combines chemistry and sediment bioassay measurements with in situ studies: Chemistry and bioassay estimates are based on laboratory measurements with field collected sediments. In situ studies may include, but are not limited to, measures of resident organism histopathology, benthic community structure and bioaccumulation/metabolism. Areas where the three facets of the triad show the greatest overlap (in terms of positive or negative results) provide the strongest data for determining numerical sediment criteria.

Numerical approaches for the assessment of environmental impact of sediment-associated metals are based on (1) accumulation, (2) pore water concentrations, (3) solid/liquid equilibrium partition (both sediment/water and organism/water) and (4) elution properties of contaminants:

- *Background Approach:* compares the actual data with sites comprising natural or insignificant pollutant concentrations. Particularly useful are samples from deeper layers of the sediment sequence at a given site, for example from drill holes, since this material is derived from the same catchment area and usually is similar in its substrate composition. Nonetheless, standardization with respect to grain size distribution is indispensable.
- *Pore water Approach:* based on the experience, that the composition of interstitial waters is the most sensitive indicator of the reactions that take place between pollutants on particles and the aqueous phase which contacts them. There is the advantage of a direct recovery and analysis of water-borne constituents. But there are several disadvantages, mainly arising from the sampling and sample preparation, which need considerable precaution, such as for exclusion of oxygen.
- *Equilibrium Approaches:* these approaches are related to the broad toxicological basis of food and water quality data – a very important advantage. On the other hand, there are the effects of sample preparation, e.g. the drying procedures; the separation techniques, for example filtration or centrifugation; there are strong effects of grain size composition and the influence of suspended matter concentration in the aquatic system, which is even more important, if the kinetics of sorption and desorption are too slow for equilibrium to be achieved in a given time of interactions. Unlike non-ionic organic chemicals, KD-values of metals are not only correlated to organic substances but also with other sorption-active surfaces; therefore, the equilibrium partition approach exhibits strong limitations for metallic elements.
- *Remobilization:* short-term effects may be studied from water/sediment-suspensions, medium-term effects from experiments using tanks, and long-term effects by applying chemical extractants, either single or in sequence. Field observations often do not show clear effects, as has been demonstrated for the release of metals from anoxic sediments during oxidation. Such implications for future criteria development, particularly important for dredging and management of dredged material, will be discussed on the basis of experience from metal speciation studies on soils and sediments.

### 11.5.1 Limitations of Available Test Procedures

From the data in Sections 11.3 and 11.4 it has become clear that considering the mobilization of heavy metals from sediments there are two master variables: redox potential and pH, the latter directly influenced by the former. For the practice of criteria development and application, available test procedures should be checked regarding their relevance for predicting middle- and long-term behaviour of metal-bearing sediment matrices under changing redox and pH-conditions.

**Pore Water Chemistry.** As mentioned above, the composition of pore water is a highly sensitive indicator for reactions between chemicals on solid substrates and the aqueous phase which contacts them. Particularly for fine-grained material the large surface area related to the small volume of its entrapped interstitial water ensures that minor reactions with the solid phase will be indicated by major changes on the composition of the aqueous phase. While the direct recovery and analysis of water-borne constituents can be seen as a major advantage of this approach there are several disadvantages, particularly arising from the sampling and sample preparation, which are not yet routine procedures, and are usually involving considerable precautionary measures such as exclusion of oxygen.

**Experimental Multi-Chamber Device.** While the actual focus in sediment criteria is on individual pollutant concentrations and their biological effects, inclusion of data on potential matrix variations by redox and pH changes has not been considered as yet to any significant extent. The only exception is the initiative of the United States Environmental Protection Agency to introduce the acid volatile sulphide (AVS) procedure as a measure for the non-availability of critical metals at a surplus of reactive sulphide ions (DiToro et al. 1990). Such effects have been investigated experimentally with the multi-chamber device, which has been described in Fig. 11.2. Metal transfer processes observed in the experimental device usually cannot be measured under field conditions, due to the complexity of natural systems and short residence time of released metals in the water phase. Extrapolation of such findings, therefore, into quantitative models seems to be nearly impossible, and this may also restrict their incorporation into criteria.

**Elutriate Tests.** To estimate short-term chemical transformations, the interrelations between solid phases and water have been increasingly subjected to laboratory studies. The advantage of such experiments is that especially important parameters can be directly observed and particularly unfavourable conditions simulated. The US Army Corps of Engineers and the US Environmental Protection Agency have developed an elutriate test that is designed to detect any significant release of chemical contaminants in dredged material. This test involves the mixing of one volume of the dredged sediment with four volumes of the disposal site water for a 30-min shaking period. If the soluble chemical constituent in water exceeds 1.5 times the ambient concentration in the disposal site water, special conditions will govern the disposal of the dredged material.

**Sequential Extraction.** In connection with the problems arising from the disposal of solid wastes, particularly of dredged materials, chemical extraction sequences have been applied which are designed to differentiate between the exchangeable, carbonatic, reducible (hydrous Fe/Mn oxides), oxidizable (sulphides and organic phases) and residual fractions. The undisputed advantage of this approach with respect to the estimation of long-term effects on metal mobilities lies in the fact, that rearrangements of specific solid "phases" can be evaluated prior to the actual remobilization of certain proportions of an element into the dissolved phase. One of the widely applied extraction sequences of Tessier and co-workers (1979) has been modified by various authors (Table 11.7).

**Table 11.7.** Sequential extraction scheme for partitioning sediment samples (Kersten and Förstner 1986, 1987)

| | Extractant | Extracted component |
|---|---|---|
| Exchangeable | 1 M NH4OAc | Exchangeable ions |
| Carbonatic | 1 M NaOAc, pH 5 w/ HOAc | Carbonates |
| Easily reducible | 0.01 M $NH_2OH$ HCl w/ 0.01 M $HNO_3$ | Mn-oxides |
| Moderately reducible | 0.1 M oxalate buffer pH 3 | Amorphous Fe-oxides |
| Sulphidic/organic | 30% $H_2O_2$ pH 2 w/ 0.02 M $HNO_3$ extr. w/ 1 M $NH_4OAc$ – 6% $HNO_3$ | Sulphides together with organic matter |
| Residual | Hot $HNO_3$ conc. | Lithogenic material |

Partition of metals was determined in a sample from Hamburg harbour , which was pretreated in different ways (Kersten et al. 1985): (1) The EPA Standard Elutriate Test, 1:4 sediment/site water for 30 min (see preceding); (2) freeze-dried sample; (3) oven-drying at 60°C. It was observed that oxidation had a great effect on regulating the chemical form of cadmium and other trace metals (Fig. 11.20). Compared to the original sample *A*, which was extracted under an argon atmosphere, there was a typical change from oxidizable phases, which were mainly Cd-sulphide, to easily reducible forms, *B*, upon application of the shaking/bubbling test. During freeze-drying – which is commonly assumed to represent a relative smooth mode of sample pretreatment – transformation to *C*, carbonatic and exchangeable forms, takes place. This effect is further enhanced during oven-drying at 60°C, *D*.

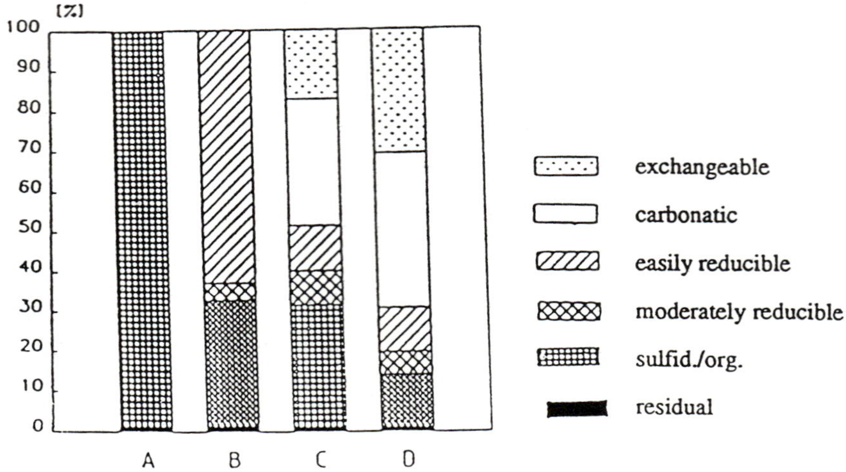


**Fig. 11.20.** Partitioning of Cd in anoxic mud from Hamburg Harbour in relation to the pretreatment procedure (Kersten et al. 1985)

Chemical forms of metals have been studied in a sediment sample from Hamburg harbour during three succeeding redox cycles in 210 days. Figure 11.21 indicates, that after each oxidation step the percentage of the sulphidic/organic fraction of zinc decreased continuously, whereas the dissolved zinc contents increased – an indication of the weakening of zinc binding intensity.

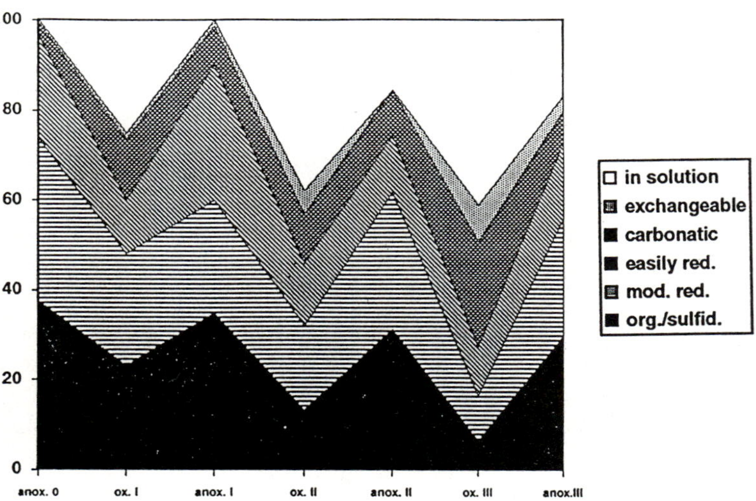


**Fig. 11.21.** Transformation of Zn forms in succeeding redox cycles (Calmano et al. 1993)

In Table 11.8, an example is given for the possibilities of standardizing the data from elution experiments with respect to numerical evaluation. An "elution index" for sediment samples from various rivers in West Germany is based on the metal concentrations exchangeable with 1 N ammonium acetate at pH 7. These metal fractions are considered to be remobilizable from polluted sediments at a relative short term under more saline conditions, for example in the estuarine mixing zone. Comparison of the release rates from oxic and anoxic sediments clearly indicates, that the oxidation of samples gives rise to a very significant increase in the mobilization of the metals studied. This effect was particularly important for Cd. When proceeding further in the extraction sequence, more long-term effects can be estimated, but generally with a reduction of prognostic accuracy (Table 11.8).

**Table 11.8.** Elution-index for selected river sediments, as determined from exchangeable proportions (1 M ammonium-acetate). Calculated relative to background data from Elbe river sediments (Förstner et al. 1990). These values are multiplied by a factor of 100

| | Neckar river | Main river | Rhine river | Elbe river | Weser river |
|---|---|---|---|---|---|
| Copper | 0.2 | - | 1 | 1 | - |
| Lead | 1 | 1 | 2 | 1 | 1 |
| Zinc | 7 | 9 | 28 | 36 | 9 |
| Cadmium | 30 | 30 | 230 | 30 | - |
| Total oxic | 38 | 40 | 261 | 68 | 10 |
| (Anoxic) | 0.5 | 0.3 | 8 | >4 | 4 |

### 11.5.2 Calculating Long-term Acid Production and Buffer Capacities

Criteria for the prognosis of the middle- and long-term behaviour of metals should include the abilities of sediment matrices for producing acidity and for neutralizing such acid constituents. Important acid-producing oxidation reactions leading to a decrease of pH in low buffered sediments are listed in Table 11.9.

**Estimating Acid Producing and Consuming Capacities.** Experimental approaches for calculating APC and ACC for sulphidic mining residues have been summarized by Ferguson and Erickson (1988). A test described by Sobek et al. (1978) involves the analysis of total pyritic sulphur; potential acidity is then subtracted from neutralizing potential which can be obtained by adding a known amount of HCl, heating the sample and titrating with standardized NaOH to pH 7. Bruynesteyn and Hackl (1984) calculated APC from total sulphur analysis; here, acid producing capacity was then subtracted from acid consuming

**Table 11.9.** Important acid-producing oxidation reactions in aquatic systems (Calmano et al. 1993)

| Elements | | Reactions | f [a] |
|---|---|---|---|
| Inorganic | S | $H_2S + 2\,O_2 = SO_4^{2-} + 2\,H^+$ | 2 |
| | S | $S^- + 3/2\,O_2 + H_2O = SO_4^{2-} + 2\,H^+$ | 2 |
| | S, Fe | $FeS + 9/4\,O_2 + 3/2\,H_2O = FeOOH + SO_4^{2-} + 2\,H^+$ | 2 |
| | S, Fe | $FeS_2 + 15/4\,O_2 + 5/2\,H_2O = FeOOH + 2\,SO_4^{2-} + 4\,H^+$ | 4 |
| | Fe | $Fe^{2+} + 1/4\,O_2 + 5/2\,H_2O = Fe(OH)_3 + 2\,H^+$ | 2 |
| | N | $NH_4^+ + 2\,O_2 = NO_3^- + H_2O + 2\,H^+$ | 2 |
| | N | $NO_x + 1/4(5\text{-}2x)O_2 + 1/2\,H_2O = NO_3^- + H^+$ | 1 |
| Organic | N | $R\text{-}NH_2 + 2\,O_2 = R\text{-}OH + NO_3^- + H^+$ | 1 |
| | S | $R\text{-}SH + H_2O + 2\,O_2 = R\text{-}OH + SO_4^{2-} + 2\,H^+$ | 2 |

[a] acid formation coefficient

capacity, obtained by titration with standardized sulphuric acid to pH 3.5. The APC-relationships of sediments are more complex than that in sulphidic ores because the APC from organic matter must be considered; time-scale plays a major role in sediments.

The acid neutralization capacity of a natural aquatic system is composed of the acid consuming capacity of the solids ACC(s) and that of the dissolved phase ACC(aq):

$$ACC = ACC(s) + ACC(aq) \qquad (11)$$

The ACC(aq) of the dissolved phase – in aquatic systems is always much smaller than ACC(s) – may be described by the following Eq. (12):

$$ACC(aq) = [HCO_3^-] + 2\,[CO_3^{2-}] + 2\,[S^{2-}] + [HS^-] + [NH_3] - [H^+] \qquad (12)$$

A similar rule shall apply to the ACCs of the solids at pH > 5:

$$ACC(s) = 2[CaO] + 2[MgO] + 2[Na_2O] + 2[K_2O] + 2[MnO] + 2[FeO] - 2[SO_3] - 2[P_2O_5] - HCl] \qquad (13)$$

From the two parameters ACC and APC an effective acid producing capacity $APC_{eff}$ can be calculated. This may be helpful in practice for the specification of sediment quality criteria. The $APC_{eff}$ of a sediment suspension may be defined as:

$$APC_{eff} = V/W\,([H^+]_e - [H^+]_o) \qquad (14)$$

where $APC_{eff}$ = effective acid producing capacity
$V$ = suspension volume
$W$ = solid mass
$[H^+]_{o,e}$ = $[H^+]$ before and after oxidation of the suspension.

For example, for an anoxic Elbe River sediment with an original $pH_o = 7.03$ and a $pH_e = 3.29$ after an oxidation time of 35 days an $APC_{eff} = 5.12$ mmol $kg^{-1}$ could be calculated. Such simple test procedures allow better statements of the sediment quality than simple considerations of limiting values.

**Geochemical Characterization of the Potential Metal Mobility.** The most efficient fixation process within anoxic sediments for trace metals is production of free sulphide during anaerobic organic matter degradation and reduction of sulphate (study of heavy metal associations with sulphide and carbonate formations in anoxic sediments, therefore, provides insight into early diagenetic processes; Berner 1981, Morse and Mackenzie 1990).

Whereas the ability of the sediment to produce free sulphide is determined by the sulphate reduction rates, the ability to remove all produced free $HS^-$ is given by the reactive metal – predominantly reducible $Fe^{3+}$ concentrations – available to form sulphide minerals (*Available Sulphide Capacity* [ASC]; Williamson and Bella 1980). Simultaneous application of standard sequential leaching techniques (BCR-Version; Ure et al. 1993) on critical trace elements and matrix components can be used for geochemical characterization of anoxic, sulphide-bearing sediments in relation to the potential mobility of critical trace metals (Kersten and Förstner 1991):

| | | | |
|---|---|---|---|
| Step 1 | Exchangeable ions | 0.5 g sediment – 5 h | $Ca^{2+}$ and $Fe^{2+}$ |
| | Carbonates | – 20 ml of acetic acid 0.11 mol $l^{-1}$ | from carbonates and phosphates |
| Step 2 | Reducible metals | 16 h – 20 ml $NH_2OH{\cdot}HCl$ 0.1 mol $l^{-1}$, $HNO_3$ pH 2 | $Fe^{3+}$ from reactive oxides |
| Step 3 | Sulphidic and organic bound metals | 10 ml hydrogen peroxide 8.8 mol $l^{-1}$(twice) - extr. 50 ml $NH_4Ac/HOAc$ pH 5 | S from easily oxidizable sulphides (FeS) |

For determining the acid producing capacity (*maximum APC*) in anoxic sediments, both the FeS pool (*actual APC*) and the maximum ferrous sulphide (worst case: pyrite)-producing capacity upon disposal has to be taken into consideration. The latter is given by the *sulphide binding capacity* (ASC, see preceding), which were determined in our example of Hamburg harbour mud (Table 11.10) using the Fe-concentrations in step 2 of the sequential leaching results. The stoichiometry of the oxidizable S- and Fe-fractions of step 3 indicate that the ferrous sulphide extracted in this step was in the FeS form. The sum of both the ASC and the *actual APC* gives the *maximum APC* for the samples as shown in Table 11.10. The acid consuming capacity value for these samples was more simply determined from the Ca-concentration released from reactive carbonates by the Na-acetate solution of step 1, which has to be multiplied by a factor of 0.5 to account for the stoichiometric ratio between pyrite and calcite within the redox reaction:

$$FeS_2 + 2\ CaCO_3 + 15/4\ O_2 + 1/2\ H_2O = FeOOH + 2\ SO_4^{2-} + 2\ Ca^{2+} + 2\ CO_2$$

The negative balance between the ACC and APC indicates, that the mud sample from Hamburg harbour has a significant acidification potential.

**Table 11.10.** Balance between APC and ACC values for an anoxic sediment sample from Hamburg Harbour (Kersten and Förstner 1991)

| Compound | Function | Value | Parameter ($mmol\ kg^{-1}$) | |
|---|---|---|---|---|
| Ca in step 1 | base potential | (x 0.5 =) | ACC | 90 |
| Fe in step 2 | ASC | (335) | | |
| S in step 3 | sulphide sulphur | (+ 85 =) | max. APC | 420 |
| Balance | | | ACC - APC = | -330 |

### 11.5.3 Assessing Long-term Mobility of Metals in Sediments by Titration Experiments

With regard to prediction of long-term effects of sediment-bound metals, chemical extraction procedures are of limited value because they usually involve neither reaction-mechanistic nor kinetic considerations. This lack can be avoided, e.g. by an experimental approach, originally been used by Patrick et al. (1973) and Herms and Brümmer (1978), where sediments can be treated in a circulation system under controlled intensification of significant release parameters such as pH-value, redox-potential, and temperature. Our experimental design (Schoer and Förstner 1987) includes an ion-exchange system for extracting and analysing the released metals at an adequate frequency, and compares sequential extraction results before and after treatment of the sample in the circulation apparatus. Individual metal species are released at different time intervals. Taking into account both element contents released during the 10 weeks experiments (equivalent to several thousand years of solid/water interaction) and those extrapolated from extraction "pools", concentrations can be calculated for different scenarios.

While these extrapolations have been made from pH 5 conditions, titration curves from investigations on a wide spectrum of metal-bearing waste materials (Obermann and Cremer 1992) suggest, that pH 4 may be more appropriate for long-term predictions of potential metal releases from contaminated sediments. In this pragmatic approach, which is outlined in Fig. 11.22, the pH is automatically adjusted to 4 during the time period of 24 h; apart from the release rates of metals, which can be determined from samples taken at different time intervals, the sum curve (diagram of discrete acid additions would be valuable!) of acid consumption provides information on the potential changes of the matrix composition during acidification and the availability of buffer capacity at different time-scales.

The acid consumption curve in Fig. 11.22 reflects slow long-term metal release from sediments. Because calcite dissolution was fast, the acid consumption in the first stage increased drastically within a short period of time. Cation release and alumosilicate dissolution were dominant factors consuming acid in the later stage. The reactions can be treated as:

$$SO{=}Me + H^+ = SO{=}H_2 + Me^{2+}, \text{ and}$$

$$Al_2O_3 + 6\,H^+ = 2\,Al^{3+} + 3\,H_2O$$

where SO= and Me are reaction groups in solid phase and metal ion, respectively. The reactions of hydrogen ion with metal and alumosilicate are delayed due to the complex sediment structure and matrix. For example, special penetration of hydrogen ion is required for reacting with cation on clay minerals coated with organic matter or biofilms. Rates of reactions can be estimated by measurements of metal concentrations in solution.

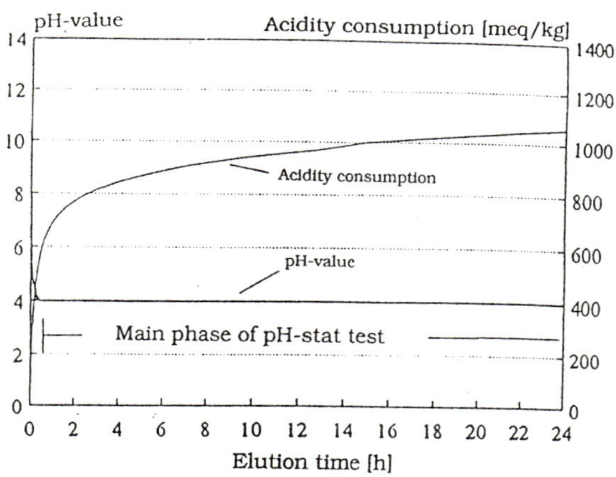


**Fig. 11.22.** Patterns of a suspension titration (Obermann and Cremer 1992). The initial phase, during which the target-pH-Value is not attained, is followed by a 50-fold longer main phase with highly stable pH-conditions.

### 11.5.4 Discussion: Prognostic Tools for Long-term Behaviour of Metals in Sediments

Regarding the potential release of metals from contaminated sediments changes of redox and pH conditions are of prime importance. In practice, therefore, characterization of sediment substrates with respect to their buffer capacity is an indispensable step for the prognosis of middle- and long-term processes of metal mobilization.

*(1) Description of Basic Processes.* Experimental studies in the multi-chamber device and reactor cell provided information on the metal transfer between different sedimentary matrices and on interrelations between mobilization and readsorption mechanisms. Further studies should focus on the role of iron sulphides in the large-scale changes in metal species, i.e. solid/solution interactions and

bioavailability of trace metals in aquatic systems. The combination of matrix and trace element data in sediments, extracted simultaneously with different reagents, can provide additional information, which can be used to establish models on chemical interactions occurring under characteristic borderline conditions, with respect to sulphide, chloride and complexing agents (Wallmann et al. 1993).

Reliability of long-term prognoses on the impact of metal-bearing sediments is restricted due to the dominance of non-linear and delayed processes in redox-controlled systems. While measurements of both actual acid producing and consuming capacities can be undertaken with available methods (Sect. 11.5.2), a possible "split" or translocation of their inventories between solid and dissolved phases renders such parameters difficult to be predicted in a long-term perspective. It seems, that calculation of maximum acid producing capacity (APC) involving "available sulphide capacity", i.e. reducible Fe(III)-concentration (Sect. 11.5.2), gives a relative safe basis for such considerations, because iron generally is less mobile than some sulphur species. From data on long-term acid producing potential of anoxic sediments, appropriate management techniques for dredged materials can be derived (Sect. 11.6).

*(2) Definition of Parameters.* The pragmatic $pH_{stat}$-approach (presented in Sect. 11.5.3) for evaluating the behaviour of solid waste subsequent to landfilling involves the "spot test" of lowering the pH value and – in addition – provides data on the buffer capacity of the solid matrix. Direct assessment of the acid producing capacity can be performed by ventilation with air or oxygen; our data, however, have shown that these experiments should be designed for relative long time periods, i.e. for several weeks duration.

Although the three parameters forming the framework of future criteria for assessing long-term effects from metal-bearing sediments – acid producing potential, acid consuming capacity and metal concentration – are closely interrelated, regulatory activities will focus on the concentration term, mainly because its priority function as effect parameter in biological systems. In this respect, however, the long-term prognostic approach is different from the assessment of the short-term impact by metal-contaminated sediments, which is preferentially done by applying bioassays on sediment pore water and elutriates (recent developments are directed towards solid phase bioassays). Long-term prediction, i.e. extrapolations of experimental or field data to a hypothetical situation, necessarily implies some type of numerical method. From the available approaches (Sect. 11.5), the group of elution techniques generally seems to be best suited for this purpose; within this group, long-term predictive capacity is predominantly reflected by the more rigid extractants, in particular acid solutions. Guideline values based on acid extraction data have been proposed for soil and solid waste materials. The most common pH level of about 4 was primarily aimed to simulate conditions of acid precipitation; however, as has been shown from sediment and solid waste data, internal processes such as oxidation of sulphides can by far be more effective in the lowering of pH.

*(3) Criteria for Dredging and Up-Land Deposition of Sediments.* In dredged material management, two different target areas for combined matrix/metal criteria can be distinguished: (1) sediment resuspension and (2) dredged material disposal. With respect to "resuspension of aquatic sediments", which involves more short-term effects than the disposal of dredged material, special emphasis should be posed on the factor "available metal species". Within certain categories of acid producing and consuming capacities, guideline values for individual metals should be based on elutriate data, preferentially at pH 4, for better comparison with other solid matrices (e.g. Swiss Ordinance for Waste Materials, which includes two categories of limit values: *Inert-/Construction Materials* and *Residue Material Deposits*, the latter mainly comprising pretreated products of municipal solid waste incineration; Anonymous 1990).

Environmental impact of sediment deposits is influenced by the internal chemical conditions rather than by the concentration and extractability of metals. Priority, therefore, should be given to the optimization of long-term chemical stability ("geochemical engineering").

*(4) Outlook.* Above discussion has shown that potential mobility, bioavailability and toxicity of metals in sediments strongly depend on redox conditions. From earlier chemical extraction procedures, as proposed by Tessier et al. (1979) to more recent SEM/AVS ratio determinations (Di Toro et al. 1990, 1992), metal speciation on is a major characteristic describing its potential mobility in sedimentary environments. The inherent link between the parameters to describe "metal speciation" is the characterization of chemical stability and bioavailability of metal compounds in sediments under certain redox conditions.

In a sulphidic anoxic environment, even of the sediment is heavily polluted by metals, organism are considered to be still safe due to the strong fixation of metal ions by $S^{=}$ or $HS^{-}$ as source of acid volatile sulphide. Such heavily polluted sediment, however, can behave as a time bomb which is triggered by only one factor – redox increase to a critical point, i.e. by exposure to oxygen-rich overlying water or directly to air. Once this situation occurs (a possible pathway also is oxygen transfer via plant roots), toxic metals in the sediments will be release to the water phase or transformed into more bioavailable species (Calmano et al. 1992).

At the moment, research on long-term effects of redox variations on metal behaviour in sediments is mostly based on thermodynamic considerations. Future research should emphasize studies on the kinetics of metal species transformations, hydrogen ion production and metal release as affected by changing redox conditions. Additional important aspects involve the bridging of the gaps between numerical criteria approaches, as reflected by matrix composition and metal mobility, and biological approaches. Assessment for bioavailability of metal species is still based on results of water phase and elutriate, mostly using acute toxicity data (Ankley et al. 1990; Ahlf et al. 1992); initial data on solid phase bioassays suggest good perspectives for future applications in sediment quality assessment. It may well be, that for such systems, which are much less disturbed than artificial sediment elutriates, relationships between matrix conditions – as reflected, e.g. by redox indices – and metal species bioavailability may be found, which may serve as a more solid basis for the interpretation of results from bioassays, eventually with respect to chronic toxicity.

## 11.6 Remediation Procedures

### 11.6.1 Treatment of Dredged Sediments

Dredging is a human activity, which can well compete with highly dynamic natural sedimentary processes such as landslides. In the harbours around the North Sea, approximately 100 million $m^3$ of sediment has to be dredged annually – about 10 times the average annual sediment discharge of Rhine river. Typical problems with these sediments are:

- Increasing volumes of dredged materials
- High concentrations of toxic substances
- Problems have been concentrated mainly at the mouth of large rivers and in coastal areas

In Rotterdam Harbour, at the mouth of Rhine river, the volume of sediment, which has to be dredged annually, increased from less than 1 million $m^3$ in 1920 to more than 10 million $m^3$ in 1980. During the last 80 years, the cadmium concentrations in the river sediment increased by a factor of more than 50.

**Assessment and Selection of Remediation Technology.** Beside the costs of the remediation technique, the major questions relate to the contaminant loss pathways. Contaminant loss can occur through one or more pathways. The example of a confined disposal facility indicates, that the potential pathway for contaminant loss include surface run-off, effluent, seepage, leachate, dust and uptake by plants and animals. Predictive techniques for estimating contaminant losses comprise to categories – a priori techniques and techniques based on pathway-specific laboratory testing:

- A priori techniques are suitable for planning-level assessments
- Techniques that use pathway-specific test data provide state-of-the-art loss estimates (generally more advanced)

The various types for sediment remediation can be subdivided according to the mode of handling "in place" or "excavation", or in relation to technology "containment" or "treatment". Important containment techniques include "in situ capping" and "confined disposal facility". Regarding in-place-treatment, biological

**Table 11.11.** Technology types for sediment remediation (Anonymous 1993)

| | In place | Excavated |
|---|---|---|
| Containment | In situ capping | Beneficial use |
| | Contain/fill | Capping/confined aquatic disposal |
| | | Commercial landfills |
| | | Confined disposal facility |
| Treatment | Bioremediation | Chemical |
| | Chemical | Biological |
| | Immobilization | Extraction |
| | | Immobilization |
| | | Physical separation |
| | | Thermal treatment |

processes may be applied. Excavated sediments – apart from physical separation – can be treated to immobilize pollutants, mainly metals.

A more general conceptual scheme related to excavated sediment material has been proposed by the TNO, the Netherlands scientific technological organization (Van Gemert et al. 1988). "A-" and "B-" techniques are distinguished: "A" is for large-scale concentration techniques like mechanical separation; "A"-techniques are characterized by low costs per unit of residue, low sensitivity to variations, and they may be applied in mobile plants. "B"-techniques are decontamination procedures, which are especially designed for relatively small-scale operations. They involve higher operating costs per unit of residue; they are more complicated, need specific experience of the operators and are usually constructed as stationary plants. "B"-techniques include biological treatment, acid leaching, solvent extraction, etc.

In a trial to rank the individual technologies, one can follow the proposals in the Draft Remediation Guidance Document prepared by the USEPA Oceans and Coastal Protection Division/Great Lakes National Program (1994). Three criteria have been used:

- *State of Development.* It is mostly good except for non-removal and treatment alternatives. Such deficiencies should be overcome by intensified research.

- *Potential Contaminant Loss.* It is particularly high for removal. This would favour in situ techniques. In situ capping could be the method of choice for areas, where maintenance dredging is not essential.

- *Cost.* It is already clear at this point, that the cost factor will mostly exclude treatment (in the closer sense) of large-volume contaminated sediments. The only possibility would be the reduction of the volume by mechanical pretreatment, then chemical/biological techniques could be used potentially.

**Table 11.12.** Ranking of remediation technologies (Anonymous 1994). Also See Text

| Technology | State of development | Potential contaminant loss | Cost |
|---|---|---|---|
| Non-removal | Low/Medium | Low | Low/Medium |
| Removal | High | High | Low |
| Transport | High | Low | Low |
| Pretreatment | Medium | Medium | Low/Medium |
| Treatment | Low | Low /Medium | High |
| Disposal | High | Low/Medium | Low/Medium |
| Effluent/leachate Treatment | High | Low | Low |

It is quite obvious, that technological options are more restricted for dredged sediments than for other waste materials in most cases. In particular, remediation techniques in the closer sense often are economically unacceptable because of the large volume of contaminated sediments.

### 11.6.2 Geochemical Engineering – Application to Contaminated Sediments

"Geochemical engineering" (Salomons and Förstner 1988) applies geochemical principles (such as concentration, stabilization, solidification, and other forms of long-term, self-containing barriers) to determine the mobilization and biological

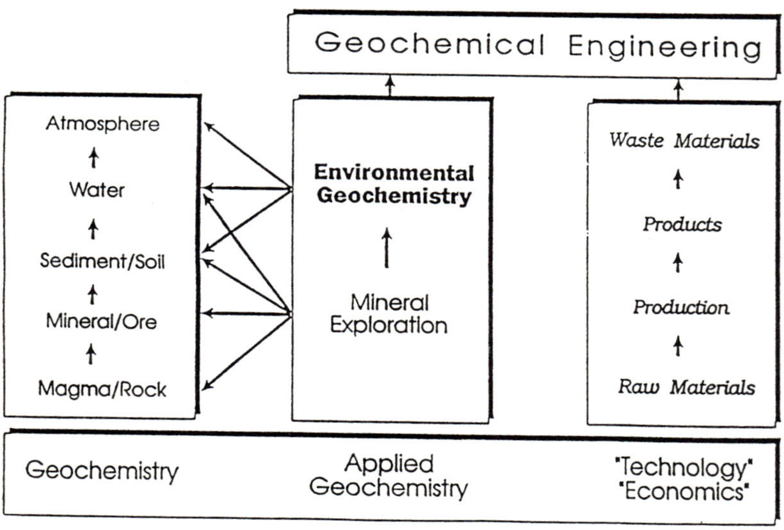


**Fig. 11.23.** New developments in environmental geochemistry

availability of critical pollutants (Fig. 11.23). Including new experience from impact evaluations related to capacity controlling properties, the mobility concept of environmental geochemistry can be implemented into waste management practice by different ways of optimizing barrier systems.

Inclusion of the time factor moves beyond a traditional chemical approach. It also transcends the civil engineering approach in waste management, which usually devotes little attention to long-term emissions from waste disposal sites. "Because we have become accustomed to considering the filling period as the most important phase in landfill operation, we have forgotten that subsequent to the active working period there is the infinitely long time in which the site has to function as a depository for all materials unwanted in the biosphere" (Stief 1987).

In modern waste management, the fields of geochemically oriented technology include:

- The study of material fluxes within and between the anthroposphere and "geospheres"
- The optimization of elemental distribution at high-temperature processes
- The selection of favourable milieu conditions for the deposition of large-volume wastes
- The selection of additives for the solidification and stabilization of waste materials
- The development of test procedures for long-term prognoses of pollutant behaviour in all kinds of waste depositories

As shown from the examples of large-mass wastes – dredged material, mining residues and municipal solid waste, long-term immobilization of critical

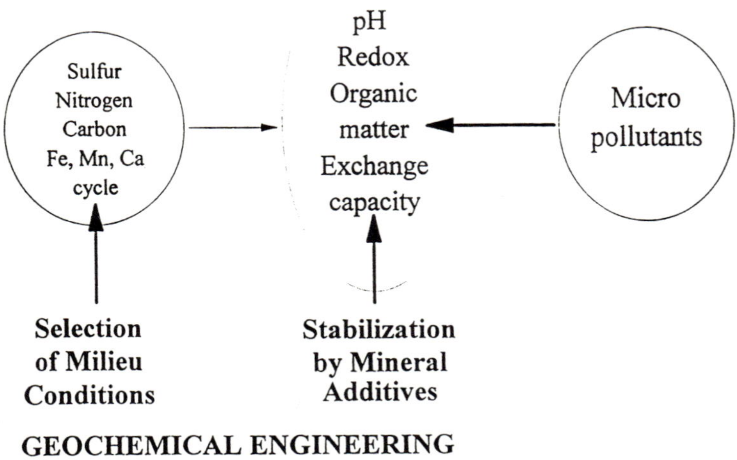


**Fig. 11.24.** Interference of geochemical engineering methods with biogeochemical cycles

pollutants can be achieved by promoting less soluble chemical phases, i.e. by thermal and chemical treatment, or by providing respective milieu conditions. Figure 11.24 schematically indicates, that selection of appropriate environmental conditions predominantly influence the geochemical gradients, whereas chemical additives are aimed to enhance capacity controlling properties in order to bind (or degrade!) micropollutants. In general, microscale methods, e.g. formation of mineral precipitates in the pore space of a sediment waste body, will be employed rather than using large-scale enclosure systems such as clay covers or wall constructions. A common feature of geochemically-designed deposits, therefore, is their tendency to increase overall stability in time, due to the formation of more stable minerals and closure of pores, thereby reducing water permeation.

### 11.6.3 Storage Under Permanent Anoxic Conditions

Regarding the various containment strategies is has been argued that upland containment (e.g. on heap-like deposits) could provide a more controlled management than containment in the marine environment. However, contaminants released either gradually from an imperfect impermeable barrier (also to groundwater) or catastrophically from failure of the barrier could produce substantial damage (Kester et al. 1983). On the other hand, near-shore marine containment (e.g. in capped mound deposits), offers several advantages, particularly with respect to the protection of groundwater resources, since the underlying water is saline and inherent chemical processes are favourable for the immobilization or degradation of priority pollutants.

In a review of various marine disposal options, Kester et al. (1983) suggested that the best strategy for disposing of contaminated sediments is to isolate them in a permanently reducing environment. Disposal in capped mound deposits above the prevailing sea-floor, disposal in subaqueous depressions, and capping deposits in depressions provide procedures for contaminated sediment (Bokuniewiscz 1982). In some instances it may be worthwhile to excavate a depression for the disposal site of contaminated sediment which can be capped with clean sediment. This type of waste deposition under stable anoxic conditions, where large masses of polluted materials are covered with inert sediment became known as "subsediment-deposit". The first example was planned for highly contaminated sludges from Stamford Harbour in the Central Long Island Sound following intensive discussions in the US Congress (Morton 1980).

Under subsediment conditions there is a particular low solubility of metal sulphides, compared to the respective carbonate, phosphate, and oxide compounds (Fig. 11.25). One major prerequisite is the microbial reduction of sulphate. Thus, this process is particularly important in the marine environment, whereas in the anoxic freshwater milieu there is a tendency for enhancing metal mobility due to the formation of stable complexes with ligands from decomposing organic matter. Marine sulphidic conditions, in addition, seem to repress the formation of

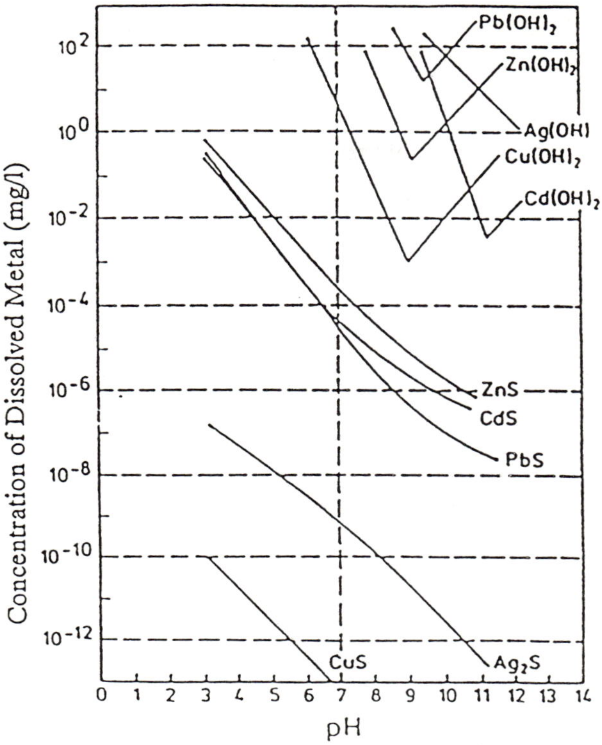


**Fig. 11.25.** Solubility of metal sulphides and metal oxides (Ehrenfeld and Bass (1983)

monomethyl mercury, one of the most toxic substances in the aquatic environment, by a process of disproportionation into volatile dimethyl mercury and insoluble mercury sulphide (Craig and Moreton 1984). There are indications that degradation of highly toxic chlorinated hydrocarbons is enhanced in the sulphidic environment relative to oxic conditions (Sahm et al. 1986; Kersten 1988).

### 11.6.4 Chemical Stabilization by Additives

In general, solidification/stabilization technology is considered a last approach to the management of hazardous wastes. The aim of these techniques is a stronger fixation of contaminants to reduce the emission rate to the biosphere and to retard exchange processes. Most of the stabilization techniques aimed for the immobilization of metal-containing wastes are based on additions of cement, water glass (alkali silicate), coal fly ash, lime or gypsum (Malone et al. 1982; Wiedemann 1982; Goumans et al. 1991).

Laboratory studies on the evaluation and efficiency of stabilization processes were performed by Calmano et al. (1986). As an example Fig. 11.26 shows acid titration curves for Hamburg harbour mud without and after addition of limestone and cement/fly ash stabilizers. Best results are attained with calcium carbonate, since the pH-conditions are not changed significantly upon addition of $CaCO_3$. Generally, maintainance of a pH of neutrality or slightly beyond, favours adsorption or precipitation of soluble metals (Gambrell et al. 1983). On the other hand, it can be expected that both low and high pH-values will have unfavourable effects on the mobility of heavy metals.

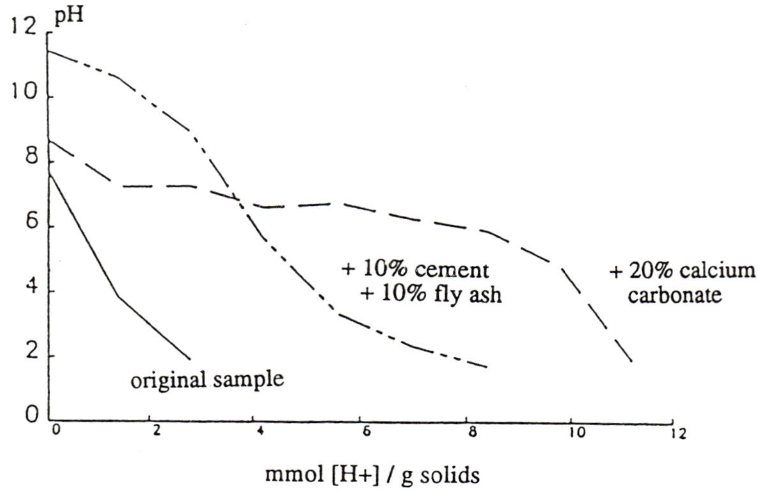


**Fig. 11.26.** Effect of calcium carbonate and cement/fly ash additives on chemical stabilization of fine-grained sediment from Hamburg harbour (Calmano et al. 1986)

## 11.7 Summary and Outlook

The enormous problems arising from the historic thoughtless dissipation of chemicals in the environment has become obvious in the last 10 years, and especially with the opening of eastern Europe. The detection of extremely toxic chemicals at Love Canal, Lekkerkerk and Georgswerder are only a few of the harsh lessons. After solving a few spectacular cases with intensive efforts, it now seems that financial restrictions may inhibit the use of many newly developed remediation techniques (such as chemical extraction, high-temperature incineration, and some biological procedures) to sanitize contaminated land on a large scale (i.e. for areas of hundreds of square kilometres. Although this relates primarily to parts of eastern Europe, the tens of thousands of old landfill sites which must be excavated, treated and recultivated will soon strain even more prosperous economies. It may well be, that economic considerations will strengthen the popularity of geochemically engineered solutions in these situations, as well as for large waste masses such as mine residues, dredged materials and filter ashes described here. But even if traditional engineering continues to dominate practical solutions the specific potential of geochemistry to provide instruments for long-term assessment of processes should be used and expanded.

Management of contaminated sediments requires – similar to other waste materials – a holistic approach. In addition to the common predictive techniques for estimating contaminant losses – a priori techniques and pathway-specific tests – the interactive nature of various parameters has to be recognized. This means, that particular emphasis should be posed on the evaluation of the two data sets *driving force/geochemical gradient* parameters (mostly redox-sensitive components, which may be critical for the "acid producing capacity") and *capacity controlling properties* (with a prevalence of "carbonate content"). Assessment of these parameters and their long-term borderline conditions, as well as the potential links between the two groups *driving force/capacity controlling properties* and assessment of individual parameters should be an integrated part of the wider management scheme. The analytical and experimental parameters should always be related to the potential remediation options for a specific sediment problem.

Remediation techniques on contaminated sediments generally are much more limited than for most other solid waste materials, except for mine wastes. The widely diverse contamination sources in larger catchment areas usually produces a mixture of pollutants, which is more difficult to treat than an industrial waste. Even if one has procedures at hand to reduce 9 priority pollutants below the guideline values – number 10, for example for mercury or PCB – this may render the whole business unrealistic. There is a long retention time for sediments in larger catchment areas. Improvement at the source may need decades to become effective in the sediments at the lower reaches and harbours close to the river mouth. For most sediments from maintenance dredging, there are more arguments in favour of "disposal" rather than "treatment". Considering the worldwide dredging activities – more than 1 billion $m^3$ per year – only a very small percentage of these materials can undergo "treatment" in the closer sense – solvent extraction, bioremediation, thermal desorption, vitrification, etc. Mechanical separation of less strongly contaminated fractions, however, may be an useful step prior to final storage of the residues.

**Acknowledgements**. I wish to record my particular gratitude to my collaborators at the Section of Environmental Protection Technology, Technical University of Hamburg-Harburg, who have provided significant portions of the material for this review. With regard to the present subject, I received much inspiration by Dr. Wim Salomons during the last 10 years and more recently by both Dr. Gerrit Hekstra and Dr. William Stigliani as well as from numerous colleagues the "Chemical Time Bomb Project". I am particularly grateful to the National of Water Research Institute at Burlington, especially to Dr. Rod Allan, who encouraged me to present the theme for a lectureship in honour to Professor Richard A. Vollenweider.

## References

Ahlf W (1983) The River Elbe: behaviour of Cd and Zn during estuarine mixing. Environ Tech Letts 4:405-410

Ahlf W, Munawar M (1988) Biological assessment of environmental impact of dredged material. In: Salomons W, Förstner U (eds) Chemistry and biology of solid waste – dredged material and mine tailings. Springer, Berlin Heidelberg New York pp 127-142.

Ahlf W et al. (1992) Mikrobielle Biotests mit Sedimenten. Schr Reihe Verein WaBoLu/Berlin 89:427-435

Allan RJ (1974) Metal contents of lake sediment cores from established mining areas: an interface of exploration and environmental geochemistry. Geol Surv Can 74-1/8:43-49

Ankley GT, Katko A, Arthur JW (1990) Identification of ammonia as an important sediment-associated toxicant in the Lower Fox River and Green Bay, Wisconsin. Environ Toxicol Chem 9:313-322

Anonymous (1979) Suggested guidelines for methods of operation in surface mining of areas of potentially acid-producing materials. West Virginia Surface Mine Drainage Task Force. WV Dept Nat Resour, Charleston, WV

Anonymous (1986) Leitbild für die schweizerische Abfallwirtschaft (Guidelines for the Waste Management in Switzerland), Schriftenreihe Umweltschutz No 51. Eidgenössische Kommission für Abfallwirtschaft, Bundesamt für Umweltschutz, Bern/Switzerland

Anonymous (1993) ARCS Overview of risk assessment and modeling in the assessment and remediation of contaminated sediments. Prepared by ASCI for the US Environmental Protection Agency, Great Lakes National Program Office, Chicago, Illinois

Anonymous (1990) Technische Verordnung über Abfälle (TVA). Der Schweizerische Bundesrat (Swiss Federal Parliament), SR 814.015, Dec 10, 1990. Bern/Switzerland

Anonymous (1994) Remediation guidance document. USEPA Oceans and Coastal Protection Division and Great Lakes National Programme Office. Draft, Aug 20, 1993

Baccini P (ed)(1989) The landfill – reactor and final storage. Lecture Notes in Earth Sciences 20. Springer, Berlin Heidelberg New York, 439 pp

Belevi H, Stämpfli DM, Baccini P (1992) Chemical behaviour of municipal solid waste incinerator bottom ash in monofills. Waste Management Research 10:153-167

Benjamin MM, Hayes KL, Leckie JO (1982) Removal of toxic metals from power-generated waste streams by adsorption and co-precipitation. J Water Pollut Control Fed 54:1472-1481

Berner RA (1981) A new geochemical classification of sedimentary environments. J Sediment Petrol 51:359-365

Berner RA, Rainwell R (1984) C/S method for distinguishing freshwater from marine sedimentary rocks. Geology 12:365-368

Bernhard M, Brinckman FE, Sadler PJ (Editors)(1986) The importance of chemical "speciation" in environmental processes. Dahlem Konferenzen, Life Sciences Research Report 33. Springer, Berlin Heidelberg New York, 763 pp

Binkley D et al. (1989) Acidic deposition and forest soils. Springer, Berlin Heidelberg New York, 146 p

Bokuniewicz HJ (1982) Submarine borrow pits as containments for dredged sediments. In: Kester DR, Ketchum BH, Duedall IW, Parks PK (eds) Dredged material disposal in the ocean. John Wiley & Sons, New York, pp 215-227

Breemen N van (1975) Acidification and deacidification of coastal plain soils as a result of periodic flooding. Soil Sci Soc Amer Proc 39:1153-1157

Breemen, N van (1987) Effects of redox processes on soil acidity. Neth J Agric Sci 35:271- 279

Breemen N van (1988a) Effects of seasonal redox processes involving iron on the chemistry of periodically reduced soils. In: JW Stucki, BA Goodman, U. Schwertmann (eds) Iron in soils and clay minerals. D Reidel Publishing Company, Dordrecht, The Netherlands, pp 197-809.

Breemen N van (1988b) Long-term chemical, mineralogical, and morphological effects of iron-redox processes in periodically flooded soils. In: Stucki JW, Goodman BA, Schwertmann U (eds) Iron in soils and clay minerals. D Reidel Publ Co, Dordrecht, The Netherlands, pp 825-841

Breemen N van, Mulder J, Driscoll CJ (1983) Acidification and alkalinization of soils. Plant Soil 75:283-308

Breemen N van, Driscoll CT, Mulder J (1984) Acidic deposition and internal proton sources in acidification of soils and water. Nature 307:599-604

Brinkman R (1970) Ferrolysis, a hydromorphic soil forming process. Geoderma 3:199- 206

Brinkman R (1979) Ferrolysis, a soil-forming process in hydromorphic conditions. Agricultural Research Reports 887. PUDOC, Wageningen.

Bruynesteyn A, Hackl RP (1984) Evaluation of acid production potential of mining waste materials. Miner Environ 4:5-8

Calmano W (1988) Stabilization of dredged mud. In: Salomons W, Förstner U (eds) Environmental management of solid waste: dredged materials and mine tailings, pp 80-98. Springer, Berlin Heidelberg New York

Calmano W, Förstner U, Kersten M, Krause D (1986) Behaviour of dredged mud after stabilization with different additives. In: Assink JW, Van Den Brink WJ (eds) Contaminated soil. Martinus Nijhoff Publ, Dordrecht, pp 737-746

Calmano W, Ahlf W, Förstner U (1988) Study of metal sorption/desorption processes on competing sediment components with a multi-chamber device. Environ Geol Water Sci 11:77-84

Calmano W, Hong J, Förstner U (1992) Einfluß von pH-Wert und Redoxpotential auf die Bindung und Mobilisierung von Schwermetallen in kontaminierten Sedimenten. Vom Wasser 78:245-257

Calmano W, Hong J, Förstner U (1993) Binding and mobilization of heavy metals in contaminated sediments affected by pH and redox potential. In: Christensen E et al. (eds) Proc 1st Intern Spec Conf on Contaminated Aquatic Sediments: Historical Records, Environmental Impact, and Remediation. Water Sci Technol 28:223-235

Calmano W, Förstner U, Hong J (1994) Mobilization and scavenging of heavy metals following resuspension of anoxic sediments from the Elbe River. In: Alpers CN, Blowes DW (eds) Environmental geochemistry of sulfide oxidation. American Chemical Society, Washington DC, ACS Symp Ser 550:298-321

Carignan R, Rapin F, Tessier A (1985) Sediment pore water sampling for metal analysis: a comparison of techniques. Geochim. Cosmochim. Acta 49:2493-2497

Chapman PM (1986) Sediment quality criteria from the sediment quality triad: an example. Environ Toxicol Chem 5:957-964

Church TM (1986) Biogeochemical factors influencing the residence time of microconstituents in a large tidal estuary, Delaware Bay. Marine Chemistry 18:393-406

Craig PJ, Moreton PA (1984) The role of sulphide in the formation of dimethyl mercury in river and estuary sediments. Mar Pollut Bull 15:406-408

Darby DA, Adams DD, Nivens WT (1986) Early sediment changes and element mobilization in a man-made estuarine marsh. In: Sly PG (ed) Sediment and water interactions. Springer, Berlin Heidelberg New York, pp 343-351

Davison W (1988) Interactions of iron, carbon and sulphur in marine and lacustrine sediments. In: Fleet AJ, Kelts K, Talbot MR (eds) Lacustrine Petroleum Source Rocks. Spec Publ 40, Geological Society of America, Boulder, CO, pp 131-137

Davison W, Lishman JP, Hilton J (1985) Formation of pyrite in freshwater sediments. Implications for C/S ratios. Geochim Cosmochim Acta 49:1615-1620

DiToro DM et al. (1990) Toxicity of cadmium in sediments: the role of acid volatile sulfide. Environ Toxicol Chem 9:1487-1502

DiToro DM et al. (1992) Acid volatile sulfide predicts the acute toxicity of cadmium and nickel in sediments. Environ Sci Technol 26:96-101

Drever JI (1982) The geochemistry of natural waters. Prentice-Hall, Englewood Cliffs New York

Duinker JC (1980) Suspended matter in estuaries: adsorption and desorption processes. In: Olausson E, Cato I (eds) (1980) Chemistry and biogeochemistry of estuaries. John Wiley & Sons Chichester, New York, pp 121-151

Duinker JC (1986) Formation and transformation of element species in estuaries. In: Bernhard M, Brinckman FE, Sadler PJ (eds) The importance of chemical "speciation" in environmental problems. Dahlem-Konferenzen Life Sciences Research Report 33, pp 365-384. Springer, Berlin Heidelberg New York

Duinker JC, Nolting RF (1978) Mixing, removal and mobilization of trace metals in the Rhine estuary. Neth J Sea Res 12:205-223

Duinker JC, Nolting RF, Michel D (1982) Effects of salinity, pH, and redox conditions on the behaviour of Cd, Zn, Ni and Mn in the Scheldt estuary. Thalassia Jugoslavica 18:191-202

Eaqub M, Blume HP (1982) Genesis of a so-called ferrolysed soil of Bangladesh. Z Pflanzen- ernähr Bodenkde. 145:470-482

Edmond JM, Spivack A, Grant BC, Hu M-H, Chen Z, Sung C, Zeng X (1985) Chemical dynamics of the Changjiang Estuary. Cont Shelf Res 4:17-36

Ehrenfeld J, Bass J (1983) Handbook for evaluating remedial action technology plans. Municipal Environ Res Lab Cincinnati. EPA-600/2-83-076. Aug 1983

Elbaz-Poulichet F, Martin JM, Huang WW, Zhu JX (1987) Dissolved Cd behaviour in some selected French and Chinese estuaries, consequences on Cd supply to the ocean. Marine Chemistry 22:125-136

Emerson S, Jahnke R, Heggie D (1984) Sediment-water exchange in shallow water estuarine sediments. J Mar Res 42:709-730

Ferguson KD, Erickson PM (1988) Pre-mine prediction of acid mine drainage. In: Salomons W, Förstner U (eds) Environmental management of solid waste - dredged material and mine tailings. Springer, Berlin Heidelberg New York, pp 24-43

Förstner U (1981) Trace metals in fresh waters (with particular reference to mine effluents). In: Wolf KH (Ed) Handbook of strata-bound and stratiform ore deposits, vol 9. Elsevier, Amsterdam, pp 271-303

Förstner U (1985) Chemical forms and reactivities of metals in sediments. In: Leschber R, Davis RD, and L'Hermite P (eds) Chemical methods for assessing bio-available metals in sludges and soils. Elsevier Applied Science, London, pp 1-30

Förstner, U (1989) Contaminated sediments. Lecture Notes in Earth Sciences No 21. Springer, Berlin Heidelberg New York, 157 pp

Förstner U (1993) Metal speciation – an overview. Intern J Environ Anal Chem 51:5-27

Förstner, U, Kersten, M (1988) Assessment of metal mobility in dredged material and mine waste by pore water chemistry and solid speciation. In: Salomons W, Förstner U (eds) Chemistry and biology of solid waste – dredged material and mine tailings. Springer, Berlin Heidelberg New York, pp 214-237

Förstner U, Kersten M, Wienberg R (1989) Geochemical processes in landfills. In: Baccini P (ed) The landfill – reactor and final storage. Lecture Notes in Earth Sciences 20. Springer, Berlin Heidelberg New York, pp 39-81

Förstner U, Schoer J, Knauth H-D (1990) Metal pollution in the tidal Elbe River. Sci Total Environ 97/98:347-368

Förstner U et al. (1990) Sediment criteria development – contributions from environmental geochemistry to water quality management. In: Heling et al. D (eds) Sediments and environmental geochemistry. Springer, Berlin Heidelberg New York, pp 311-338

Froelich PN et al. (1979) Early oxidation of organic matter in pelagic sediments of the eastern equatorial Atlantic: suboxic diagenesis. Geochim Cosmochim Acta 43:1075-1090

Gambrell RP, Reddy CN, Khalid RA (1983) Characterization of trace and toxic materials in sediments of a lake being restored. J Water Pollut Control Fed 55:1271-1279

Gendron A, Silverberg N, Sundby B, Lebel J (1986) Early diagenesis of cadmium and cobalt in sediments of the Laurentian Trough. Geochim Cosmochim Acta 50:741-747

Giesy J.P., Hoke R.A. (1990) Freshwater sediment quality criteria: Toxicity bioassessment. In: Baudo R, Giesy J, Muntau H (eds) Sediments: chemistry and toxicity of in-place pollutants. Lewis Publ. Chelsea/Michigan, pp. 265-348

Goumans JJJM, Van der Sloot HA, Aalbers ThG (eds)(1991) Waste materials in construction. Studies in environmental science 48. Elsevier, Amsterdam, 672 pp

Hawkes HE, Webb JS (1962) Geochemistry in mineral exploration. Harper & Row, New York, 415 pp

Herms U, Brümmer G (1978) Löslichkeit von Schwermetallen in Siedlungsabfällen und Böden in Abhängigkeit von pH-Wert, Redoxbedingungen und Stoffbestand. Mitt Dt Bodenk Ges 27:23-43

Hines ME, Lyons BWM, Armstrong PB, Orem WH, Spencer MJ, Gaudette HE (1984) Seasonal metal remobilization in the sediments of Great Bay, New Hampshire. Mar Chem 15:173-187

Holmes CW, Slade EA, McLerran CJ (1974) Migration and redistribution of zinc and cadmium in marine estuarine systems. Environ Sci Technol 8:255-259

Hong J, Förstner U, Calmano W (1994) Effects of redox processes on the acid-producing potential and metal mobility in sediments. In: Hamelink J, Landrum PF, Bergman HL, Benson WH (eds) Bioavailability – physical, chemical, and biological interactions. Lewis Publ, Boca Raton, pp 119-141

Hunt CD, Smith DL (1983) Remobilization of metals from polluted marine sediments. Can J Fish Aquat Sci 40:132-142

Irmer U, Weiler K, Wolter K (1986) Untersuchungen zur Schwermetall- und Schwebstoffdynamik im Elbeästuar – Mögliche Folgerungen für die Überwachungspraxis. Vom Wasser 67:111-123

Irmer U, Knauth H-D, Weiler K (1988) Einfluß des Schwebstoffregimes auf die Schwermetall- verschmutzung der gezeitenbeeinflußten Elbe bei Hamburg. Z Wassere Abwasser Forsch 21:236-240

Kersten M (1988) Geochemistry of priority pollutants in anoxic sludges: cadmium, arsenic, methyl mercury, and chlorinated organics. In: Salomons W, Förstner U (eds) Chemistry and biology of solid waste – dredged material and mine tailings. Springer Berlin Heidelberg New York, pp 170-213

Kersten M (1989) Mechanismus und Bilanz der Schwermetallfreisetzung aus einem Süßwasser- watt der Elbe. Dissertation Technische Universität Hamburg-Harburg.

Kersten M, Förstner U (1986) Chemical fractionation of heavy metals in anoxic estuarine and coastal sediments. Water Sci Technol 18:121-130

Kersten M, Förstner U (1987) Effect of sample pretreatment on the reliability of solid speciation data of heavy metals – implications for the study of early diagenetic processes. Mar Chem 22:299-312

Kersten M, Förstner U (1991) Geochemical characterization of the potential trace metal mobility in cohesive sediment. Geo-Marine Letts 11:184-187

Kersten M et al. (1985) Freisetzung von Metallen bei der Oxidation von Schlämmen. Vom Wasser 65:21-35.

Kester DR, Ketchum BH, Duedall IW, Park PK (eds)(1983) Wastes in the ocean. Vol 2: Dredged-material disposal in the ocean. Wiley, New York, 299 pp

Kittrick JA, Fanning DS, Hossner LR (eds)(1982) Acid sulfate weathering, SSSA special publ 10, Soil Sci Soc Amer Madison, WI, pp 234

Knauth H-D, Schroeder F, Irmer U (1986) Investigations on the heavy metal variability in the River Elbe estuary upstream and downstream of Hamburg Harbour. In: Ernst WHO (ed) Proc 2nd Intern Conf Environ Contam, Amsterdam. CEP Consultants Edinburgh, pp 186-188

Knauth H-D,Schwedhelm E, Sturm R, Weiler K, Salomons W (1989) The importance of physical processes on contaminant behaviour in estuaries. GKSS Report, Research Centre Geesthacht/Germany

Krebs J, Belevi H, Baccini P (1988) Long-term behavior of bottom ash landfills. Proc 5th Intern Solid Wastes Exhibition and Conf, ISWA 1988, Copenhagen

Lee GF, Plumb RH (1974) Literature review on research study for the development of dredged material disposal criteria. US Army Corps of Engineers, Dredged Material Research Program, report D-74-1. Vicksburg MS 1974, 145 pp

Lichtensteiger T, Brunner PH, Langmeier M (1988) Klärschlamm in Deponien. EAWAG Project No 30-681. EC-COST

Luoma SN, Davis JA (1983) Requirements for modeling trace metal partioning in oxidized estuarine sediments. Mar Chem 12:159-181

Maaß B, Miehlich G (1988) Die Wirkung des Redoxpotentials auf die Zusammensetzung der Porenlösung in Hafenschlickspülfeldern. Mitt Dtsch Bodenkunde Ges 56:289-294

Maaß B, Miehlich G, Gröngröft A (1985) Untersuchungen zur Grundwassergefährdung durch Hafenschlick-Spülfelder. II. Inhaltsstoffe in Spülfeldsedimenten und Porenwässern. Mitt Dtsch Bodenkundl Ges 43/I:253-258

Malone PG, Jones LW, Larson RJ (1982) Guide to the disposal of chemically stabilized and solidified waste. Report SW-872, Office of Water and Waste Management. Washington DC: US Environmental Protection Agency

Meyer JS, Davidson W, Sundby B, Oris JT, Laurén DJ, Förstner U, Hong J, Crosby DG (1994) The effects of variable redox potentials, pH, and light on bioavailability in dynamic water-sediment environments. In: Hamelink J, Landrum PF, Bergman HL, Benson WH (eds) Bioavailability – physical, chemical, and biological interactions. Lewis Publ, Boca Raton, pp 155-170

Moore JN, Luoma SN (1990) Hazardous wastes from large-scale metal extraction. Environ Sci Technol 24:1278-1285

Morse JW, Mackenzie FT (1990) Geochemistry of sedimentary carbonates. Elsevier Publ Co, New York

Morton RW (1980) "Capping" procedures as an alternative technique to isolate contaminated dredged material in the marine environment. In: Dredge spoil disposal and PCB Contamination: Hearings before the Committee on Merchant Marine and fisheries. House of Representatives, Ninety-Sixth Congress, 2nd Session, on exploring the various aspects of dumping of dredged spoil material in the ocean and the PCB contamination issue, Mar 14, May 21, 1980. USGPO Ser No 96-43, Washington DC, pp 623-652

Müller G (1979) Schwermetalle in den Sedimenten des Rheins – Veränderungen seit 1971. Umschau Wiss Tech 79:778-783

Nipkow F (1920) Vorläufige Mitteilungen über Untersuchungen des Schlammabsatzes im Zürichsee. Z Hydrol 1:101-123

Obermann P, Cremer S (1992) Mobilisierung von Schwermetallen in Porenwässern von belasteten Böden und Deponien: Entwicklung eines aussagekräftigen Elutionsverfahrens. Landesamt für Wasser und Abfall Nordrhein-Westfalen, Düsseldorf/-Germany, vol 6, 127 pp

Patrick WH, Williams BG, Moraghan JT (1973) A simple system for controlling redox potential and pH in soil suspensions. Soil Sci Soc Amer Proc 37:331-332

Peiffer S (1989) Biogeochemische Regulation der Spurenmetallöslichkeit während der anaeroben Zersetzung fester kommuler Abfälle. Dissertation Universität Bayreuth, 197 pp

Plant JA, Raiswell R (1983) Principles of environmental geochemistry. In: Thornton I (ed) Applied environmental geochemistry. Academic Press, London, pp 1-39

Prause B, Rehm E, Schulz-Baldes M (1985) The mobilisation of Pb and Cd from contaminated dredge spoil after dumping in the marine environment. Environ Technol Letts 6:261-266

Prenzel L (1985) Verlauf und Ursache der Bodenversauerung. Z Dt Geol Ges 136:293-302

Reynoldson RB (1987) Interactions between sediment contaminants and benthic organisms. In: Thomas RL et al. (eds) Ecological effects of in situ sediment contaminants. Hydrobiologia 149:53-66

Sahm H, Brunner M, Schobert SM (1986) Anaerobic degradation of halogenated aromatic compounds. Microbial Ecol 12:147-53

Salomons W (1980) Adsorption processes and hydrodynamic conditions in estuaries. Environ. Technol Letts 1:356-365

Salomons W (1985) Sediment and water quality. Environ Technol Letts 6:315-326

Salomons W (1993) Non-linear responses of toxic chemicals in the environment: a challenge for sustainable development. In: ter Meulen GRB et al. (eds) Chemical time bombs, Proceedings of the European State-of-the-art Conference on Delayed Effects of Chemicals in Soils and Sediments, Veldhoven/The Netherlands, 2-5 Sept 1992, pp 31-43

Salomons W, Eysink W (1981) Pathways of mud and particulate trace metal from rivers to the southern North Sea. In: Nio SD, Schuettenhelm RTE, Weering TCE (eds) Holocene marine sedimentation in the North Sea Basin. Spec Publ Intern Assoc Sedimentologists 5:429-450.

Salomons W, Förstner U (1984) Metals in the hydrocycle. Springer, Berlin Heidelberg New York

Salomons W, Förstner U (eds)(1988a) Chemistry and biology of solid waste: dredged materials and mine tailings. Springer, Berlin Heidelberg New York,, 305 pp

Salomons W, Förstner U (eds)(1988b) Environmental management of solid waste: dredged materials and mine tailings. Springer, Berlin Heidelberg New York, 396 pp

Salomons W, De Rooij NM, Kerdijk H, Bril J (1987) Sediments as a source of contaminants. In: Thomas RL et al. (eds) Ecological effects of in situ sediment contaminants. Hydrobiologia 149:13-30

Santschi PH (1982) Application of enclosures to the study of oceanic chemistry. In: Grice GD, Reeve (eds) Marine mesocosms. Biological and chemical research in experimental ecosystems. Springer, Berlin Heidelberg New York, pp 63-80

Schoer J, Förstner U. (1987) Abschätzung der Langzeitbelastung von Grundwasser durch die Ablagerung metallhaltiger Feststoffe. Vom Wasser 69:23-32

Sholkovitz ER (1976) Flocculation of dissolved organic and inorganic matter during the mixing of riverwater and seawater. Geochim Cosmochim Acta 40:831-845

Sholkovitz ER (1978) The flocculation of dissolved Fe, Mn, Al, Cu, Ni, Co and Cd during estuarine mixing. Earth Planet Sci Lett 41:77-86

Sholkovitz ER, Copland D (1981) The coagulation, solubility and adsorption properties of Fe, Mn, Cu, Ni, Cd, Co and humic acids in a river water. Geochim Cosmochim Acta 45:181-189

Sobek AA, Schuller WA, Freeman JR, Smith RM (1978) Field and laboratory methods applicable to overburden and mine spoils. Report EPA-600/2-78-054. US Environmental Protection Agency, Washington DC

Stief K (1987) Zukünftige Anforderungen an die Deponietechnik und Konsequenzen für die Sickerwasserbehandlung. In: Deponiesickerwasserbehandlung. UBA Materialien 1/87. Erich Schmidt Verlag, Berlin, pp 27-36

Stigliani WM (1991) Chemical time bombs: definition, concepts, and examples. Executive report 16 (CTB basic document). IIASA Laxemburg, Austria, 23 pp

Stigliani WM (1992) Chemical time bombs, predicting the unpredictable. In: Chemical time bombs. European State-of-the-Art Conference on Delayed Effects of Chemicals in Soils and Sediments. Veldhoven, The Netherlands, Sept 2-5, pp 12

Stumm W, Morgan JJ (1981) Aquatic chemistry. John Wiley & Sons, New York

Swift RS (1977) Soil organic matter studies. IAEA Vienna. pp 275-281

Tabatabei MA (1987) Physico-chemical fate of sulfate in soils. J Amer Phys Chem Assoc 37:34-38

Tent L (1987) Contaminated sediments in the Elbe estuary: Ecological and economic problems for the Port of Hamburg. In: Thomas RL et al. (eds) Ecological effects of in situ sediment contaminants. Hydrobiologia 149:189-199

Tessier A, Campbell PGC, Bisson M (1979) Sequential extraction procedure for the speciation of particulate trace metals. Anal Chem 51:844-851

Ure AM, Quevauviller Ph, Muntau H, Griepink B (1993) Speciation of heavy metals in soils and sediments, an account of the improvement and harmonization of extraction techniques undertaken under the auspices of the BCR of the Commission of the European Communities. Intern J Environ Anal Chem 51:135-151

Van Gemert WJT, Quakernaat J, Van Veen HJ (1988) Methods for the treatment of contaminated dredged sediments. In: Salomons W, Förstner U (eds) Environmental management of solid waste – dredged material and mine tailings. Springer, Berlin Heidelberg New York, pp 44-64

Wallmann K, Kersten M, Gruber J Förstner U (1993) Artifacts in the determination of trace metal binding forms in anoxic sediments by sequential extraction. Intern J Environ Anal Chem 51:187-200

Wangersky PJ (1986) Biological control of trace metal residence time and speciation: a review and synthesis. Marine Chemistry 18:269-297

Wiedemann HU (1982) Verfahren zur Verfestigung von Sonderabfällen und Stabilisierung von verunreinigten Böden. Ber Umweltbundesamt 1/82. Erich Schmidt Verlag, Berlin

Wiles CC, Barth E, de Percin P (1988) Status of solidification/stabilization in the United States and factors affecting its use. In: Wolf K, Van den Brink WJ, Colon FJ (eds) Contaminated Soil '88. vol 1, Kluwer Academic Publ, Dordrecht, The Netherlands pp 947-956

Williamson KJ, Bella DA (1980) Estuarine sediments: successional model. J Environ Engin Div ASCE 106:695-710

Windom H, Wallace G, Smith R, Dudek N, Maeda M, Dulmage R, Storti F (1983) Behaviour of copper in southeastern United States estuaries. Mar Chem 12:183-193

Wollast R (1988) The Scheld Estuary. In: Salomons W, Bayne BL, Duursma EK, Förstner U (eds) Pollution of the North Sea – an assessment. Springer, Berlin Heidelberg New York, pp 184-193

Züllig H (1956) Sedimente als Ausdruck des Zustandes eines Gewässers. Schweiz Z Hydrol 18:7-143

Zwolsman JJG, Van Eck GTM (1993) Behaviour of trace metals in the Scheldt Estuary, S.W. Netherlands. Neth J Aquatic Ecol 27:287-300

# 12 How to Manage Accumulated Contaminants

*H. Eijsackers*

## 12.1 Introduction

In the preceding chapters various aspects of biogeodynamics of pollutants in soils and sediments have been described. Due to continuing immissions from anthropogenic activities different contaminants have accumulated in soils and sediments and become bound to the soil. Changes in environmental conditions may change the binding capacities of soils, resulting in a non-linear, possibly massive, release of these contaminants again leading to polluting loads.

The main question is under which circumstances and to what extent such release may occur, in order to be preventive. Therefore a practical issue will be whether it is possible to manage these contaminants so as to avoid a recontamination by unloading. The various management options in this perspective are, firstly, to keep the contaminants immobilized, and secondly, to allow the contaminants to be released in such a way (flux or rate) that the natural system can cope with them. As a third option we can leave the system as it is and let it 'adapt' to the contaminants, which implicitly means that we accept further impacts. As this third option is not within the scope of this book on biogeodynamics, Chapter 12 will concentrate on the first two approaches mentioned.

The mechanisms of immobilization and release are governed by the intrinsic characteristics of the various contaminants (Sect. 12.2) and environmental conditions in connection with the major geochemical cycles (Sect. 12.3). The most important ways to change environmental conditions are:

- Manipulating them
- Supplementing the system with various binding materials
- Changing the land use

Manipulation of the environmental conditions is described in Section 12.3.1, supplementing the system in Section 12.4, and the different ways to change land use, i.e. to manage contaminated soils and sediments in Section 12.5.

### 12.1.1 Type of Contaminants

The importance of the various management options changes with the characteristics of the organic and inorganic contaminants. Organic contaminants can be split into three main groups: the easily degradable organic compounds, the more complex and resistant chlorinated organo compounds and the very resistant (hardly/undegradable) complex compounds like the drins, dioxins and furans. Inorganic compounds comprise mainly the heavy metals and metalloids, although there is some discussion in the Netherlands now on whether exotic heavy elements as well as accumulating nutrients (e.g. phosphate) should be included too. Nieboer and Richardson (1980) divide metals into two major groups according to their physico-chemical behaviour. Metals forming oxygen bonds are biologically important and can act as macronutrients. Metals forming nitrogen and sulphur bonds are toxic at a biochemical and enzymatic level.

## 12.2 Contamination Pattern

The different contamination situations can be characterized by their scale, and horizontal and vertical dispersal/distribution pattern (Table 12.1).

In a well-defined contaminated area, like a very localized soil contamination, management is relatively easy. Soil can be dug out, removed and treated ex situ in special thermal, extractive or biological treatment plants. If the contaminant has penetrated to deeper soil layers, it becomes more complicated. Floating layers on the groundwater table (like petrol) can be easily pumped off. Homogeneously

**Table 12.1.** Classification of contamination patterns according to scale, horizontal and vertical dispersal

| Scale | | Horizontal border line | | Vertical distribution | |
|---|---|---|---|---|---|
| Small | Large | Confined | Gradient | Upper soil | Deeper soil |
| X | | | X | X | |
| X | | X | | X | |
| X | | | X | | X |
| X | | X | | | X |
| | X | X | | X | |
| | X | | X | X | X |

distributed contamination in a confined area can be treated in situ with venting, extraction, biotreatment or a combination of these techniques.

Management becomes more difficult with a heterogeneous distribution pattern, which unfortunately is rather common. Alternating layers of sandy and clayey material or clay lenses hamper transport and lead to fingered flow patterns through the soil. These flow patterns are complicated to model and it is difficult to dimension pump rates and fluxes precisely. As a consequence, treatment approaches have to be over-dimensioned. Typical but rather extreme examples of heterogeneous distribution are TRI and PER solvents used in dry-cleaning. These have been found in concentrated amounts in very confined soil spots at 50–100 m below ground level. These spots are very hard to locate and to treat with in situ techniques. Far greater volumes of soil have to be treated and large amounts of water (flushing) or air (venting) have to be circulated.

Larger confined areas with controlled borders like waste dumps, car dumps etc. can be reasonably well kept under control. The rather complicated composition of the contaminants, and the dynamic nature of the degradation, leaching and volatilization processes within waste dumps (see Boels and Fleming 1993) complicate proper management. Continuous monitoring, also in the long term (several decades), is therefore advised to indicate the first slight emissions and so to prevent unforeseen adverse emissions from a dump.

Even more complicated is the control of diffusely contaminated areas like the surroundings of major point sources emitting to the atmosphere, e.g. heavy metal smelters which emit contaminants over many square kilometres (e.g. Avonmouth in the UK and Sudbury in Canada), or huge mine tailings which influence the water catchment areas downhill (Salomons, Chap. 1). These types of sources are concentrated in mining areas like the Dirty Triangle (the German, Polish, Czech border areas) leading to mass contamination over very large areas (ter Meulen-Smidt, Chap. 7). Also areas with high industrial activity and resulting long-term accumulation are diffusely contaminated; these also include harbours and harbour sludges (Förstner, Chap. 11). Here the problem should be approached regionally and combine the containment of a whole area and the monitoring at the borders with treatment of hotspots in the area itself. Deposited dredged harbour sludges can be contained too.

But also areas which have been heavily manured or treated with persistent contaminants like heavy metals and POPs (Persistent Organic Pollutants e.g. DDT and dieldrin) will become so heavily contaminated in the long term that special management will be necessary. In the Netherlands already over 20 000 ha have become phosphate saturated and this area is still increasing. For this type of diffusely contaminated large-scale area, ex situ or in situ remediation techniques do not help, rather special management schemes for land and groundwater have to be developed. These should primarily improve, reinforce and control natural degradation and leaching processes, physico-chemical as well as biological. This kind of 'eco' approach, in line with natural biogeochemical processes to which organisms are adjusted, demands for very long-term treatment (many decades).

## 12.3 Biogeochemical Processes as Management Tools

When trying to manipulate biogeochemical processes we have to realize that we are interfering not only with different ecosystems but also with major chemical element cycles and related physico-chemical processes. As Salomons and Förstner (in earlier chapters) have sketched, these processes and the consequences for the extensive (un)loading of contaminants are mentioned only briefly here.

To sum up: the (im)mobilization of chemical elements and organic compounds is governed by the elementary environmental conditions, pH, redox potential (Eh), organic matter content (OM) and cation exchange capacity (CEC). By changing these conditions the rates and patterns of a number of major biogeochemcial processes are also changed, and as a consequence also the availability of a great number of other organic and inorganic compounds.

Therefore, Section 12.3 will concentrate on environmental conditions and the physico-chemical and biological processes which influence the fate of contaminants. The biogeochemical processes also have different temporal and spatial scales, which makes it very complicated to fully assess the outcomes of a particular management approach. Since a good basic understanding of these processes and their temporal and spatial scales is an essential prerequisite to adequately managing contaminated soils and sediments, these processes will be discussed as well.

### 12.3.1 Manipulating Environmental Conditions

Changing the physical conditions can be achieved by manipulating temperature or redox conditions. Soil aeration or saturation will lead to aerobic and anaerobic conditions, respectively. In areas with high water-levels – lowlands, river banks and estuarine areas – this can be manipulated well. This can also be regulated in peaty moor soils with high water-tables. Moreover, in catchment areas, specific conditions could be created in shallow parts of the catchment at different intervals downstream, for instance, for treatment of mining effluents. For storage areas for contaminants or contaminated material (e.g. harbour sludge) this kind of manipulation could be used as well. In the course of settling of sludge material the top layers will change from anaerobic to aerobic conditions. Maintenance of anaerobic conditions is important to prevent mobilization and consecutive release of toxic elements present in the contaminated waste. An exception must be made for a few elements as these become more mobile under anaerobic conditions. Even in hotspots around specific local pollution point sources, creation of saturated soil conditions could be considered.

Manipulating soil temperature is at present only used in in situ remediation techniques like soil venting or extensive ex situ treatment like land farming.

Because of the specific energy content of soil material considerable energy input is necessary for an effective temperature increase. Only by preventing plant cover, heating by solar irradiation of the upper 0–10 cm layer of a bare soil can be promoted. This not only accelerates biological but also chemical processes and leads to increased mobility of the contaminants. Recent research has made clear that also at lower temperatures in deeper soil and groundwater layers biodegradation processes go on, be it at lower, but still considerable, rates. With land farming, the alternate or consecutive manipulation of temperature, redox and pH can speed up biological degradation processes. Biodegradation of organic compounds is a combination of subprocesses, each with their own set of optimal conditions, partly aerobic and partly anaerobic. Optimum conditions should not only be created for the breakdown of the primary compound, but also for further degradation of resulting breakdown products (metabolites), which may be more mobile or toxic than the primary compound. Otherwise accumulation of some of these breakdown products may occur, leading to unwanted secondary contamination.

These environmental conditions also influence, however, the behaviour and metabolism of organisms. Lowering of pH increases mobility of heavy metals (e.g. Pb) but at the same time decreases mobility and metabolism of soil organisms (e.g. earthworms); the final result in the case of Pb is neutral (Bengtsson and Rundgren 1992). Another example is the impact of environmental conditions on the degradative capacities of various micro-organisms; chemical degradation rates may be increased with increasing pH but microbial activity decreased above a certain optimum pH (see Doelman, Chap. 2). Organic matter in general will bind chemical elements. Finely degraded organic particles – dissolved organic carbon (DOC) – act as carriers of adsorbed contaminants and, hence, enhance leaching of the contaminants into deeper soil layers and groundwater aquifers. Redox conditions (Eh) can, depending on the chemical element, work both ways and either strengthen or weaken the binding to soil.

### 12.3.2 Impacts on Biodegradative Potentials

The environmental conditions mentioned above play a major role in the various biodegradation processes. Micro-organisms react very sensitively to changes in for example temperature or pH in a sometimes almost digital way; the difference between full degradative activity and zero being within a tenth of a pH-unit (see Doelman, Chap. 2).

With respect to biodegradation potential, should be distinguished:

- The easily degradable organic compounds
- The more complex and resistant chlorinated organo compounds
- The very resistant (hardly/undegradable) complex compounds like the drins, dioxins and furans

The first group is more or less similar to compounds of normal organic matter.Biodegradation will not pose a problem; many different micro-organisms are capable of breaking down these types of compounds and degradation will take weeks to months, sometimes even years. A compound which has not been degraded after several years is usually unavailable; i.e. the compound is dispersed in a heterogeneous or even concentrated pattern, or the micro-organisms cannot reach the compound or the environmental conditions are unsuitable.

For the second group, the environmental conditions must be suitable and proper micro-organisms must be available. Sometimes anaerobic conditions have to be followed by aerobic or vice versa because the degradation of different metabolites require different conditions (see Fig. 2.7 in Doelman, Chap. 2). Recent research by Harkness et al. (1993) has shown that PCBs are aerobically degradable in situ: 37–55% of the lightly chlorinated PCBs were totally degraded. Rhee et al. (1989) have shown that also anaerobic degradation in clay encapsulations is possible, with 53% degradation of mainly mono- to pentachlorobiphenyls.

The third group really pose a problem: very intensive treatment technologies combining pretreatment with gamma or ultrasonic radiation with biodegradation by very specialized micro-organisms under finely tuned conditions in bioreactors seems to be the best approach so far.

### 12.3.3 Impacts on Geochemical Binding Capacities

The chemical stability of minerals (the conditions by which specific metal salts can be formed) can be conveniently plotted on a pH-Eh diagram. For a more extensive overview on binding conditions and related extraction procedures, refer to Pickering (1981). Different chemical 'species' react differently with respect to changing pH and Eh conditions: increased mobility is shown in Fig. 12.1. Similar plots can be made for different types of environment and environmental impacts: acidification, desiccation, and inundation. In this way different management practices can be combined with their consequences for release of elements and binding. However, these management practices have different temporal scales. Changes in a slow process will work out only in the long term, but so will countermeasures! The buffering system of most soils consists of a carbonate, an exchange, an Al and a silicate buffer (Fig. 12.2). Each of these systems has its own effectiveness (the amount of protons $H^+$ causing a certain change in pH) and timescale. The carbonate buffer is very effective and works in the short term; the Al (or the silicate) buffer works in the long term. Consequently, countermeasures like liming will be effective in the short term but do not last very long and have to be repeated frequently, whereas impacts on the Al system have a longer timescale.

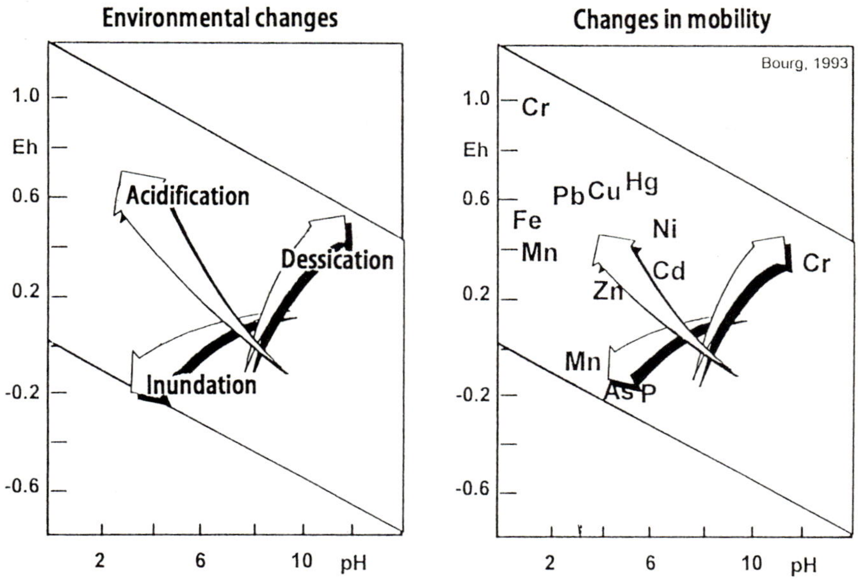


**Fig. 12.1.** pH-redox diagram for environmental changes and the resulting change in mobility of metals

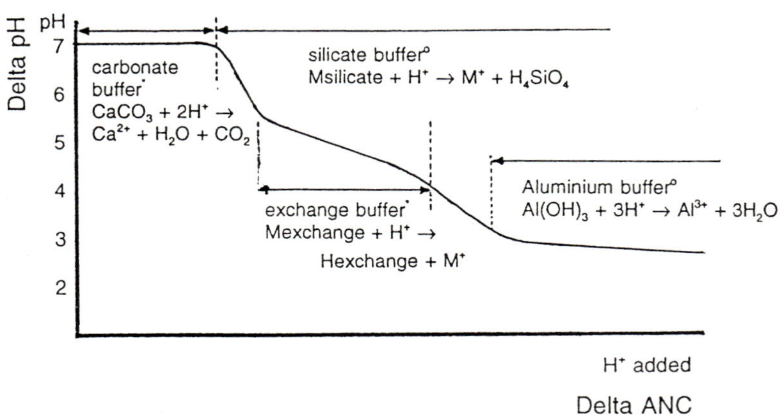


**Fig. 12.2.** Relation between the amount of $H^+$ added and the rsulting pH-decrease for the soil buffer system

The chemical occurrence of compounds (their chemical speciation) is of importance to the ease in which their (im)mobility is changed (Fig. 12.3). Different binding forms are more or less sensitive to various changes in conditions: e.g. molybdene is easily changed through the organic matter system, whereas chromium is hard to control as it is mostly present in a residual form.

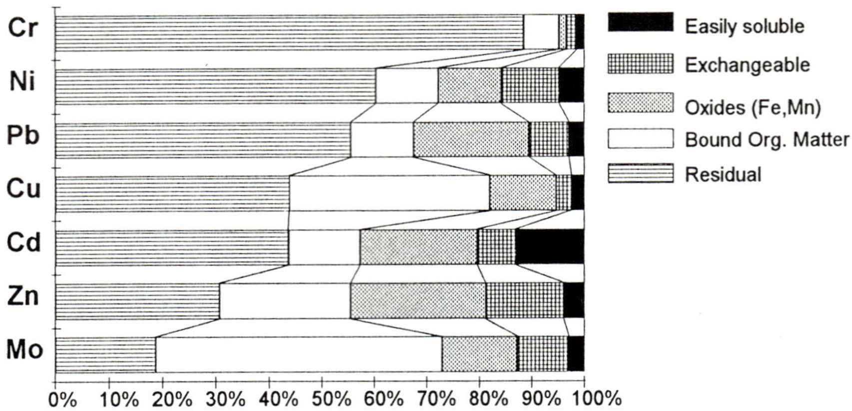


**Fig. 12.3.** Binding forms for heavy metals in soils

Following are a number of options for proper management of contaminated areas to accommodate these various binding forms, impacts of environmental conditions and time-scales:

- One can supplement the systems with certain compounds, so as to raise the binding capacity, adding carbonate (liming) to counteract acidification, ferro compounds to bind phosphate, or organic matter to increase organic binding.
- One can change the environmental conditions; however, this is in fact not a separate option as it either adds materials to change conditions or changes in land-use patterns. Examples are changing pH by liming and CEC by clayey material. However, also redox conditions must be considered. In soil redox can be changed by influencing the aeration of the soil through drainage or inundation, or adding large amounts of organic material, the decomposition of which will use all locally available oxygen. In this case we are in fact applying the third option.
- In land-use change or active soil management we combine both activities, for instance, in changing from agricultural to forestry use, or from roughlands in river banks and marine estuaries to agricultural lands.

These different options will be further described in the following sections.

## 12.4 Supplementing the System to Increase the Immobilization Capacity

Supplementing the system is possible in various ways and for different purposes:

- Materials can be added with a calcareous character or a high pH to influence the acidity: lime, gypsum, cement, wood ash, coal fly ash (the last two types also have many adsorptive sites).
- Specific binding conditions can be influenced by adding concentrated $Fe^{2+}$, manganese or aluminium compounds: concentrated $Al^{3+}$ or $Al^{3+}/Si(OH)_4$.
- The CEC or, in general, the adsorptive capacity can be raised by mixing clayey soils or specific clays or treated clay minerals (montmorillonite, illite), alkali silicate, or zeolites through the soil.
- To maintain or raise the organic matter content various materials, varying from specific humic or fluvic acids, humus or peat materials to organic materials through manuring or mulching can be added.

Depending on the system, its scale and whether it is built up of new (depots) or has already existed for a longer period, it is possible to add these materials at the influx (depots) or mix them with the upper material layers. Larger scale and more diffusely contaminated areas can be treated in this way when soil management is common practice. Examples are liming and deep-ploughing (to mix sandy soils with deeper clayey or peaty deposits) in agricultural areas. Even complete turnover of the soil profile is possible by bringing up deeper soil layers as done in Dutch sandy bulb growing areas. However, it is not feasible to apply these methods on very large (national) scales or to prescribe them at international (EU) levels.

### 12.4.1 Liming

Supplementing with lime to improve productivity has been a common practice in agronomy for many decades now. It has as well been used to resist acidification in forest and lakes this past decade.

However, there is an essential difference between agricultural soils, forestry and natural soils, and sediments; this is related to their homogeneity and the mixing of supplements like lime through the upper layer. Agricultural soil is both horizontally and vertically homogeneous (the cultivated zone) as a result of regular tilling. Liming is mostly maintained at a yearly cycle. Forest soils can be homogeneous but most will be heterogeneous, horizontally as well as vertically. The various soil horizon layers have different and specific binding capacities. Sediments of lakes and other water bodies are relatively homogeneous, partly through continuous mixing of the upper layer, which results in a gradual shift from sediment-rich water down to saturated sediment. Sediments as such can show microstratigraphy (Doelman, Chap. 2) with an aerobic layer of a few millimetres followed by a saturated deeper zone which is anaerobic. Moreover, when the water has decreased (by circulation or treatment) in contamination, there can be serious post-delivery of contaminants.

As a consequence of these differences in profile the effectiveness of liming (or binding materials in general) in the long term will differ among agriculture, forestry and natural soils, and sediments.

In agriculture the finely grained lime particles added are intended to dissolve quickly. The particles are homogeneously distributed and further dispersed into the soil profile by cultivating. In forestry, liming is intended to be a long-term treatment. Hesterberg (1993) points out an essential difference between agricultural and 'environmental' liming. By providing coarser particles the liming effect will last longer as the particles gradually break apart and release lime. Moreover, coarser particles are cheaper as production costs are for a great part related to grinding. Besides coarse particles, other slow-release compounds can be used. In a long-term liming programme in Sweden, not only lime and dolomite but also wood ash, peat ashes and slag products have been tested (Andersson and Persson 1988).

In natural, non-cultivated soils one of the problems is to achieve a homogeneous dispersal over an area. Solution of lime particles and further leaching by infiltrating rain and groundwater is the main vertical transport mechanism. Due to the heterogeneous composition of the abiotic soil material there are facilitated transport pathways in the soil, resulting in 'fingered flow'. Biological structures like earthworm burrows and root channels too will facilitate and confine transport to specific parts. As a consequence some parts of the lower soil horizons will show elevated calcium levels in the longer term, whereas other parts are still almost completely devoid of Ca. To improve the mixing of lime in natural soils, Judas and Schauermann (1994) suggest earthworms as facilitators and burrowers. Rundgren (pers. comm.) could not, however, observe any effect of indigenous earthworms on the effects of liming of a few small watersheds, even after 12 years; inoculation of earthworms was suggested. This is possible, as shown by Judas and Schauermann, but the actual change inferred after a few years was limited to a few centimetres alongside the burrows of the worms.

Liming has been applied as a long-term treatment of the sediment in a great number of American, Canadian and Scandinavian lakes. Technical procedures for a proper application are now available.

After abandoning arable land and cessation of liming the pH will drop. Data from Johnston et al. (1986) on a long-term field experiment from 1880 onwards show that the rate with which cessation of liming becomes prominent depends on the system (Fig. 12.4). In a grassland system a single application led to a pH maintenance for over 60 years. In combination with $NO_3$-N fertilizer, pH first increased half a unit, whereas with $NH_4$-N the pH dropped continuously. Knoop and Traas (1992) simulated a change in abandoned arable soils with accumulated cadmium due to fertilization and manuring including cessation of liming into roughland. Thirty years after cessation of liming the amounts of mobile Cd entering the terrestrial food chain will surpass toxicity limits for moles and other small carnivorous mammals.

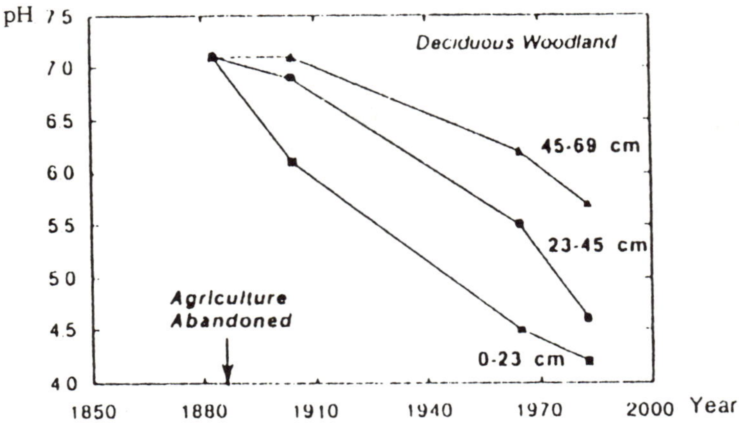


**Fig. 12.4.** Long-term pH changes in various depth intervals of an abandoned agricultural soil that evolved into woodland. Data taken from Johnston, et al. (1986)

### 12.4.2 Maintaining and Raising the Organic Matter Content

Organic matter (OM) is the primary binding component for a number of chemical elements (see Fig. 1.6 in Salomons, Chap. 1). Peat has been used to take up adverse volatile organic compounds and could similarly be used as an end filter of water from groundwater or sediment/soil sludge leachate. OM is the primary energy source for all natural plant growth and an important energy source for crop growth. As a consequence of cultivation organic matter is metabolized and OM content declines. Cultivation is combined with drainage to lower groundwater levels and increased aeration. Especially in the organic rich prairie and forest soils in the United States, but also in similar soils all over the world, OM is declining to an alarming extent (Table 12.2). In the Mediterranean areas this occurs massively too, leading to serious slope erosion (see Imeson, Chap. 5).

Breakdown of OM also results in the formation of dissolved organic carbon (DOC) which acts as a carrier of bound contaminants. DOC formation may enhance downward leaching through the soil profile but also in lateral transport over the surface. Lateral transport is enhanced by soil erosion, another consequence of OM breakdown. Displacement of soil by erosion may result in accumulation of the debris in downhill shallow areas with patchy aerobic/anaerobic conditions, ideal for mobilization of contaminants.

The long-term drawback of OM decline has been recognized by farmers and agronomists. They have tried to counteract it by addition of organic materials like pig and cattle manure, applying different straw and litter materials and using mulch fertilizers (Figs. 12.5 and 12.6). Only massive applications of manure appear to be able to maintain organic matter content in the long term. On cropped

**Table 12.2.** Organic matter content decline in soils

| Organic matter content |
|---|
| **Humus production grassland** |
| Yearly NPP 2.8 ton $ha^{-1}$ $a^{-1}$ |
| 1% undecomposed contribution to $_{org}C$-content 0.028 ton $ha^{-1}$ $a^{-1}$ |
| **Humus loss** |
| Prairie soils USA 40 years cultivation |
| sandy: 8–16 ton $_{org}C$ $ha^{-1}$(42-54%) = 0.2–0.4 ton $_{org}C$ $ha^{-1}$ $a^{-1}$ |
| fine: 10–26 ton $_{org}C$ $ha^{-1}$(36-54%) = 0.225–0.65 ton $_{org}C$ $ha^{-1}$ $a^{-1}$ |
| **Loss percentages other soils** |
| USA grassland 25 years: 56% $_{org}C$ |
| USA virgin forest 3 years: 33% $_{org}C$ |
| Canada grassland 60 years: 50-60% $_{org}C$ |
| Canada grassland 60-80 years: 0-15 cm 57% and 40–80 cm 20% $_{org}C$ |
| European peaty soils100 years: several metres |

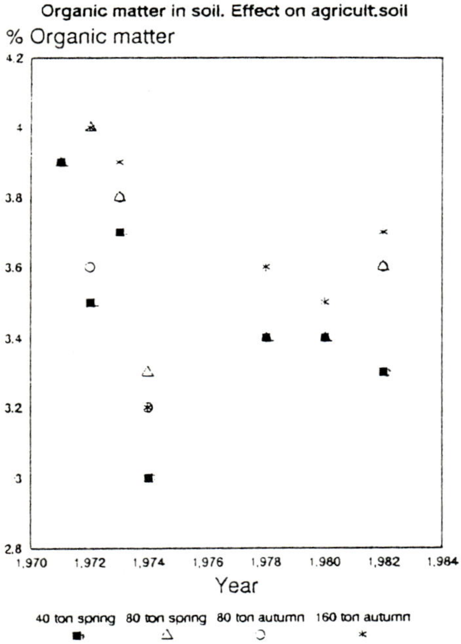


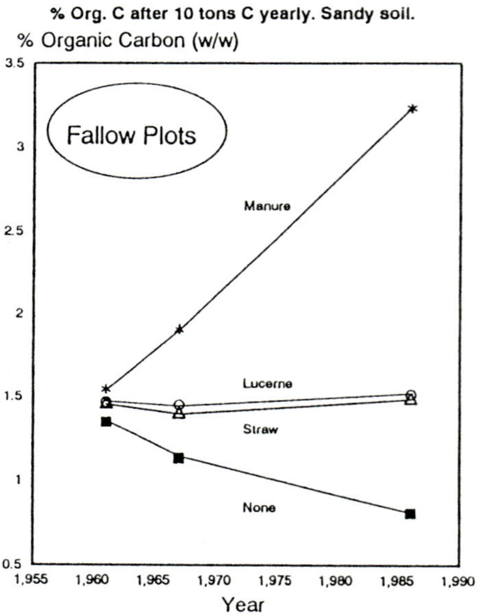


**Fig. 12.5.** **Fig. 12.6.**
Change in soil organic matter content following addition of pig manure (Fig. 12.5) or different forms of organic matter (Fig. 12.6) on fallow plots of arable soil

soil with OM removal by the crop, a massive addition of 160-ton cattle slurry $ha^{-1}$ $a^{-1}$ resulted only in maintenance of OM content, not an increase. On bare soil this amount caused a positive increase, addition of straw and lucerne only maintained the content. Also, within complete cropping systems, massive suppletion of manure was necessary to maintain OM content. A long-term Dutch experiment comparing: (1) only fertilizer application; (2) combined with ley grassland which was mulched; and (3) extra application of organic manure; showed maintenance of the OM level only for the latter system over a 20-year cropping cycle (Fig. 12.7). This was mainly due to the extra suppletion of manure, totalling 13 500 kg $ha^{-1}$ in 8 years (Kooistra et al. 1989).

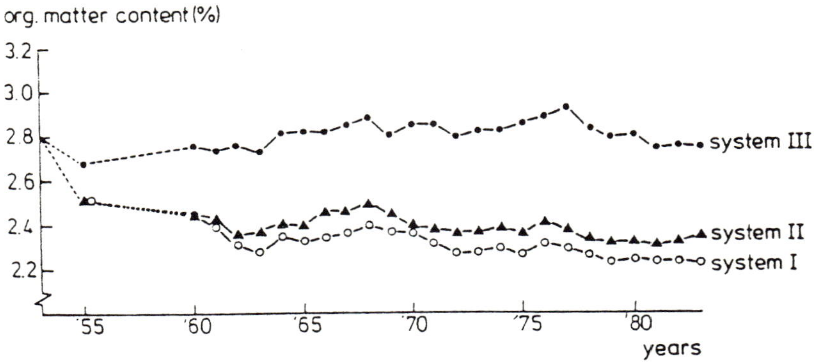


**Fig. 12.7.** Course of organic-matter contents of the soils of systems *1–3* during the period 1952–1984. The organic-matter content of the soils of each pair of systems is significantly different (P<0.001)

From these experiences one can conclude that given the problems of maintaining OM content, not to speak of raising it, one has to be most careful not to stimulate OM mineralization, as the secondary consequences for mobilization of contaminants may be serious.

### 12.4.3 Other Binding Compounds

Suppletion of other compounds has been less common, although certain 'raw' fertilizers also add other compounds than just N, P and K to the soil. Direct treatment of sediments to bind phosphate by injecting $Fe^{2+}$ has been tested and technical equipment developed (Quaak et al. 1993). Further dispersal by bioturbation and in shallow lakes by wind turbulence will readily occur, but will also limit the duration of the treatment as non-treated sediment comes to the surface and release of contaminants starts again. In peaty sediments rich in organic content the iron reduction is accompanied by a loss of the phosphate adsorption ability (Keizer et al. 1993). Turbidity could also indirectly increase aerobic mineralization of OM to $CO_2$, thereby decreasing pH and changing the equilibrium between Ca-$PO_4$ and $PO_4^{3-}$ in favour of $PO_4^{3-}$ (Driscoll et al. 1993). This application of ferric compounds has also been tested for heavily manured sandy soils by injecting $Fe(OH)_3$-solution in field margins next to drainage ditches to prevent leaching of phosphate to surface water. Results are not yet available (Schoumans, pers. comm.).

Application of concentrated $Al^{3+}$ or $Al^{3+}/Si(OH)_4$ is still in the experimental phase (Gerhard, pers. comm.). Promising is the use of an aggregated combination of Al molecules: polynuclear $Al_{13}$. The excellent adsorptive capacities, also in pH ranges as low as pH 4.7 (Furrer 1993) and the ease of forming this Al-aggregate under semi-natural conditions (Furrer et al. 1992) makes it an adsorbent which can be successfully applied in defined spots.

A more diffuse application will depend on the role of Al in soil geochemistry. Leaching of contaminants to lower soil horizons, either due to a decrease of pH or to OM in the upper horizons, may result in further leaching in lower horizons despite an unaffected pH there. The adsorption systems in these lower horizons, $Al(OH)_3$ in equilibrium with gibbsite, are less effective than OM/humic systems, as shown for Zn (Hesterberg 1993). This may be counteracted for other contaminants by binding with ferric compounds which in such a manner get taken up in Fe-rich B-horizons after being leached from the upper soils, as observed by Bergkvist et al. (1989).

## 12.5 Managing Environmental Impacts in Changed Land-use and Soil Management Systems

### 12.5.1 Land-use Changes

Land use is changing gradually and permanently. In the Netherlands each year a slight percentage of the total area assumes a new function. In the long term these changes are massive, also on a European scale. In western Europe this is due to the spreading of living, traffic and industrial areas combined with the intensification of agriculture. In eastern Europe it is the consequence of the change from big estates to the "colchoze" system to the most recent privatizing in agricultural areas, each system with its typical land use and management system (Varallyay, pers. comm.). In the European Union (former EC) the intensification of agriculture has led to large areas with too low production levels; large "set-aside" schemes have been developed.

Within these set-aside options the most relevant land-use changes now seem to be:

- Abandoning agricultural areas (which includes cessation of liming) and natural succession to either roughland or forest
- Changing arable land in forest areas by planting specific species/cultivars
- Inundation of past drained soils (mainly limited to polder areas like in the Netherlands)
- Drainage of areas with a high water-table to improve aeration and, hence, organic matter breakdown and soil productivity (occurring in peat areas, forested areas on soils rich in organic matter, and in marshlands in estuarine areas)

For a further description see ter Meulen-Smidt in Chapter 7.

So far, very few studies have been carried out on the consequences of contaminant mobility due to land-use changes. A few examples of these studies and some potential mechanisms are described in this section.

Abandonment of arable soils and a change to roughland has been modelled by Knoop and Traas (1993), see Section 12.4.1. Their model included the development of a litter layer which will influence the behaviour of organically bound contaminants.

Planting forest trees will result in a changed litter layer, and consequently changes in environmental conditions. P. Römkens (pers. comm.) observed a distinct drop in pH, accompanied by an increase in dissolved organic carbon, in a series of forests of different age planted on former arable sandy/loamy soils (see Chap. 1). DOC facilitates transport of organic matter bound to heavy metals, and of organic components. De Vries et al. (1994) modelled the pH change for arable land to be planted with coniferous Douglas fir stands. Both for sandy and clayey soils, pH and acid-binding capacity will change within several decades. Restoring the pH of the soil has proven to be rather difficult, not with respect to the very upper soil layers after superficial application of for instance lime but to deeper soil layers, as has been described in Section 12.4.1. Therefore, maintaining the pH of the soils as well as possible is to be advised.

Another major consequence of land-use change, as mentioned above, is a change in litter quality and OM breakdown and OM cycling. Inadequate maintenance of organic matter of soils has resulted in alarmingly decreased OM contents (Table 12.2). This may result in mobilization of organic-matter bound compounds, and destruction of soil structure and increased soil erosion. Increased wind and water erosion may lead to further dispersal of chemical compounds, especially for those adsorbed to mineral particles. Wind-blown pesticides, for a large part adsorbed to soil or dust particles, comprise a larger percentage of total dispersal and occur over a longer distance than previously thought (Gosselink et al. 1993). Kuhnt (1993) and Imeson et al. (1993) call attention to water erosion of pesticide-loaded particles downhill and concentration in lower shallower areas. Accumulation of pesticides and local anaerobic conditions due to waterlogging in these areas may lead to mobilization.

Restorating or raising the OM status of soils is a hard and long task. Proper maintenance of organic matter is the best option for proper treatment of accumulated contaminants with land-use change schemes.

In inundation and drainage the primary force is the change of redox conditions, which mobilizes accumulated contaminants. Dredged contaminated sediments will first change from anaerobic to aerobic conditions, especially when temporarily stored on the land. Raising the groundwater level will reverse the situation to anaerobic. As a consequence, elements mobilizing under aerobic conditions as well as elements mobilizing under anaerobic conditions will be released. A similar process may occur when contaminated soil from a riverbank is dug out to create new side channels for the sake of nature area development and deposited again elsewhere under saturated conditions. Inundation of river banks as such is described by Japenga and Salomons (1993).

### 12.5.2 Controlled and Integrated Soil Management

For a controlled and integrated soil management system, two major options seem to be most promising:

- Manipulating the organic matter cycling through management of the plant cover
- Creating aerobic, anaerobic or mixed redox conditions by managing the run-off or groundwater system

As addition of materials like pig manure or cover crops (lucerne) seems to be less effective (see Sect. 12.4.2), the selection of a proper plant cover (vegetation) is of crucial importance. It contributes to the litter layer and thus to the OM content; moreover, it protects the soil against wind and water erosion. Litter composition and resulting humus quality also have a distinct influence on the binding capacities of Cd, Cu, Pb and Zn (Krosshavn et al. 1993; Ernst and Peterson 1994). To provide the highest amount of leaf and litter material with the lowest depletion of the soil nutrient status, it is generally advisable to select species with a high shoot-root ratio. These kinds of fast and luxuriously (exuberantly) growing plants potentially accumulate heavy metals. Between different species and cultivars there are great differences with respect to accumulation (Muys et al. 1992). One has to realize that the application of plants as specific accumulators of contaminants – as a management option developed by Baker et al. (1991) but previously mentioned by Ernst et al. (1983) – is in this context less useful. Taking away the above ground parts which have accumulated high amounts of contaminants is only possible in small defined areas. Further, by doing so, the organic material which should add to the litter layer and OM content is taken away too and the OM content will decrease instead of increase. The choice of plant species is further important for the behaviour of contaminants in the soil, as illustrated for Zn with Norway spruce and silver birch by Bergkvist et al. (1989). With similar inputs the impacts of these species on soil processes will be very different (Fig. 12.8).

Muys et al. (1992) have developed a management system which combines the impact of cultivar choice, soil type, acidification and impact of soil organisms.

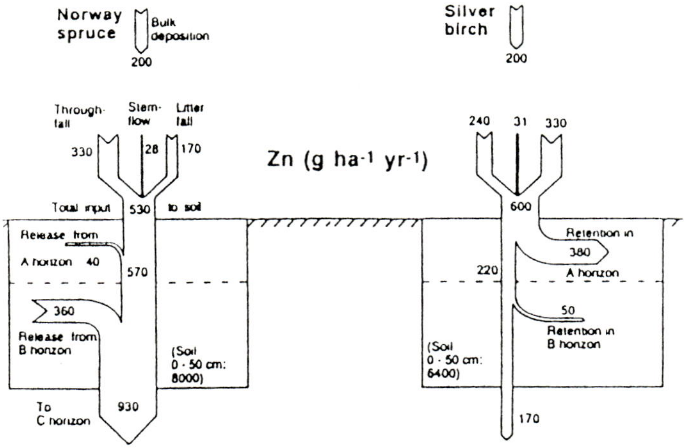


**Fig. 12.8.** Effect of tree type on the flow of zinc in two forest systems. Bergkvist et al. (1989)

By applying the proper tree species or cultivar, soil horizons can be changed in a geochemically more desirable way, for instance, a Mull type soil. These Mull types have a higher pH, are intensively mixed through the soil horizons, partly due to a rich soil life (including deep-burrowing earthworms), and a resulting homogeneous soil profile coupled with a good CEC. This approach has been worked out by De Vos (pers. comm.) in a management system to control harbour sludge deposits. These slurries are covered by mats of Salix cultivars (Salimat®, see Muys et al. 1992). The Salix readily develops and spreads out, covering the soil horizon and preventing erosion. In addition it improves maturing of the soil by adding litter to the surface. Selection of the proper cultivars prevents much of the contaminants being removed from the soil into the plant and in this way transferred to the soil surface. Harvesting of the timber could provide some release but in a controlled way. Further testing of these cultivars is needed. As a further management option, pumping of the sludge into the deposits can be combined with forced aeration of the watery sludge mixture. See *A 1–4* in Fig. 12.9 which gives an artist's impression of such a system. After diagenesis (ageing) of the sludge the upper layer can be cultivated to stimulate aerobic biodegradation *B*, and planted *B–D*, using the Salimat-system.

Creating desirable redox conditions could be realized by such unrefined measures as drainage and inundation, but also by a fine tuning of groundwater management or the management of water run off in a watershed area. Groundwater management will be particularly possible in lowland areas. By pumping with a defined quantity at specific locations, groundwater level, flow direction and rate will be controlled. At the outlet – whether this is a lowland polder area or a

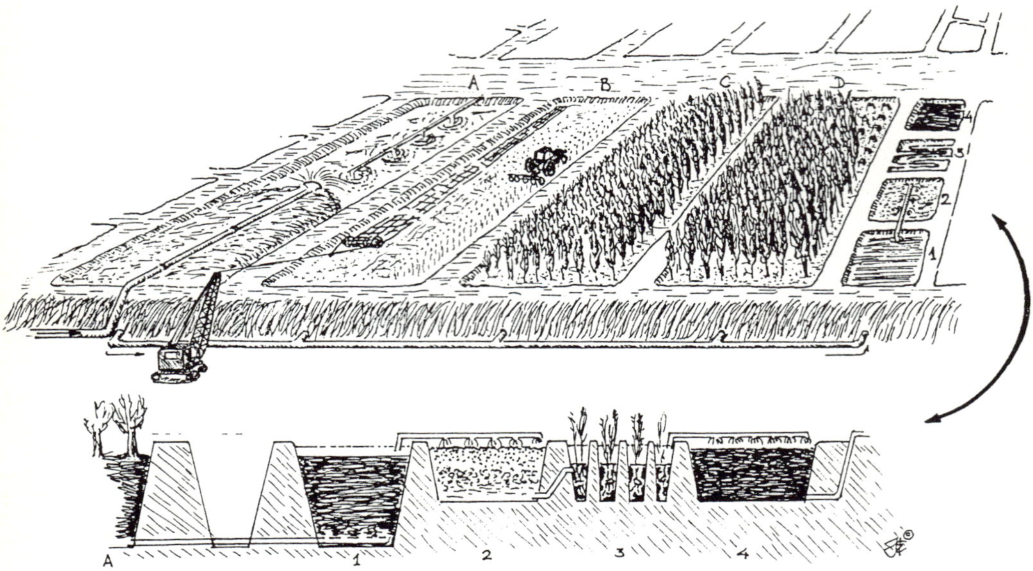


**Fig. 12.9.** Treatment of pumped groundwater via *1* a combination of an anaerobic upwell reactor unit, followed by *2* an aerobic trickling filter, *3* a halophytes-filter and finally, *4* a humus/peat adsorptive end unit

(small) watershed – the quality of the discharged water can be monitored. The pumped groundwater can then be treated either preventively or when some residual contaminants are measured. To this end, pumping, monitoring and post-treatment should be combined in one facility. Also, for defined areas like waste dumps or sludge deposits groundwater management will be a manageable option. Treatment under aerobic as well as anaerobic conditions can be realized (Fig. 12.9) by *1* a combination of an anaerobic upwell reactor unit, followed by *2* an aerobic trickling filter, *3* a halophytes-filter and finally, *4* a humus/peat adsorptive end unit. This sequence is chosen because anaerobic dechlorination is a necessary first step in the breakdown of a number of important organic contaminants, mostly followed by further aerobic breakdown processes. The halophytes filter functions in this sequence as an extra anaerobic step; moreover, with very active microbial activity due to the rhizosphere micro-organisms. Halophytes are plant species filtering out the contaminants via their roots and storing these in their plant parts below or above ground level. According to Abdul-Shafy et al. (1994) vascular plants can act as scavengers of metals from municipal waste water while still maintaining a healthy status. But also for organic compounds these halophytes could be useful, as the microbial rhizosphere (the microbes occurring especially directly around the roots) are very effective in the breakdown of a great number of complex organic compounds. The final treatment water could be reused, while the final residues are concentrated in the humus filter material. When the filter capacity is used up the peat can be dug out and the final contaminant residues can be disposed by burning (if necessary at very high temperatures). The heat generated could be used for pumping, warming the bioreactors and general maintenance. Creation of such treatment subunits and of anaerobic soil conditions in the deposits themselves can have negative impacts on plant growth and the species selected should temporarily be able to withstand anaerobic conditions.

The kind of extensive treatment sketched above can be realized on different spatial scales:

- The previously mentioned very local scale of waste dumps and sediment deposits which are bordered by barriers and ideally sealed at the bottom and the top (Polanska 1993)
- The regional scale of a specific industrial area for which a confined groundwater management system can be designed and realized
- The regional scale of a geographical region with a specific groundwater system

Examples of these last two categories can be found both in lowland (polder) areas and upland watersheds. Polder can be managed by pumping the water from the surrounding ring canal – which is normal practice to maintain drainage – and transporting it to a control, monitoring and treatment station where possibly contaminated drainage water is monitored and cleaned. For watersheds this could be realized by a cascade of treatment areas (anaerobic mini-marshes, aerobic peat catchment plots, etc.) at the base part of the catchment. Such a system, at least parts of it, have been developed for a few mining areas in Asia by Salomons (see Chap.1).

One has to be aware, however, that most of these treatment provisions are at present still in a pilot stage. They have yet to be developed in further detail. Further development of the different processes (modules) as described in Fig. 12.9 seems to be the best route. as the basic mechanisms and general operation procedures are now available in principle. Combining and scaling up these different modules will nevertheless cost considerable effort before integrated and controlled management systems are put into practice.

## References

Abdul-Shafy HI, Hegeman W, Teiner A (1994) Accumulation of metals by vascular plants. Env Manag and Health 5:21-24

Andersson F, Persson T (eds)(1988) Liming as a measure to improve soil and tree condition in areas affected by air pollution. Results and experience of an ongoing research programme, Report 3518. National Swedish Environmental Protection Board, Solna, Sweden

Baker AJM, Reeves RD, McGrath SP (1991) In situ heavy metal decontamination of polluted soils using crops of metal accumulating plants – a feasibility study. In: Hinchee RF, Olfenbuttel RF (eds) In situ bioreclamation. Butterworth, Boston, pp 600-605

Baker G (1994) Ecology and management of earthworms in agricultural soil with particular reference to southern Australia. ISEE 5, Ohio
Bengtsson G, Rundgren S (1992) Seasonal variation of lead uptake in the earthworm *Lumbricus terrestris* and the influence of liming and acidification. Arch Environ Contam Toxicol 23:198-205
Bergkvist B, Folkeson I, Berggren D (1989) Fluxes of Cu, Zn, Pb, Cd, Cr, and Ni in temperate forest ecosystems. Water Air Soil Pollution 17:217-286
Boels D, Fleming G (1993) Chemical time bombs from landfills: appraisal and modelling. Land degradation and rehabilitation 4:399-406
Breemen N van, Mulder J, Driscoll CT (1983) Acidification and alkalinization of soils. Plant and Soil 75:283-308
Brussaard L, Bouwman LA, Geurs M, Haasink J, Zwart KB (1990) Biomass, composition and temporal dynamics of soil organisms of a silt loam soil under conventional and integrated management. Neth J Agric Sci 38:283-302
Driscoll CT, Effler SW, Auer MT, Doerr SM, Penn MR (1993) Supply of phosphorus to the water column of a productive hardwater lake: controlling mechanisms and management considerations. In: Boers PCM, Cappenberg ThE, Raaphorst W van (eds) 1993 Proceedings of the third international workshop on phosphorus in sediments. Hydrobiologia 253:61-72
Ernst WHO, Peterson PJ (1994) The role of biomarkers in environment assessment. 4. Terrestrial plants. Ecotoxicology 3:180-192
Ernst WHO, Verkleij JAC, Vooijs R (1983) Bioindication of a surplus of heavy metals in terrestrial ecosystems. Environ Monit Assess 3:297-305
Francis AJ (1982) Effects of acidic precipitation and acidity on soil microbial processes. Water Air and Soil Pollution 18:375-394
Furrer G (1993) New aspects on the chemistry of aluminium in soils. Aquatic Sciences 55/4:281-290
Furrer G, Trusch B, Müller C (1992) The formation of polynuclear $Al^{13}$ under simulated natural conditions. Geochim Cosmochim Acta 56:3831-3838
Gosselink GRA, Ferdinandus HNM, Zee SEATM van der (1993) De verplaatsing van aan grond gehechte bestrijdingsmiddelen als gevolg van winderosie. Rapport LUW Wetenschapswinkel, Wageningen, The Netherlands, 75 pp
Harkness MR, McDermott JB, Abrancowicz DA, Salvo JJ, Flanagan WP, Stephens ML, Mondello FJ, May RJ, Lobos JH, Carroll KM, Breman MJ, Bracco AA, Fish KM, Warner GL, Wilson PR, Dietrich DK, Lin DT, Morgan CB, Gately HL (1993) In situ stimulation of aerobic PCB-biodegradation in Hudson river sediments. Science 257:503-507
Hesterberg D (1993) Effects of stopping liming on abandoned agricultural land. Land Degradation and Rehabilitation 4:257-267
Imeson AC, Perez-Trejo F, Verstraten JM (1993) The erodibility of soils relevant to the triggering of chemical time bombs. In: Meulen GRB ter, Stigliani WM, Salomons W, Bridges EM, Imeson AC (eds) Chemical time bombs. Proceedings of the European state-of-the-art conference on delayed effects of chemicals in soils and sediments. Veldhoven, The Netherlands, 2-5 September 1992. Foundation for Ecodevelopment, Hoofddorp, pp 61-66
Japenga J, Salomons W (1993) Dyke-protected floodplains: a possible chemical time bomb ? Land degradation and rehabilitation 4:373-380
Johnston AE, Goulding KWT, Poulton PR (1986) Soil acidification during more than 100 years under permanent grassland and woodland at Rothamsted. Soil Use and Management 2:3-10
Judas M, Schauermann J, Meiwes KJ (1994) The inoculation of *Lumbricus terrestris* in a limed spruce forest. ISEE 5, Ohio

Keizer P, Galas J, Sinke A, Joode P de (1993) Phosphate immobilization in peaty sediments. In: Boers PCM, Cappenberg ThE, Raaphorst W van (eds) 1993 Proceedings of the third international workshop on phosphorus in sediments. Hydrobiologia 253:374

Knoop J, Traas ThP (1993) Acidification, changing land use, and cadmium mobilization: a modelling approach. In: Meulen GRB ter, Stigliani WM, Salomons W, Bridges EM, Imeson AC (eds) Chemical time bombs. Proceedings of the European state-of-the-art conference on delayed effects of chemicals in soils and sediments. Veldhoven, The Netherlands, 2-5 September 1992. Foundation for Ecodevelopment, Hoofddorp, pp 107-118

Kooistra MJ, Lebbink G, Brussaard L (1989) The Dutch programme on soil ecology of arable farming systems 2. Geogenesis, agricultural history, field site characteristics and present farming systems at the Lovinkhoeve experimental farm. Agriculture, Ecosystems and Environment 27:361-387

Krosshavn M, Steinnes E, Varskog P (1993) Binding of Cd, Cu, Pb and Zn in soil organic matter with different vegetational background. Water Air and Soil Pollution 71:1-2

Kuhnt G (1993) Soil-bound residues of pesticides. In: Meulen GRB ter, Stigliani WM, Salomons W, Bridges EM, Imeson AC (eds) Chemical time bombs. Proceedings of the European state-of-the-art conference on delayed effects of chemicals in soils and sediments. Veldhoven, The Netherlands, 2-5 September 1992. Foundation for Ecodevelopment, Hoofddorp, pp 87-94

Muys B, Lust N, Granval Ph (1992) Effects of grassland afforestation with different tree species on earthworm communities, litter decomposition and nutrient status. Soil Biol Biochem 24:1459-1466

Nieboer E, Richardson DHS (1980) The replacement of the nondescript term 'Heavy metals' by a biologically and chemically significant classification of metal ions. Environ Pollut 1B:3-26

Pickering WF (1981) Selective chemical extraction of soil components and bound metal species. CRC Crit Rev in Anal Chem 12:233-266

Polanska K (1993) Quaternary aquifer pollution in the Notec Valley, Poland. Land Degradation and Rehabilitation 4:393-398

Quaak M, Does J van der, Boers P, Vlugt J van der (1993) A new technique to reduce internal loading by in-lake phosphate fixation in shallow lakes. In: Boers PCM, Cappenberg Th E, Raaphorst W van (eds) 1993 Proceedings of the third international workshop on phosphorus in sediments. Hydrobiologia 253:337-344

Rhee GY, Bush B, Brown MP, Kane M, Shane L (1989) Anaerobic biodegradation of polychlorinated biphenyls in Hudson river sediments and dredged sediments in clay encapsulation. Wat Res 23:957-964

Vos B de (1994) De aanpak van primaire natuurtechnische milieubouw op baggerslibgronden. Groene Band 93, 36 pp

Vries W de, Kros J, Salm C van der (1994) Long-term impacts of various emission deposition scenarios on Dutch forest soils. Water Air and Soil Pollution 75:1-35

# 13 Global Perspectives and Risk Assessment

*William M. Stigliani*

## 13.1 Introduction

This chapter presents some thoughts on the function of land as a receptor and emitter of toxic substances. The former function is an anthropogenic imposition on the land and its soils. Soils did not evolve to serve the role as a storage bin for toxic persistent, chemicals, at least not at the levels of input experienced in most industrialized areas over the last 100 years. The latter function of soil as an emitter is mostly determined by natural processes in the soil. One perspective on this function is that it serves to purge the soil of toxic materials, and so can be viewed as a useful function from the point of view of improving soil quality. On the other hand, in the process of purging itself of polluting substances, such substances may be passed on to vegetation, or possibly transferred to groundwaters. Since such transfers are guided by the natural laws of soil chemistry and physics, human control over the outcomes are limited. In light of the above, major questions arise regarding the management of these soils; i.e., the range of available options for insuring their long-term sustainability, and controlling soil emissions so that they do not degrade groundwaters, or toxify flora and fauna beyond acceptable limits.

## 13.2 Land as a Receptor and Emitter of Environmentally-Harmful Chemicals

From a technical point of view, it is relatively easy to improve the quality of air contaminated by industrial pollution. When the emissions of most air pollutants are reduced (greenhouse gases being a notable exception), their concentrations in the atmosphere will decrease in proportion to the decrease in emissions on a time scale of days to a few weeks. The reason for this behavior is that such air pollutants have short residence times in the atmosphere. Through processes of dry and wet deposition, the atmosphere quickly cleanses itself of heavy metals, volatile organics, and acidic substances. Without this cleansing action, these pollutants would quickly build up to levels that would be intolerable for humans and most other living species. But there is a downside to this rapid atmospheric broom-sweeping. Because of the law of mass balance, the pollutants swept out of the atmosphere must be deposited somewhere, and often, it is the soils that are the receptors. Unlike the atmosphere, however, well-buffered soils do not quickly pass on the received wastes to other compartments of the biosphere, namely lower soil horizons, groundwaters and vegetation. Typically, a large fraction of the chemicals is bound to the soil in a phase that is not bioavailable, and the mobilized portion of the waste load is released slowly. Thus, we may consider one important function of soils to be their ability to protect soil flora and fauna, groundwaters, and vegetation from rapid transfers of pollutants originally generated in the atmosphere and via other sources (e.g., application of agrochemicals).

This discussion raises two important questions. The first is: how can we improve soil quality? The situation is not so clear-cut as it is for improving air quality. Unlike the case for air, if we reduce the input of a particular pollutant to the soil, the concentration of that pollutant may not decrease appreciably for a long period of time. The beneficial effect of decreasing the input is merely to slow down the rate of increase in concentration of that pollutant in the soil, but not to reduce the concentration. Figure 13.1 shows the example of historical trends in air emissions of cadmium in the Rhine Basin, compared to the calculated increases in cadmium concentration in a typical agricultural soil in the basin. One may observe that despite a ten-fold decrease in air emissions over the last several decades, agricultural soils are now more polluted than ever, although the rate of increase in concentration of cadmium has slowed down considerably.

The second question is: can we improve soil quality (i.e., decrease or stabilize the concentration of pollutant in the soil) without polluting the groundwaters and vegetation beyond acceptable standards? Going back to the principle of mass balance, soils can't simultaneously improve in quality *and* protect groundwaters and vegetation. In fact, by adsorbing and immobilizing a large fraction of pollutant inputs, soils protect other compartments of the biosphere by becoming more polluted themselves. Therefore, the concept of "improved soil quality" has to be formulated in the context of soils and their impacts on groundwaters, vegetation, soil flora and fauna, and perhaps other components as well.

The answers to these questions are not simple. Firstly, we need to have a clear objective – do we want to improve soil quality by reducing concentrations of pollutantsin the soil? If so, then we must assess the potential impacts on the receptors of emissions from the soil. Do we want to contain the pollutants in the soil? If so, then we must understand the risks associated with the potential mobilization of the stored pollutants at some point in the future. In all of these issues we need to be guided by the principles of soil science, which can specify the dynamic behavior of accumulated toxic materials in the soils, and mechanisms

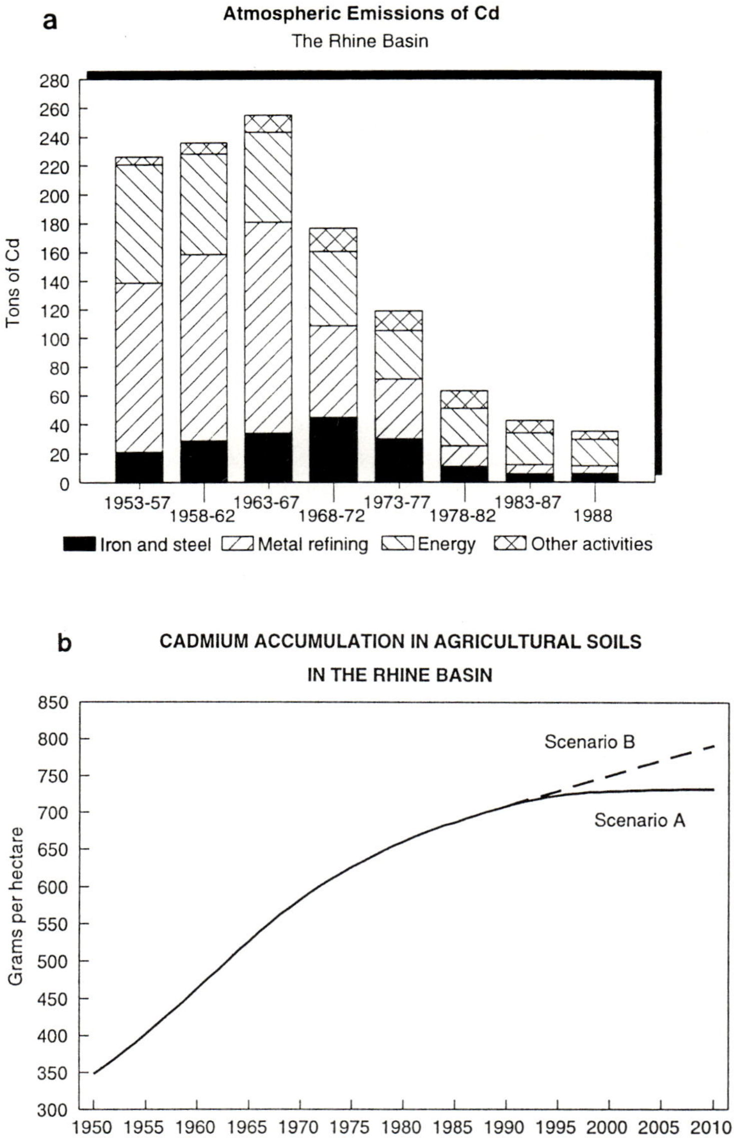


**Fig. 13.1. a** Atmospheric emissions of Cd in the Rhine Basin, compared to **b** the build up of Cd in a typical agricultural soil of the Basin, 1955-1988. *Scenario A* refers to case in which all Cd is removed from phosphate fertilizer; *scenario B* is case where Cd remains in phosphate fertilizer. Source: Stigliani et al. (1994)

that can trigger their release. In order to reduce these rather global questions to a practical level, we will discuss, in the following section, the example of the build up of cadmium in agricultural soils of the Rhine Basin and problems associated with the safe management of these soils.

## 13.3 Cadmium Pollution in Agricultural Soils of the Rhine Basin: A Practical Example

The manner in which a well-buffered soil behaves with respect to the input of a pollutant can often be described as shown in Fig. 13.2. For any given point on the curve, the vertical axis shows the amount of pollutant adsorbed in the soil (stored phase) and the horizontal axis shows the amount in soil solution (mobilized phase). The slope of the line at a particular point, $\Delta Q/\Delta C$, defines the *binding capacity* of the soil for a given pollutant. Thus, at the beginning of the pollution phase the binding capacity is quite high (see $\Delta Q_1/\Delta C_1$), and as pollution progresses it declines (see $\Delta Q_2/\Delta C_2$), until at "maximum adsorption capacity", the binding capacity goes to zero. At that point the soil is said to be saturated with zero retention of pollutant. Any pollutant inputs to the soil go quickly out of the soil. There are cases of high contamination, where soils are completely saturated with a given pollutant, but this is not the general case. It is more likely that pollutant concentration standards for soils, crops, or groundwaters would be exceeded long before such a saturation would occur.

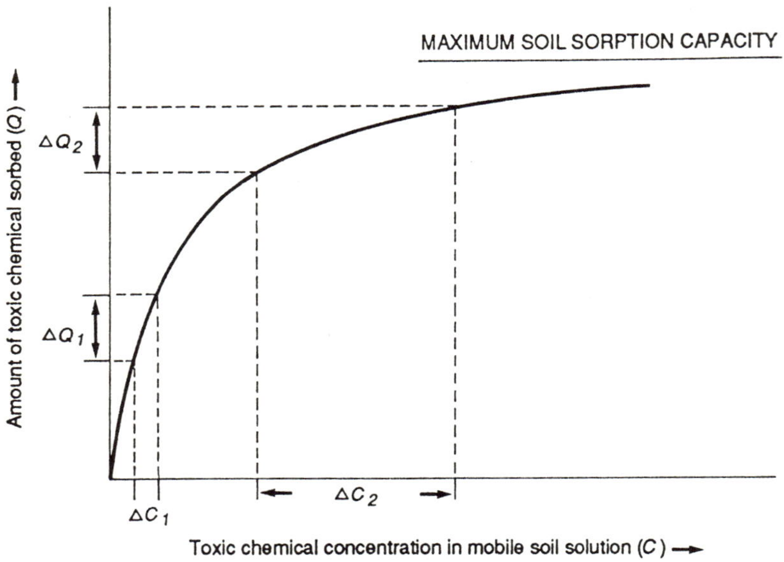


**Fig. 13.2.** Generalized adsorption-mobilization curve for toxic chemical in soil. Source: Hesterberg et al. (1992)

A practical example of the behavior displayed in Fig. 13.2 is shown in Fig. 13.3. The figure, based on a simple model of crop uptake and human intake, shows the weekly human ingestion of cadmium for the period from 1950 to 2010 as a function of soil pH by a hypothetical person receiving his or her entire food supply from crops grown in the Rhine Basin (Stigliani and Jaffe, 1993). The intakes are plotted against the World Health Organization (WHO) suggested limits of cadmium ingestion of 400 to 500 µg/week. Some of the implications of this plot with respect to changing soil pH are discussed later in this section and in Section 3.4, but for the moment let us focus on the case where the pH of the soil has been held at around 6.0. Under the assumptions of the model, intake over the 60 year period is estimated to have increased from about 120 µg/week in 1950 to about 265 µg/week in 2010. The reason for the increase over this time (referring to Fig. 13.2) is that the cadmium inputs have moved further up the adsorption-mobilization curve, resulting in a greater fraction of the inputs partitioned to the mobilizable fraction. This is the fraction that is responsible for crop uptake, and thus the percentage of cadmium ending up in the crops increases as pollution increases. Given certain assumptions about inputs, it is possible to estimate the number of years that would be required to reach a level of crop ingestion of 400 µg/week, the lower limit of the WHO standard. Assuming that when the soil reaches this point of contamination it would no longer be suitable for food production, it is possible to define a "useful lifetime" of the soil as the time it would take to reach the WHO limit. Assuming further that the cadmium inputs to the soil remain constant at the level of inputs in the year 2010 depicted in the Fig. 13.2, then a "useful lifetime" of about 100 years is estimated, corresponding to a cadmium soil content of about 1200 g Cd/ha.

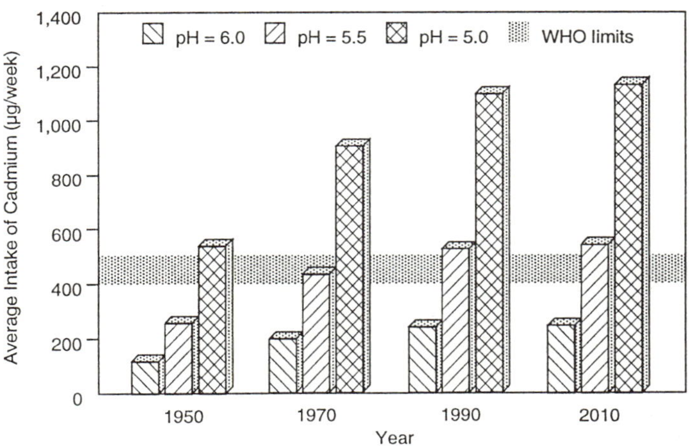


**Fig. 13.3.** Estimated average dietary intake of Cd from crops grown in the Rhine Basin as a function of time and changes in soil pH. It is assumed that the total food intake is from crops grown in the Basin. Source: Stigliani and Jaffe (1993)

The major inputs of Cd to agricultural soils are from atmospheric deposition, phosphate fertilizers, in which Cd is present as an unwanted impurity, and sewage sludge. There are three major outputs from the soil system: (1) crop uptake; (2) leaching to groundwaters; and (3) erosion of surface layers. The net inputs are simply the sum of all inputs minus the sum of all outputs. Figure 13.4 shows an estimation of net inputs over time. *Scenario A* refers to a policy in which all cadmium is removed from the phosphate fertilizer, and *scenario B* refers to a policy in which the Cd is left in the fertilizer. The fact that *scenario A* is approaching zero means that the remaining input (from atmospheric deposition) is nearly balanced by the output of cadmium from the system. If all the cadmium were removed from phosphate fertilizer, then the useful lifetime of the soil is calculated to be in excess of 1500 years.

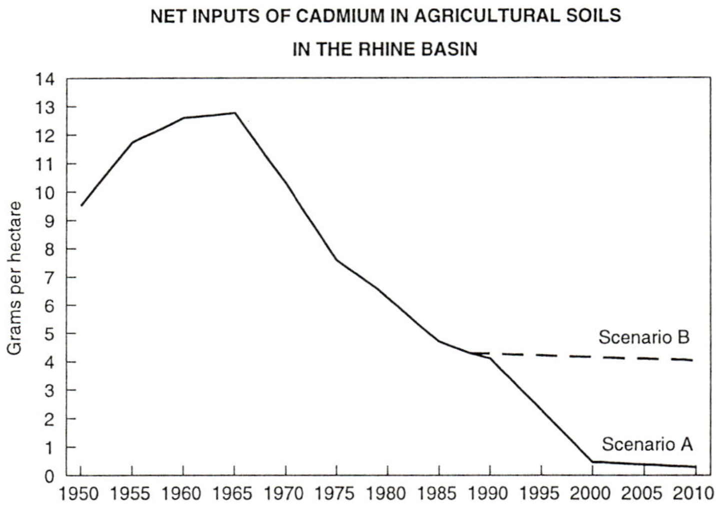


**Fig. 13.4.** Net inputs of Cd in agricultural soils in the Rhine basin. Source: Stigliani et al. (1994)

It is possible to estimate the costs of reducing the inputs of Cd, and compare these costs with the cost of doing nothing to reduce inputs. According to Klepper and Mahlau (1992), the most cost effective means of reducing the cadmium load to agricultural soils is by the removal of cadmium from phosphate fertilizer. For example, the cost of removing one kg of Cd from large coal-fired industrial furnaces is in the range of \$2500 to \$3700. For removing one kg of Cd from thermal zinc refineries, the cost is about \$1240 to \$1600. In contrast, the cost of removing one kg of cadmium from phosphate fertilizer is about \$170. Compared to the price of phosphate fertilizer spent by farmers, the cadmium removal cost is quite low, only 0.38–1.51% of the total price.

One way of calculating the total cost of doing nothing to reduce the cadmium content in phosphate fertilizer is to calculate the change in the value of the agricultural land as the concentration of cadmium in the soils begins to approach the end of itsuseful lifetime. Using certain economic estimations presented elsewhere (Forsell and Stigliani, 1994), Fig. 13.5a shows the trend in the price of agricultural land over a 100 year period. It shows that near the end of the useful lifetime of the land, its value decreases exponentially; in the last year declining by nearly 50% per year. For comparison, Fig. 13.5b shows the changes in land price over the same 100 year period, but with the useful lifetime of the soil extended to 200 years. It shows the great benefit of doubling the useful lifetime, since after 100 years the decrease in the value of the land is well under 1% per year. In summary, these crude calculations suggest that the long-term benefits of removing cadmium from phosphate fertilizer far outweigh the small additional costs to the farmer in the unit price of fertilizer. It is perhaps even fair to say that the long-term sustainability of the agricultural land in the Rhine Basin depends on a deliberate policy for cadmium removal from the fertilizer.

Thus far, we have discussed soil contamination assuming that the chemical conditions in the soil remain constant. The effect of the contamination has been to lower incrementally, the ratio of the adsorbed to the mobilized fraction, thus slowly increasing mobilization. But, as discussed in other chapters of this book, cerain changes in the soil's chemical conditions can trigger the release of stored toxic chemicals. Among the most important of these conditions are: pH, redox potential, organic matter content, and salinity.

Figure 13.6 is similar to Fig. 13.2, except that it shows the effect on soil emissions when changing chemical conditions shift the adsorption-mobilization curve to a new curve with a lower maximum adsorption capacity. Such a shift would (according to thermodynamic principles) cause a release of toxic chemical in order to bring the system to a new state of chemical equilibrium. The quantity of pollutant mobilized in the process of changing equilibrium states is $\Delta PC$ (partial capacity change for unit soil volume) or $\Delta MC$ (maximum capacity change per unit soil volume).

a

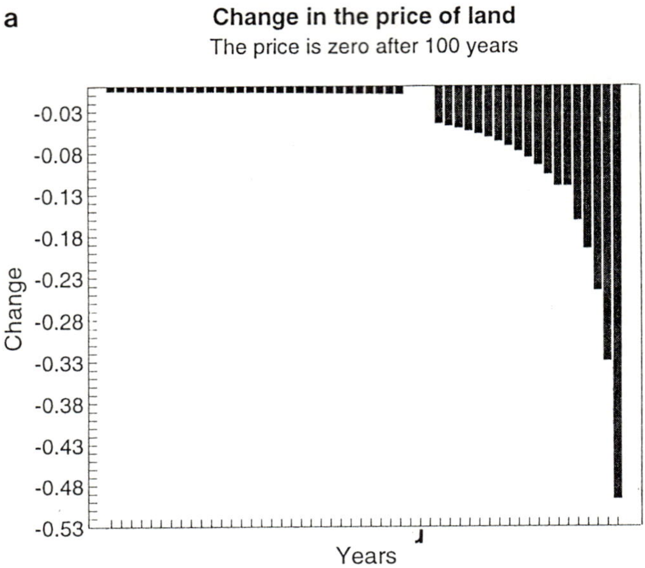


b

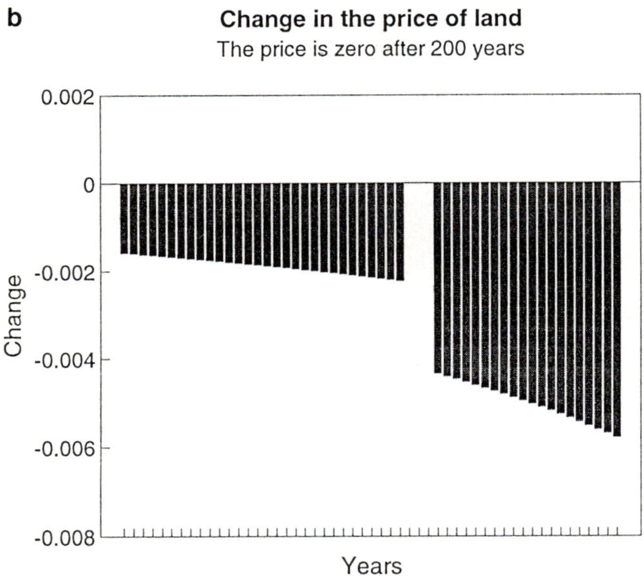


**Fig. 13.5. a** Change in the price of agricultural land in the Rhine Basin over 100 years. The price is assumed to be zero after 100 years. **b** Change in the price of agricultural land in the Rhine Basin over 100 years. The price is assumed to be zero after 200 years. Source: Forsell and Stigliani (1994)

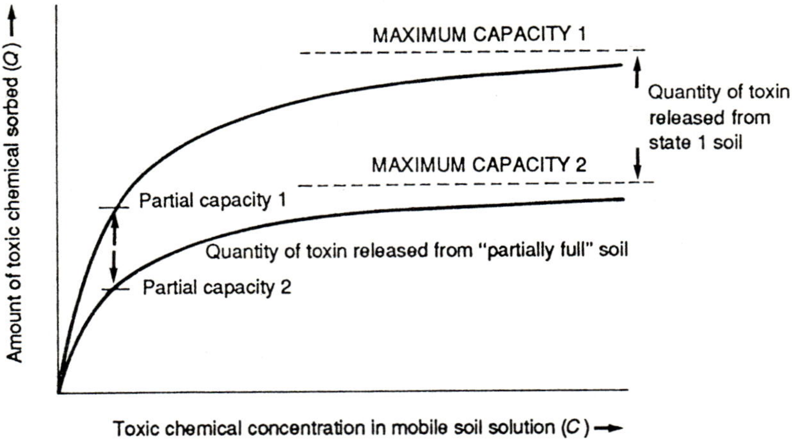


**Fig. 13.6.** Generalized adsorption-mobilization curve for a toxic chemical in soil, assuming an alteration in chemical property of soil that changes the adsorption to mobilization ratio. Source: Hesterberg et al. (1992)

Referring again to the practical example shown in Fig. 13.3, the reason why the shift to lower pH causes more cadmium to be taken up by crops, is that some portion of the adsorbed cadmium, in equilibrium with the mobile cadmium fraction at pH=6, must, at pH=5.5 or 5.0, enter the dissolved phase in order to reestablish equilibrium under the new conditions. Viewing the figure,it is obvious that the impact of changing pH on the level of cadmium ingestion increases with increasing levels of pollution. Thus, in 1950 a decline in pH from 6.0 to 5.5 would have resulted in an ingestion level below the WHO guideline, but already by 1970 such a decline would have caused a transgression of the threshold.

## 13.4 The Impact of Conversion of Agricultural Land

The problems associated with land as a receptor and emitter of toxic materials are particularly pronounced for agricultural soils. Firstly, they have been recipients of high inputs of toxic materials via atmospheric deposition, the application of agrochemicals including fertilizers and pesticides, and sewage sludge. Secondly, the pH of the soils has been maintained, through the addition of lime, at artificially high levels. The effect of these two factors has been to increase the capacity of the soil as a receptor of toxic materials, but at the same time to increase the risk of high levels of soil emissions triggered by changes in soil properties.

Currently, serious discussions are underway within the European Union (EU) concerning the reduction of agricultural production by abandoning large tracts of agricultural lands and converting them for other uses (McSharry, 1991). Conversion to forest land alone is estimated at 8.5 million ha over the period 1985-2020 (Nilsson et al., 1992). If the lands selected for conversion are no longer limed, the pH of the soils, previously maintained at a level of 6 or higher, could decline rapidly over a period of decades, resulting in an increase in mobility of trace metals in the topsoil. One consequence of abandoning agricultural lands, which are typically more polluted with heavy metals such as cadmium than are forest soils, would be the enhanced uptake of heavy metals in new vegetation. This could harm wildlife inhabiting the newly converted lands and increase leaching to groundwaters used for drinking-water supplies.

If the McSharry Plan goes into effect, it is quite likely that a significant portion of agricultural lands in the Rhine Basin will be targeted for conversion. How should the "toxic load" of stored cadmium be managed in the face of future land-use changes? The impact will depend to a great extent on the rate of pollutant leaching from the soil, and this rate will depend on the rate at which the pH changes. Table 13.1 lists the estimated time required for 40% of the soil's cadmium to leach out of the top 40 cm as a function of pH, assuming the soil's organic matter content to be 3.5%. The calculations assume that over the time periods indicated, there are no further net inputs of cadmium to the soil.

One potential problem with this plan is the possibility of broad-scale pollution by Cd in the underlying groundwater. Leaching of 40% of the Cd from the Rhine Basin means an average of about 300 g Cd/ha will be mobilized. For an agricultural area of 10 million hectares (the size of the Rhine Basin) this corresponds to 3000 tons of cadmium. If the leaching occurred in 17 years (pH = 4.0), then the annual inputs to lower soil horizons or groundwater would be around 176 tons/year.

In some areas, allowing the lands to acidify on this quick time scale may be a viable option. For a topsoil with a cadmium concentration of 0.23 mg $kg^{-1}$ (720 g $ha^{-1}$ to a depth of 20 cm), the dissolved-phase concentration is estimated to increase from 0.0005 mg $l^{-1}$ to 0.006 mg $l^{-1}$ as the soil pH declines from 6 to 4.5. Given that the primary drinking-water standards for Cd are 0.01 mg $l^{-1}$, and

**Table 13.1.** Estimated time required for a reduction in Cadmium by 40% assuming all inputs of Cadmium are discontinued

| Soil pH | Yearly leaching rate | Time required for 40% reduction (years) |
|---|---|---|
| 6.0 | 0.14% | 365 |
| 5.5 | 0.3% | 170 |
| 5.0 | 0.6% | 85 |
| 4.5 | 1.4% | 36 |
| 4.0 | 2.9% | 17 |

that mobilized cadmium will be further diluted as it moves into the saturated zone, a large-scale deterioration of groundwater quality due to an increase in cadmium concentration may not be expected. On the other hand, significant regional variations in soil cadmium concentrations exist, and heavily polluted soils that overlie shallow groundwaters could contaminate the water. A safer option would be to slow down the rate of leaching by controlling the rate of pH decline. For example, if the pH of the soil were to be lowered to 4.5 or 5.0 rather than 4.0, the rate of leaching could be reduced by one-half to one-fourth. This option would provide more assurance that the more vulnerable areas of the Rhine Basin would not be subjected to unduly high inputs of cadmium.

## 13.5 Management Options

It is clear from the above, that the most convenient option, namely, to do nothing, is not a viable one over the long term, as long as pollutant inputs to the soil exceed the outputs. The most obvious workable option, which should be taken as a first step, is to reduce inputs to the soil to a minimum. One significant barrier to achieving this step is the lack of harmonized standards for air, water, soils and sediments. For example, air quality guidelines for Europe have been established by WHO, based solely on human risk factors arising from breathing of contaminated air (WHO, 1985). The guideline for cadmium suggests that concentrations not exceed 1 to 5 ng $m^{-3}$ in rural areas, and 10 to 20 ng $m^{-3}$ in industrial areas. As noted above, even though these ranges could be quite low as far as air quality is concerned, the fact that it is the soil and not the air that accumulates the pollutants suggests that air standards should be formulated with the goal of long-term sustainability of soils in mind.

A simple analysis can be made to assess the impact of the WHO air standard for Cd on long-term soil quality, as well as the risks to humans through ingestion of agricultural crops contaminated by cadmium. Let us assume hypothetically that air quality within the specified 2 to 10 ng $m^{-3}$ range had been maintained over an approximate 40-year period from 1950 to 1990 in agricultural lands of the Rhine Basin. We can then go back and recalculate what the concentrations of Cd in the soil might have been for this range, and compare these concentratons to those calculated on the basis of actual estimated historical emissions, the results of which were presented in Fig. 13.1b. Considering rural areas alone, the WHO air concentration limits correspond approximately to European atmospheric deposition rates (wet plus dry) of 2 to 10 g Cd $ha^{-1}$ $a^{-1}$ (K. Olendrzynski, 1994, IIASA, Laxenburg, Austria, pers. comm.). The total Cd load to the soil includes, not only inputs from air, but also from phosphate fertilizer and sewage sludge. The results are shown in Fig. 13.7, which compares soil Cd concentrations based on the WHO standards with estimates of the actual concentrations

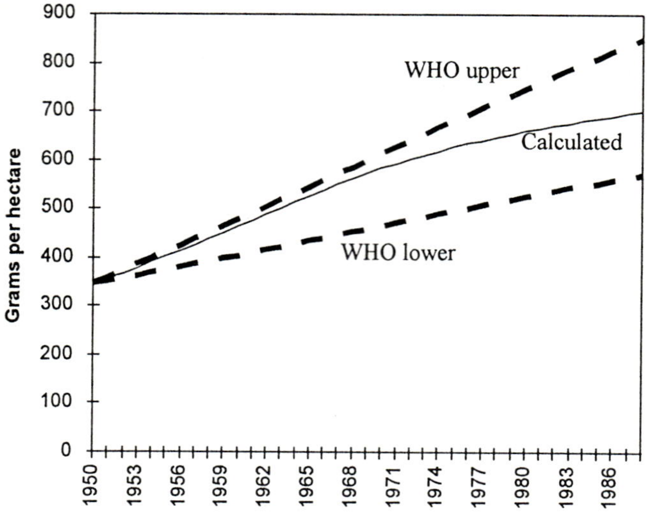


**Fig. 13.7.** Historical concentrations of cadmium in a typical agricultural soil in the Rhine Basin showing actual estimated concentrations (*calculated solid line*), and hypothetical concentrations based on high and low ranges of acceptable air concentrations of cadmium established by the WHO

(the middle *calculated* curve is thus the same as that shown in Fig. 13.1b). The top curve plots the inputs based on the high end of the WHO standard (i.e., 10 g Cd $ha^{-1}$ $a^{-1}$), and bottom curve is based on the low end of the standard (2 g Cd $ha^{-1}$ $a^{-1}$). One may observe that the actual estimated concentrations fall comfortably within the range of the WHO standard. The implication is that while the WHO standard for Cd may be appropriate for protecting human health via the inhalation pathway, it is certainly not appropriate as a standard for long-term sustainability of soils. Moreover, it is probably over the long term inappropriate for safeguarding human health when considering exposure from food intake of Cd, and not only exposure from air inhalation. In fact, in the long-term perspective the WHO air standard for Cd and the WHO standard for human ingestion (400-500 µg/week) are probably incompatible. If, as described in the preceding discussion, taking 1200 g Cd/ha as a crude threshold for crossing the WHO limit for Cd ingestion, then, extrapolating the plot in Fig. 13.7 into the future, it would take about 30 years to reach the threshold for the upper WHO value, and about 130 years for the lower value. This implies that it makes little sense to develop standards for air concentrations in isolation from considerations about soil and food quality.

After minimizing chemical inputs, the next step in managing contaminated lands is to clearly define the management goals. Cleaning the soils through natural processes such as acidification, is a possibility. For this option, however, an assessment should be made of the potential impacts to the ecological components that will be the receptors of the soil emissions. Based on the results of this assessment, it may be necessary to manipulate the decline in pH through controlled liming or other possible options in order to slow down the rate of release of the contaminants from the soil. If the strategy is to contain the contaminants in the soil, then consideration should be given to the concept of *geochemical engineering* discussed in more detail by Salomons in Chapter 1.

From the above discussion it is clear that in defining standards and management options for controlling pollution, a more holistic and consistent approach needs to be taken that consider impacts on all environmental components and exposure pathways to flora, fauna, and humans.

## References

Forsell O, Stigliani WM (1994) Costs and benefits of removing cadmium from phosphate fertilizer. Manuscript in preparation, IIASA, Laxenburg, Austria

Hesterberg D, Stigliani WM, Imeson AC (eds)(1992) Chemical time bombs: linkages to scenarios of socioeconomic development. Executive report 20 (CTB Basic Document 2) April 1992, IIASA, Laxenburg, Austria

Klepper G, Mahlau G (1992) Non-point pollution with cadmium. The Kiel Institute of World Economics, Kiel Working Paper No 532, Kiel, Germany

McSharry R (1991) European commission reforming the common agricultural policy (CAP) COM (91) 379 def, 409 def, Brussels

Nilsson S, Sallnas O, Duinker P (1992) Future forest resources of western and eastern Europe. Parthenon, Camforth, UK

Stigliani WM, Jaffe PR (1993) Industrial metabolism and river basin studies: a new approach for the analysis of chemical pollution. Research Report 93-6, IIASA, Laxenburg, Austria

Stigliani WM, Anderberg S, Jaffe PR (1994) Loading of heavy metals in the environment of the Rhine basin. Manuscript in preparation, IIASA, Laxenburg, Austria

WHO (World Health Organization) (1985) Air quality guidelines for Europe. WHO Regional Publications, European Series, WHO Regional Office for Europe, Copenhagen, Denmark

# Subject Index

*Environmental Science*

*Volumes already published*

**Biogeodynamics of Pollutants in Soils and Sediments** (1995)
W. Salomons, W. M. Stigliani (Eds.)

*Volumes in preparation*

**Heavy Metals (1995)**
W. Salomons, U. Förstner, P. Mader (Eds.)

**Sediments and Toxic Substances (1995)**
W. Calmano, U. Förstner (Eds.)

Printing: Saladruck, Berlin
Binding: Buchbinderei Lüderitz & Bauer, Berlin